SEA-DOO/BOMBARDIER PERSONAL WATERCRAFT 1988-91 REPAIR MANUAL

SELOC™

President Dean F. Morgantini, S.A.E.
Vice President–Finance Barry L. Beck
Vice President–Sales Glenn D. Potere
Executive Editor Kevin M. G. Maher, A.S.E.
Production Manager Ben Greisler, S.A.E.
Production Assistant Melinda Possinger
Project Managers George B. Heinrich III, A.S.E., S.A.E., Will Kessler, A.S.E., S.A.E., James R. Marotta, A.S.E., S.T.S., Richard Schwartz, A.S.E., Todd W. Stidham, A.S.E.
Schematics Editor Christopher G. Ritchie

Authors Joan and Clarence Coles

Brought to you by
CHILTON™ MARINE

Manufactured in USA

1020 Andrew Drive
West Chester, PA 19380
ISBN 0-89330-033-0
2345678901 8765432109

Other books
by

Seloc's OMC Stern Drive Tune-up & Repair Manual
Seloc's Marine Jet Drive Tune-up & Repair Manual
Seloc's Mercury Volume I Tune-up & Repair Manual
Seloc's Mercury Volume II Tune-up & Repair Manual
Seloc's Mercury Volume III Tune-up & Repair Manual
Seloc's Force Outboard Tune-up & Repair Manual
Seloc's Yamaha Volume I Tune-up & Repair Manual
Seloc's Yamaha Volume II Tune-up & Repair Manual
Seloc's Yamaha Volume III Tune-up & Repair Manual
Seloc's Mariner Volume I Tune-up & Repair Manual
Seloc's Mariner Volume II Tune-up & Repair Manual
Seloc's Chrysler Outboard Tune-up & Repair Manual
Seloc's OMC Cobra Stern Drive Tune-up & Repair Manual
Seloc's MerCruiser Stern Drive Volume I Tune-up & Repair Manual
Seloc's MerCruiser Stern Drive Volume II Tune-up & Repair Manual
Seloc's MerCruiser Stern Drive Volume III Tune-up & Repair Manual
Seloc's Mercury/Mariner Volume I Tune-up & Repair Manual
Seloc's Mercury/Mariner Volume II Tune-up & Repair Manual
Seloc's Mercury/Mariner Volume III Tune-up & Repair Manual
Seloc's Johnson/Evinrude Volume I Tune-up & Repair Manual
Seloc's Johnson/Evinrude Volume II Tune-up & Repair Manual
Seloc's Johnson/Evinrude Volume IIA Tune-up & Repair Manual
Seloc's Johnson/Evinrude Volume III Tune-up & Repair Manual
Seloc's Johnson/Evinrude Volume IV Tune-up & Repair Manual
Seloc's Johnson/Evinrude Volume V Tune-up & Repair Manual
Seloc's Personal Watercraft Volume I Tune-up & Repair Manual
Seloc's Personal Watercraft Volume II Tune-up & Repair Manual
Seloc's Personal Watercraft Volume III Tune-up & Repair Manual
Seloc's Volvo Penta Stern Drive Volume I Tune-up & Repair Manual
Seloc's Volvo Penta Stern Drive Volume II Tune-up & Repair Manual
Seloc's Volvo Penta Stern Drive Volume III Tune-up & Repair Manual

SAFETY NOTICE

Proper service and repair procedures are vital to the safe, reliable operation of all personal watercraft, as well as the personal safety of those performing repairs. This manual outlines procedures for servicing and repairing personal watercraft using safe, effective methods. The procedures contain many NOTES, CAUTIONS and WARNINGS which should be followed, along with standard procedures, to eliminate the possibility of personal injury or improper service which could damage the vessel or compromise its safety.

It is important to note that repair procedures and techniques, tools and parts for servicing marine engines, as well as the skill and experience of the individual performing the work, vary widely. It is not possible to anticipate all of the conceivable ways or conditions under which these engines may be serviced, or to provide cautions as to all possible hazards that may result. Standard and accepted safety precautions and equipment should be used during cutting, grinding, chiseling, prying, or any other process that can cause material removal or projectiles.

Some procedures require the use of tools specially designed for a specific purpose. Before substituting another tool or procedure, you must be completely satisfied that neither your personal safety, nor the performance of the marine engine, will be compromised.

Although information in this manual is based on industry sources and is complete as possible at the time of publication, the possibility exists that some vehicle manufacturers made later changes which could not be included here. While striving for total accuracy, Chilton Marine cannot assume responsibility for any errors, changes or omissions that may occur in the compilation of this data.

PART NUMBERS

Part numbers listed in this reference are not recommendations by Chilton Marine for any product brand name. They are references that can be used with interchange manuals and aftermarket supplier catalogs to locate each brand supplier's discrete part number.

SPECIAL TOOLS

Special tools are recommended by the marine manufacturer to perform a specific task. Use has been kept to a minimum, but, where absolutely necessary, they are referred to in the text by the part number of the tool manufacturer. These tools can be purchased, under the appropriate part number, from your local dealer or regional distributor, or an equivalent tool can be purchased locally from a tool supplier or parts outlet. Before substituting any tool for the one recommended, read the SAFETY NOTICE at the top of this page.

ACKNOWLEDGMENTS

Chilton Marine expresses appreciation to Bombardier Corporation for their assistance in the production of this manual.

FOREWORD

This is a comprehensive tune-up and repair manual for Sea-Doo/Bombardier Personal Watercraft manufactured between 1988 and 1991. Competition, high-performance, and commercial units (including aftermarket equipment), are not covered. The book has been designed and written for the professional mechanic, the do-it-yourselfer, and the student developing his mechanical skills.

Professional Mechanics will find it to be an additional tool for use in their daily work on personal watercraft because of the many special techniques described.

Boating enthusiasts interested in performing their own work and in keeping their unit operating in the most efficient manner will find the step-by-step illustrated procedures used throughout the manual extremely valuable. In fact, many have said this book almost equals an experienced mechanic looking over their shoulder giving them advice.

Students and Instructors have found the chapters divided into practical areas of interest and work. Technical trade schools, from Florida to Michigan and west to California, as well as the U.S. Navy and Coast Guard, have adopted these manuals as a standard classroom text.

Troubleshooting sections have been included in many chapters to assist the individual performing the work in quickly and accurately isolating problems to a specific area without unnecessary expense and time-consuming work. As an added aid and one of the unique features of this book, many worn parts are illustrated to identify and clarify when an item should be replaced.

Illustrations and procedural steps are so closely related and identified with matching numbers that, in most cases, captions are not used. Exploded drawings show internal parts and their interrelationship with the major component.

TABLE OF CONTENTS

1
DESCRIPTION & OPERATION

1-1 BRIEF HISTORY

The jet drive system for propelling a craft through the water arrived on the scene in the early 1960's with the jet drive boat. In the early days the jet drive system was mated only with "high performance" powerplants -- engines in the 454 cu. in. class and larger. For this reason, during the "gas crunch" of the 1970's, the jet drives were labeled as "inefficient" and as "gas hogs". In addition to these two negative terms, they earned the reputation as "bad boy" boats due to their noisey "straight" exhaust, high rpm operation, and their almost unbelievable maneuverability. These combined factors did little to enhance their image and certainly restricted their popularity.

With "new and improved" technology, personal watercraft arrived on the scene around 1968. Personal watercraft, as we know them today, were developed using the same principles as the jet boats. Only single and twin cylinder engines are used with single stage pumps (one impeller) to make the craft highly efficient.

From harbors, bays, sounds, and estuaries at sea level to high mountain lakes, personal watercraft are giving their owners hours of fun on the water.

Series Covered

All Bombardier Sea-Doo personal watercraft are covered in this manual, from the early model years thru 1991.

A 580cc rotary valve engine and an axial jet pump are used on all Sea-Doo watercraft.

Other Seloc Personal Watercraft Manuals

This manual is the second in a series of three volumes produced in 1991 covering personal watercraft. Volume I covers the Kawasaki models and Volume III covers all Yamaha personal watercraft and other craft powered with a Cuyuna engine.

1-2 PRINCIPLES OF OPERATION

One of the first lessons to be learned in any elementary physics class is Newton's basic law: "for every force, there is an opposite and equal force". This statement is the basic principle of the jet pump. Water is "sucked" and "scooped" in from under the craft by a powerful pump rotating at incredible speed, and then discharged, "blown" out, sternward in the opposite direction. In this manner the watercraft is propelled forward.

Several decals may be affixed to the stern of the hull to indicate compliance with, or exceptions granted to normal coast guard regulations.

The personal watercraft covered in this manual are equipped with either a single or twin cylinder water-cooled two-stroke engine, matched with a single stage (one impeller) jet pump. The Sea-Doo GT is equipped with a "gate" which can be lowered down over the pump outlet nozzle forcing the exhausted water back in a forward direction, thus moving the craft sternward. Personal watercraft jet pumps may be classified as "axial flow" or "mixed flow".

All Sea-Doo Personal Watercraft are equipped with an axial flow jet pump.

Axial Flow

Water in an axial pump moves on a single axis, as depicted in the adjacent illustration, thus the word "axial". In simple terms, water is ingested and discharged parallel to the axis of impeller rotation. The inside of an axial pump case is parallel to the water flow -- unlike the inside of a mixed flow pump.

EDUCATIONAL INFORMATION

Even though Bombardier Sea-Doo personal watercraft utilize only an axial flow pump, the following paragraphs are included in this chapter to acquaint the reader with a mixed flow pump -- how it differs from the axial flow pump and why.

Mixed Flow

Many experts consider the mixed flow system to the most efficient for personal watercraft application. The water flow is discharged at an inclined angle to the axis of impeller rotation, as indicated by the arrows in the accompanying illustration. The impeller blades are inclined in such a manner to cause the water to be thrown off at a 45 degree angle to the impeller shaft within the pump case. From the pump case, the water is then forced out through the outlet nozzle and steering nozzle. The vanes in the pump case help straighten the water flow through the nozzle.

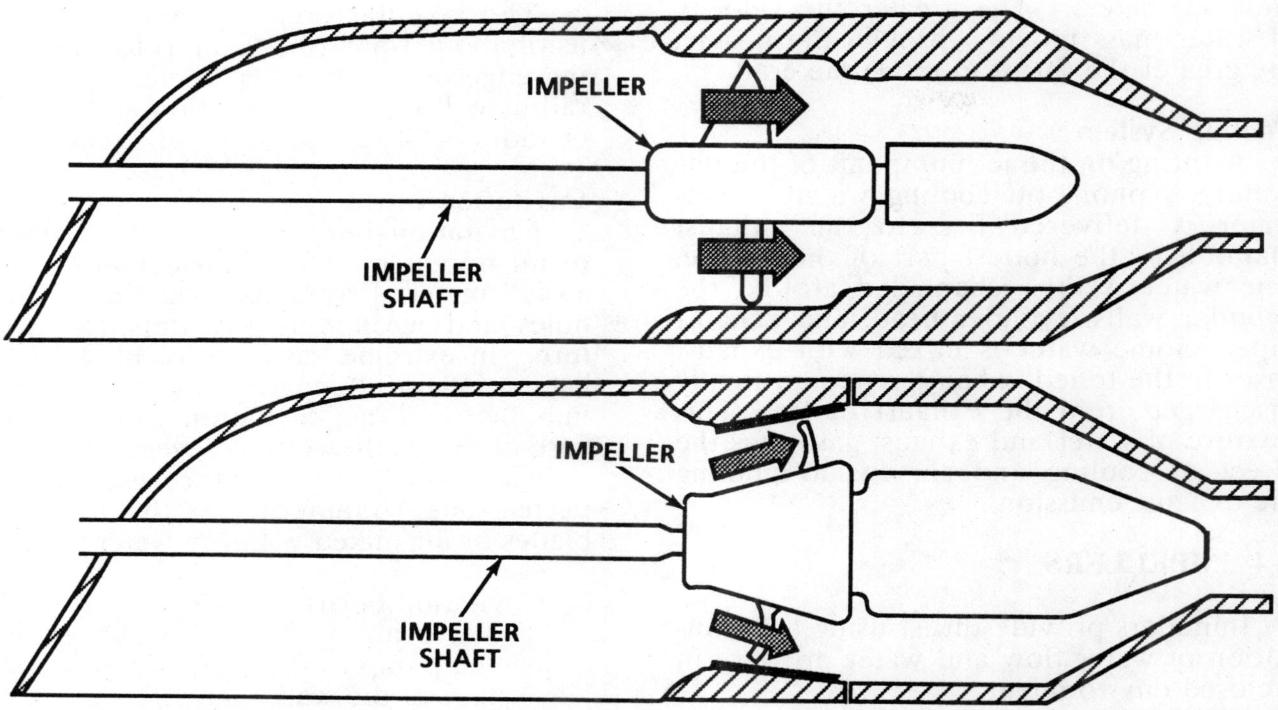

Line drawing to depict water flow through a mixed flow jet pump (top), and an axial flow pump (bottom). Notice on the mixed flow pump, how the water is directed upward at an angle to the axis of impeller rotation. Notice on the axial flow pump, how the water passes through parallel to the axis of impeller rotation.

The inside of the pump case is inclined at an angle to the water flow. Therefore, this design pump incorporates shim material to space the impeller blades at a specified distance from the pump case. Mixed flow pump efficiency decreases by 30% as blade-to-pump case clearance increases by 0.04" (1mm).

Axial or Mixed?

Axial and mixed flow pumps are intended for different applications. One is not necessarily better than the other. The axial pump can be operated at higher engine rpm and is more suitable for racing. A mixed flow pump adds thrust at low engine rpm due to the pump design. Therefore, the mixed flow pump performs more efficiently at low engine rpm. This type pump is more suitable for a craft equipped to carry extra weight.

A jet pump seldom requires service except for the following:

Impeller-to-pump case clearance becomes excessive -- mixed flow pump only.
Impeller is damaged.
The oil seal around the impeller shaft fails -- allowing water to enter the hull.

1-3 ENGINE AND JET DRIVE

The basic principle and arrangement of engine and pump are almost identical for all manufacturers.

The engine crankshaft is coupled either directly to the pump shaft or through a short driveshaft without any gear reduction. An impeller mounted on the pump shaft draws water into an opening in the hull through a suction and intake casting. Once the craft has attained forward motion the intake also serves as a "scoop" adding to the volume of water moved through the pump.

Volume and Velocity

The outlet nozzle is slightly funnel shaped. This design coupled with the capacity of the pump (impeller), causes the volume of water entering the pump to exit the nozzle with increased velocity. This principle is similar to air passing through the venturi of a carburetor.

The amount of water ejected sternward from the nozzle at high velocity is the force propelling the craft through the water. This force can actually be measured and calculated in foot pounds or

Newton meters. The greater the velocity of water mass moving through the nozzle, the greater the thrust to move the craft.

Cooling System

A fitting on the jet pump, aft of the impeller, syphons off cooling water. This water is delivered first to the exhaust manifold -- the hottest part of the engine. The water is then channeled around the cylinder walls, cylinder head, and exhaust pipe. Some water is mixed with exhaust gases in the tuned exhaust pipe and finally discharged from the exhaust outlet. This mixture of water and exhaust gases has the effect of cooling and somewhat quieting the exhaust emission.

1-4 IMPELLERS

Impellers provide thrust using a combination of water flow and water pressure in a closed environment.

The blades of an impeller overlap, thus water is trapped and forced through the impeller into the pump case and nozzle. The water moving past the impeller resists cavitation because it is under pressure. Blades in the pump case are angled exactly opposite to those of the impeller. These blades redirect the water to establish a concentrated flow through the nozzle.

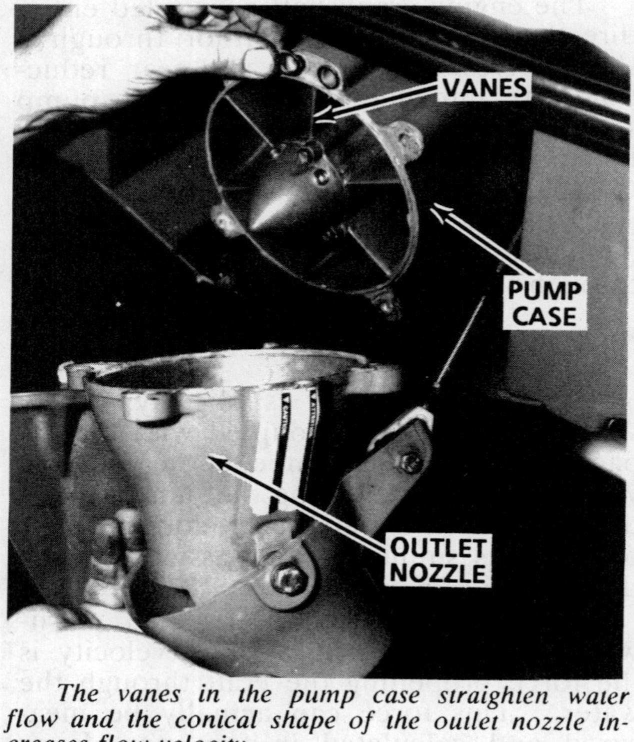

The vanes in the pump case straighten water flow and the conical shape of the outlet nozzle increases flow velocity.

The manufacturer recommends a radius of about 1/64" (0.3-0.5mm) for the leading edge of the impeller blade. A sharper radius will result in cavitation and a greater radius will reduce pump efficiency.

Cavitation Burns

Cavitation burns are the worst enemy of an impeller. These burns literally "eat away" material from the impeller, leaving holes and weakening the impeller structure. In extreme cases entire blades have been known to "depart" from the impeller hub because the cavitation burns at the base of the blade were so severe.

Cavitation burns are the result of imperfections (damage) on the impeller blades or air mixed with the water flow.

Cavitation burns may be caused by:

Wave jumping -- air is sucked into the pump as the craft leaves the surface of the water.

Bad seal between the pump case and outlet nozzle -- allowing air to enter.

Leading edge of the impeller becoming damaged.

To explain exactly what causes cavitation burns, first, let us examine the last cause. If the impeller leading edge is damaged and "mushrooms" over the blade, a low pressure area will form under the mushroomed lip.

The impeller of a jet pump is a precisely machined and dynamically balanced aluminum spiral. Significant damage to a blade will affect the balance.

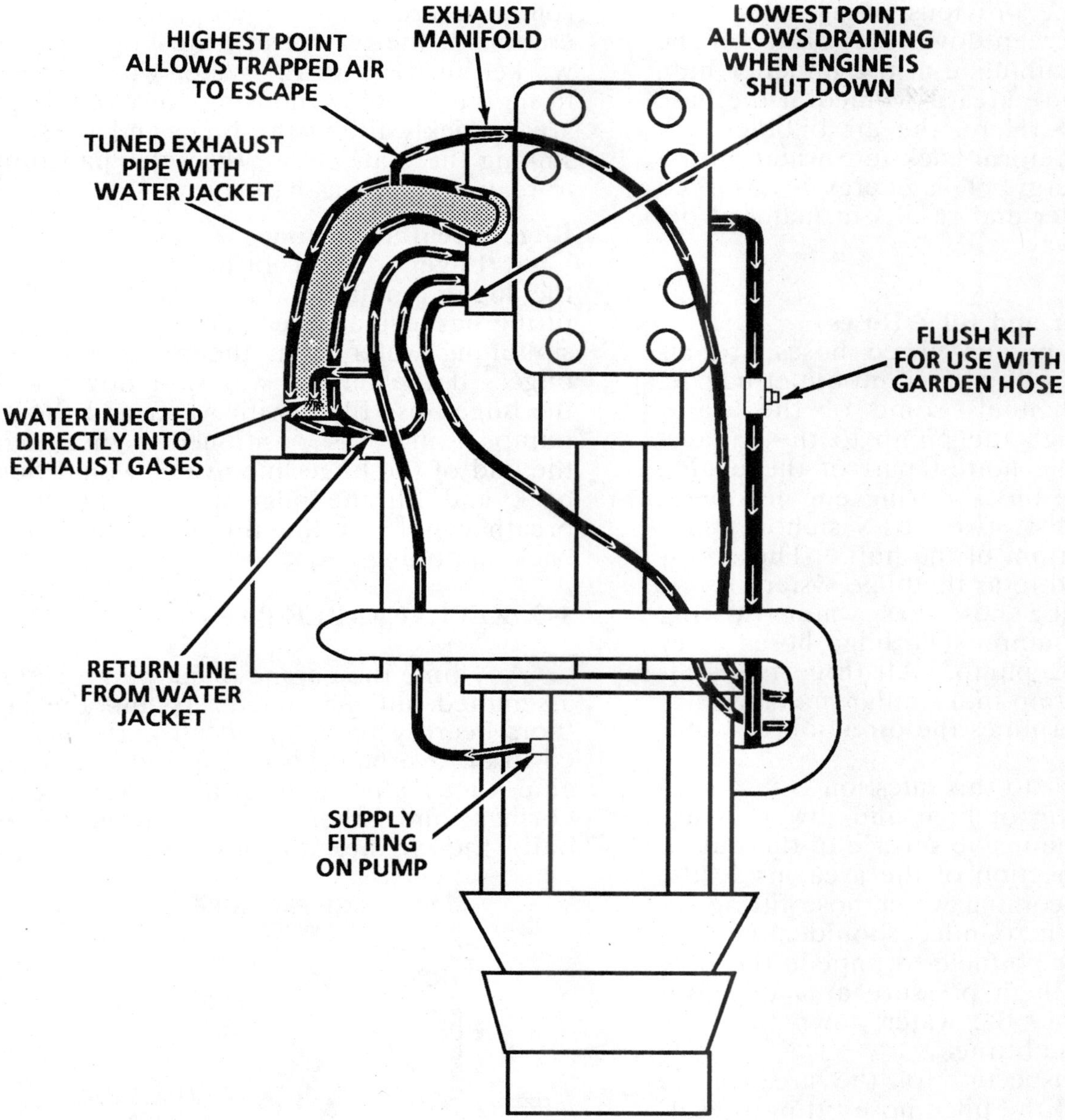

Simple line drawing to illustrate water flow from the pump, around the tuned exhaust pipe, through the engine and back to the pump. Other cooling circuits are also shown.

At atmospheric pressure, water boils at 212 degrees F. Anytime there is an imperfection on the surface of an impeller blade and water passes over that imperfection, the water pressure is lowered. If the pressure is lowered, water will boil at a much lower temperature. Therefore, air bubbles will form in the water boiling under the mushroomed lip.

The bubbles will creep down the surface of the blade and accumulate at the impeller hub. A high pressure area is formed at the base of the blade. Here, the air bubbles will collapse and reform back into water with a release of energy. This energy is absorbed by the impeller and results in material being eaten away.

The impeller should be dressed to a slightly rounded edge. It is impossible to achieve a perfectly straight sharp edge, because one side will always be convex and the other side concave. The concave side will form an area of low pressure and encourage cavitation, as described above.

The other two causes of cavitation -- defective seals and wave jumping -- occur when air is sucked in with the water flow. The same principles apply, as just described, in the previous paragraphs. Air bubbles will creep down the surface of the blade and accumulate at the impeller hub. A high pressure area is formed at the base of the blade. Here, the air bubbles will collapse and reform back into water with a release of energy. This energy is absorbed by the impeller and results in material being "eaten away".

Cooling Water and Bilge Hoses

On most pumps, three hoses are attached to the pump case and outlet nozzle. One hose channels some of the water flowing through the pump to the exhaust manifold -- the hottest part of the engine -- to cool the block during engine operation. The other two hoses siphon water out of the bottom of the hull. These hoses are referred to as the bilge system.

The cooling hose has water flowing **FROM** the pump. The bilge hoses carry water **TO** the pump. All three hoses attach to the pump in a similar manner.

What determines the direction of water flow?

The answer to this question is in a simple explanation of high and low pressure areas along the inside surface of the pump.

Close inspection of the area inside the pump at the cooling water hose fitting reveals a smooth rounded shoulder with no obstruction or obstacle to impede the flow of water. A high pressure area develops here and draws the water down the hose attached to the fitting.

Further inspection of the area inside the housing at the bilge hose fitting reveals a small protrusion around the opening. This protrusion causes disturbance to the water flow and creates an area of low pressure. This low pressure area will have the effect of emptying air and water from the hose aft with the impeller water flow -- similar to the action of a vacuum cleaner, as depicted in the accompanying illustration. If the other end of the hose is submerged in water inside the hull, the water will be "vacuumed" out. In this manner, the bilge system keeps the bilge relatively "dry".

Many an owner -- with good intentions -- thinking his action would smooth water flow and increase pump performance -- has filed the protrusion described in the previous paragraph smooth with the surrounding area of the pump. During operation of the watercraft, the bilge system worked in **REVERSE.** Water was pumped from the impeller housing into the bilge area, quickly filling the bilge and possibly sinking the craft before the rider had time to realize what was happening.

Bilge Breather Fitting

A bilge breather fitting is installed at the highest point of the bilge line. This fitting has a small breather hole to prevent siphoning water from the pump into the bilge. If the engine was shut down with the bilge hose filled with water and the jet pump fitting was positioned higher than the end of the bilge line, water would flow back and fill the bilge area. The small breather hole in the fitting prevents this back-siphoning.

1-5 REPLACEABLE PARTS

Any time the engine or jet pump is disassembled, all manufacturers and the authors heartily recommend all parts included in an overhaul kit be installed. An engine kit usually includes all gaskets, seals, O-rings, and other parts required to rebuild the engine. Pistons, rings, bearings,

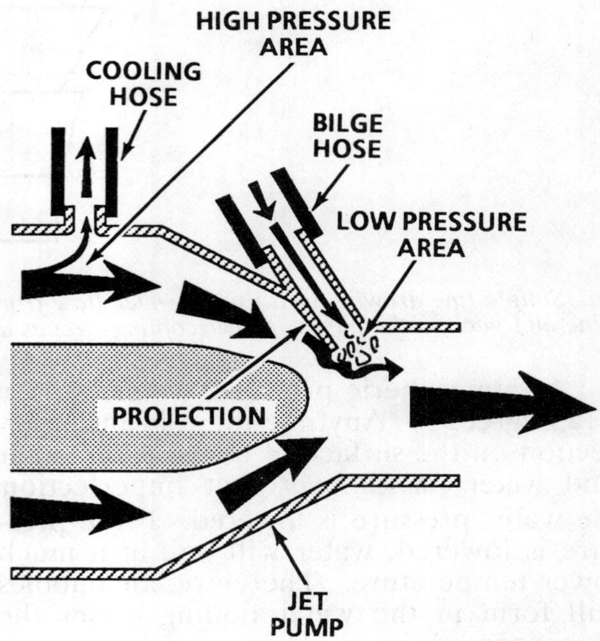

The direction of water flow through the cooling hose and the bilge hose is opposite. The cooling hose routes water from the pump to the engine. Water is "vacuumed" from the bilge through the bilge hose to the pump where it is forced out through the nozzle.

crankshafts and even blocks should be inspected to determine if they are fit for further service. The pump kits usually include all gaskets, seals, O-rings, and other parts required to restore full efficiency and "like new" condition to the jet pump unit.

1-6 DEBRIS REMOVAL AND ENGINE OVERHEATING

As may be expected, a clogged impeller will slow the craft. The volume of water passing through the nozzle and water velocity will be reduced. A clogged impeller may cause restricted cooling water through the engine.

Many times a cooling water supply hose, if not properly secured with a hose clamp, may "blow off". The engine loses its supply of cooling water and internal engine damage quickly occurs.

All Bombardier Sea-Doos are equipped with an overheat warning horn. If the horn sounds, **SHUT DOWN** the engine immediately and remove the restriction or check the cooling hose connection.

Caution must be exercised when moving the craft off a beach to avoid sucking in mud, sand, pebbles, or rocks through the intake. Such ingestion could cause damage to the impeller, the vanes in the impeller duct, or clog engine cooling passages.

The engine cooling passages must be flushed with clear water on a regular basis. This is accomplished by simply connecting a garden hose to the cooling water hose, or fitting, on the manifold, if so equipped. Such an easy maintenance practice should become a habit and will prevent accumulation of debris which could restrict engine coolant water flow and cause overheating.

Two bilge pickups actually "vacuum" the water out of the bilge and overboard through the jet pump.

1-7 REVERSE CAPABILITY

The Bombardier Sea-Doo GT is equipped with reverse capability. As mentioned earlier, "reverse" is accomplished by a gate installed in such a manner to swing down over the steering nozzle. The gate is controlled with a cable. When the gate is in the fully raised position, allowing all water from the nozzle to be discharged sternward -- the craft moves forward.

When the gate is swung down just part way, some of the discharged water moves sternward and the other portion of water is deflected downward and forward under the craft. Positioning the gate in such a manner places the jet pump in a **NEUTRAL** condition, because the water moving aft equals the water moving forward and craft is static ("dead in the water").

When the gate is moved to the full down position, virtually all discharged water from the nozzle is deflected down and forward -- the craft moves sternward.

Adjustment of the reverse cable is covered in Chapter 11, **Control Adjustments**.

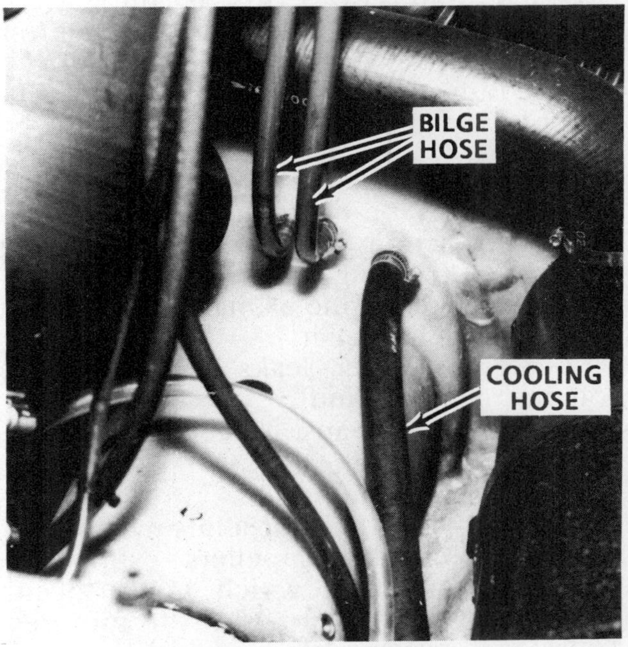

A single cooling hose and two bilge hoses are attached to the jet pump through simple fittings and hose clamps.

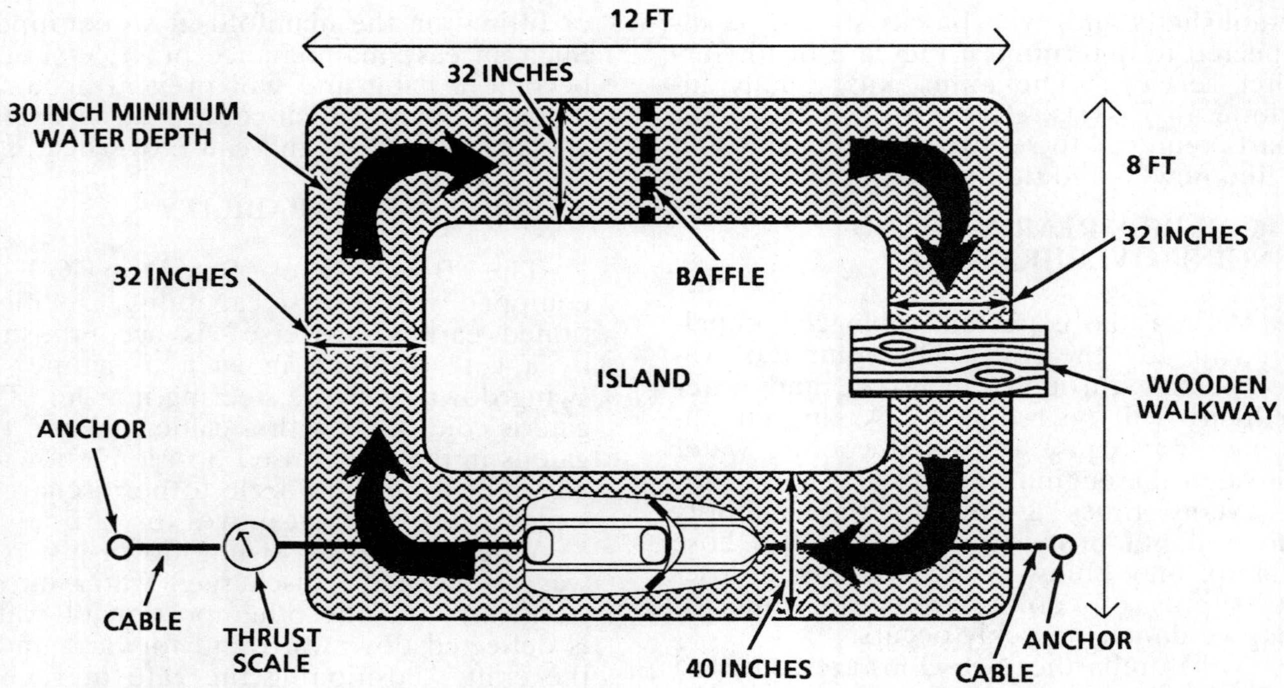

Layout of a typical test tank for testing personal watercraft. Such a tank is most valuable in verifying the completed work following service on the engine or the pump -- especially when making a timing adjustment.

1-8 TEST TANKS

Many times the performance of tune-up tasks, or adjustments are called for while the craft is in the water with the engine running. With an outboard or stern drive, the mechanic always has the boat between him/her and the wet stuff. On a personal watercraft, this is impossible without getting into the water, unless a test tank is used.

The advantages of using a test tank might be listed as:

a- Simulate actual water conditions. The craft may be operated without an exhaust extraction pipe and without using an auxiliary cooling water supply.

b- Test the engine at high rpm, placing a load on the impeller.

c- Reveal water leaks in the engine compartment and hull.

d- Uncover exhaust leaks.

e- Identify fuel leaks.

f- Detect engine noises.

g- Perform carburetor adjustments.

h- Check for bad impellers -- slippage.

i- Spot problems which might occur only after running for say 30 minutes. Therefore, prevent being stranded in the middle of a lake or far from shore if a serious malfunction should occur.

The only area a test tank cannot help diagnose is a craft handling problem.

The accompanying illustration depicts the type of test tank manufacturers use, with **APPROXIMATE** dimensions. This type of tank is partially sunk into the ground. The craft sits in a body of circulating water and the mechanic has access to both sides of the craft without getting wet and has a dry place for tools, etc.

The tank design with an island in the center allows water to circulate around the tank. With no circulation, the water will foam, and the jet pump will cavitate and lose thrust. A series of baffle plates installed on the opposite side to the craft, slow and straighten the water flow. The walls must be strong enough to hold the weight of the water and to contain the thrust of the pump, but coated to seal the walls and not scratch the sides of the craft.

The bow is secured with a stout line anchored to the edge of the tank to prevent the craft from "wandering". The stern is secured to the opposite edge of the tank to prevent the craft from moving forward. A thrust scale is sometimes attached to the stern line to determine the thrust exerted by the pump.

Seek out dealers with a test tank and attempt to use/rent their facilities to test the completed work.

2
SAFETY

2-1 INTRODUCTION

Personal watercrafting is one of the fastest growing leisure activites in the world. Riding a personal watercraft can be an exciting, exhilarating, and safe experience provided the operator exercises prudent behavior and remains in constant control of the vehicle.

Personal watercraft may be likened to a firearm. Weapons do not injure people, but the individual handling them is the true culprit. In a similar manner, if the personal watercraft is operated in an unresponsible manner the driver, passengers, and others using the same waterway may be in danger.

The greatest hazard in operating a personal watercraft stems not from the craft itself, but from the behavior of the operator. The Personal Watercraft Industry Association (PWIA) and the National Marine Manufacturers Association (NMMA) recognize there are a number of problems due to irresponsible craft operators contributing to tarnishment of the sport. Specifically, these problems range from alcohol and substance abuse to excessive noise, speed, and reckless maneuvering. Insensitive operators can disrupt an otherwise enjoyable day for other craft users through misconduct on the water. These individuals often endanger their own lives as well as the lives of others by losing full control of their craft.

Organized events, races, and exhibitions, properly planned and controlled have proven to be a spectactular event appealing to a wide range and number of participants and spectators.

Craft Classification

Any craft equipped with propulsion machinery is classified as a motorboat.

Personal watercraft are classified as "Class A inboard boats" by the Coast Guard and therefore are subject to most of the same laws and requirements as more conventional larger craft. A decal, usually affixed somewhere in the stern area, lists exceptions the Coast Guard has granted for this type of water vehicle.

Information

In 1989, several states recognized the need to impose specific regulations governing operation of personal watercraft. Rules and regulations differ from state to state and are constantly being revised. Therefore, it would be an impossible task to present current legislation on personal watercraft activites. Before the printing ink for this publication was dry, additions, deletions, and revisions would already be in effect.

For current information, the reader may call 1-800-336-BOAT for the name and address of the local Boating Law Adminis-

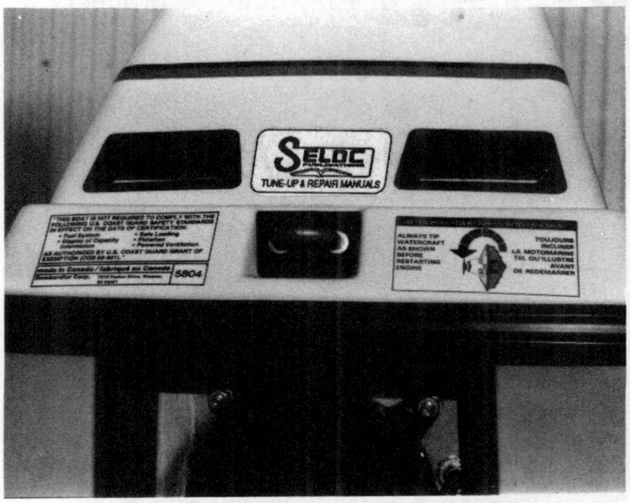

Several decals may be affixed to the stern of the hull to indicate compliance with or exceptions granted to normal Coast Guard regulations.

trator. The reader may also contact the PWIA Government Relations Representative, Washington, D.C.

Regulation Enforcement

The U.S. Coast Guard and state enforcement officers have the authority to stop any craft to check for compliance with federal or state law. They can order a craft to return to the closest dock and remain out of the water until the hazardous condition of the craft is corrected.

Examples are:

Inadequate number of Personal Flotation Devices.

Missing/Inoperative fire extinguisher.

Overloading.

Operation at night without proper lights.

Fuel leakage.

Fuel transportation (other than approved fuel tank).

Failure to meet ventilation requirements.

Failure to meet carburetor backfire flame arrestor requirements.

Excessive leakage or accumulation of water in the bilge.

The U.S. Coast Guard and state enforcement officers have the authority to stop any craft if they determine the craft is being operated in a negligent manner. Negligent operation is defined as "the failure to excercise the degree of care necessary to prevent the endangering of life, limb, or property of any person".

Examples are:

Operating while under the influence of drugs or alcohol.

Excessive speed in a congested area.

Excessive speed in stormy or foggy conditions.

Coming too close to another vessel.

Operating in a swimming area where bathers are present.

Towing water skiers where obstructions exist or where a fall might cause skier to be struck by another vessel.

Operating in the vicinity of a dam.

Cutting through a marine parade or regatta.

2-2 MINIMUM LEGAL REQUIREMENTS FOR EQUIPMENT ON BOARD CRAFT

All personal watercraft owners **MUST** provide the following equipment:

One approved Type I, II, III or IV PFD (personal flotation device) for each person on board or being towed on water skis, etc.

An efficient sound producing device.

A B-I approved hand portable fire extinquisher.

Visual distress signals for nighttime use, if operated on coastal waters.

Personal Flotation Devices (PFDs)

The Coast Guard requires an approved life-saving device be worn by each person on board. Devices approved are identified by a tag indicating Coast Guard approval. Such devices may be life preservers, or buoyant vests. Ring buoys, or buoyant cushions are not acceptable.

Life preservers have been classified by the Coast Guard into five type categories. All PFDs presently acceptable on recreational craft fall into one of these five designations. Only four of the five catagories are stipulated by legal requirements. Type V is not acceptable for use on a personal watercraft.

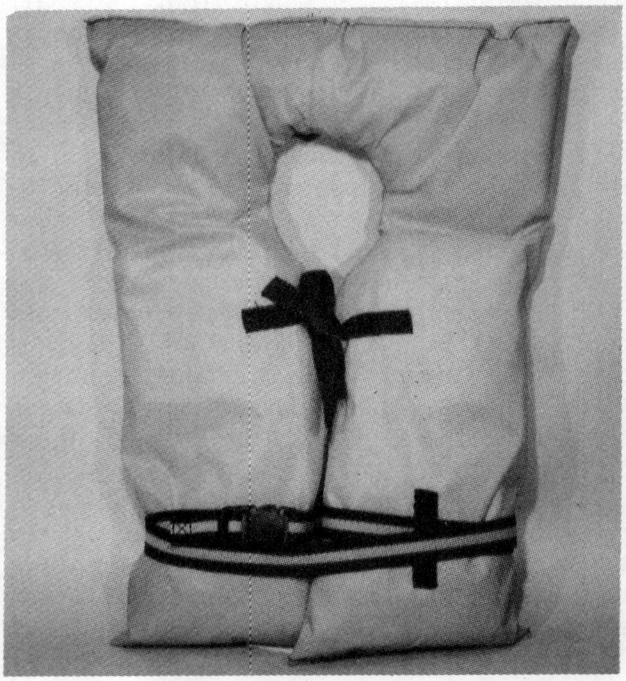

Type I PFD Coast Guard Approved life jacket. This type flotation device provides the greatest amount of buoyancy. ***NEVER*** *use them for cushions or other purposes.*

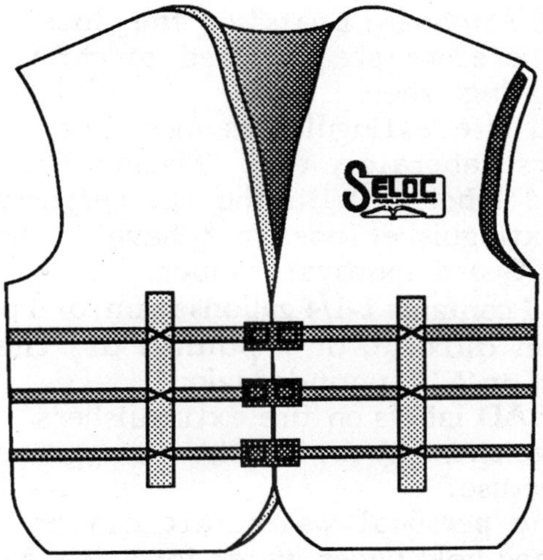

A Type III PFD requires the wearer to be active to remain upright in the water. This type device is comfortable and allows the wearer to actually swim while wearing it in the water.

A PFD **MUST** be U.S. Coast Guard approved, in good and serviceable condition, and of an appropriate size for the person intending to wear it.

It is an accepted fact that most boating people own life preservers, but too few actually wear them. There is little or no excuse for not wearing one because the modern comfortable designs available today do not subtract from an individual's boating pleasure.

PFDs may have a long serviceable life, if proper care is exercised and the device is not abused. Wipe them clean and dry before storing in a well ventilated area during the off-season. **NEVER** use a PFD as a fender. A crushed PFD loses its buoyancy and becomes useless.

Type I PFD has the greatest required buoyancy and is designed to turn most **UNCONSCIOUS** persons in the water from a face down position to a vertical or slightly backward position. The adult size device provides a minimum buoyancy of 22 pounds and the child size provides a minimum buoyancy of 11 pounds. The Type I PFD provides the greatest protection to its wearer and is most effective for all waters and conditions.

Type II PFD is designed to turn its wearer in a vertical or slightly backward position in the water. The turning action is not as pronounced as with a Type I. The device will not turn as many different type persons under the same conditions as the Type I. An adult size device provides a minimum buoyancy of 15½ pounds, the medium child size provides a minimum of 11 pounds, and the infant and small child sizes provide a minimum buoyancy of 7 pounds.

Type III PFD is designed to permit the wearer to place himself (herself) in a vertical or slightly backward position. The Type III device has the same buoyancy as the Type II PFD but it has little or no turning ability. Many of the Type III PFD are designed to be particularly useful when water skiing, sailing, hunting, fishing, or engaging in other water sports. Several of this type will also provide increased hypothermia protection.

Type IV PFD is designed to be thrown to a person in the water and grasped and held by the user until rescued. It is **NOT** designed to be worn. The most common Type IV PFD is a ring buoy or a buoyant cushion.

ONE LAST WORD

Common sense dictates the "cheap and cheerful" approach does not pay when applied to a life preserving device. Select and buy the very best. The life you save may be your own.

NOW, THESE WORDS

Due to the nature and maneuvering characteristics of personal watercraft, and from a "practical" standpoint, the following paragraph on bells, horns, and whistles hardly apply. The information is included to make the manual "complete" and because according to the U.S. Coast Guard they are required. Also, the authors feel the reader may "skipper" or "crew" on a larger craft and therefore, might be required to use the information presented here.

Bells, Horns, and Whistles

All craft must carry some means of producing an efficient sound signal and the sound must be audible for a specific distance and duration, depending on the size of the craft. For personal watercraft, an audible sound must carry for one half mile and be of four to six second duration.

Law enforcement craft are the **ONLY** boats allowed to use sirens.

Personal watercraft operators are required to sound fog and maneuvering signals when underway in fog or rain as a means of communication between craft which are not visible. This system was devised to alert nearby craft of the presence of others in the water and as a means of communicating maneuvering intentions. Signals indicating maneuvering tactics are short one second blasts.

One blast indicates the craft is turning to starboard.
Two blasts indicate the craft is turning to port.
Three blasts indicate the craft is in reverse.
Five or more blasts indicate - stay away - danger.

A prolonged blast lasting 4 to 6 seconds duration every two minutes indicates a straight line course in poor visibility.

These signals should be memorized by every personal watercraft operator, not necessarily because the operator would give them, but in the event of a potential head on collision -- with say a large powerboat -- the skipper may signal his intention to turn either to port or starboard using this system of communication.

Personal watercraft are required to carry a Type B-1 fire extinguisher aboard. The compartment is in front of the handlebars where it is readily available.

Fire Extinguishers

Personal watercraft are classified as "Class A inboard boats" by the Coast Guard and therefore are required to carry a B-I fire extinguisher.

All fire extinguishers must bear Underwriters Laboratory (UL) "Marine Type" approved labels. With the UL certification, the extinguisher does not have to have a Coast Guard approval number.

B-I contains 1-1/4 gallons foam, or 4 pounds carbon dioxide, or 2 pounds dry chemical agent, or 2-1/2 pounds Halon.

READ labels on fire extinguishers. If the extinguisher is U.L. listed, it is approved for marine use.

Any personal watercraft can be rolled over and held under water for a few seconds to extinguish the fire. Turning the craft over in the water will prevent oxygen from entering the engine compartment. Without oxygen, the fire will soon exhaust the remaining oxygen in the compartment and then go out. **DO NOT** open the engine compartment until the smoke has completely stopped, then open with great caution and have a fire extinguisher at hand.

Visual Distress Signals

Coast Guard Regulations require personal watercraft to carry night visual distress signals when used on coastal waters - even though it is illegal to operate a personal watercraft at night.

Coastal waters include: the Great Lakes, the territorial seas and those waters directly connected to the Great Lakes and the territorial seas, up to a point where the waters are less than two miles wide.

These signals assist rescuers in locating and aiding people in distress. This information is useful to every personal watercraft owner, to know how to respond to others in distress.

For night use: Pyrotechnic visual distress signaling devices (red flares, hand held or aerial) **MUST** be Coast Guard Approved, in serviceable condition and stowed to be readily accessible. If they are marked with a date showing the serviceable life, this date must not have passed.

Coast Guard Approved pyrotechnic devices carry an expiration date. This date can **NOT** exceed 42 months from the date of manufacture and at such time the device can no longer be counted toward the minimum requirements.

The only non-pyrotechnic visual distress signaling device approved by the coast guard for use at night is an electric distress light -- not a flashlight but an approved electric distress light which **MUST** automatically flash the international **SOS** distress signal (. . . - - - - . . .)four to six times each minute.

2-3 MINIMUM LEGAL REQUIREMENTS FOR REGISTRATION OF CRAFT

Federal regulations require all personal watercraft to be issued a registration number and have the number displayed in a specific manner on the craft.

The "Certificate of Number" is issued by the state and must be carried on board at all times. The Hull Identification Number (HIN) is necessary to register the craft with the state.

The registration number must be displayed as follows:

The figures must be read from left to right.

The figures must be displayed on the forward half of each side of the bow of the craft.

The figures must be in bold, block letters of good proportion and must not be less than 3 inches high.

The figures must be of contrasting color to the craft hull, or background.

The figures must be as high above the waterline as practical.

No figures other than those assigned to the craft can be displayed on the forward half of the craft.

Letters must be separated from numbers by spaces or hyphens.

Validation decals (if required by state) must be displayed within 6 inches of the registration number.

Elaborate color schemes and custom detailing are very much a part of the fun of owning a personal watercraft. **HOWEVER,** the paint design must not obstruct the registration numbers or the validation decal. Like an automobile without a license plate, sooner or later a craft with no visible registration numbers will attract the attention of the law.

Hull Identification Number

Federal regulations require all personal watercraft to have a Hull Identification Number (HIN) permanently attached to the craft in two separate locations: on the starboard side of the transom, above the waterline. The HIN must also be displayed in an unexposed location, which is usually left to the discretion of the manufacturer.

2-4 SAFETY PRACTICES

Riding Techniques

Of all watercraft, large and small, the personal watercraft are the most responsive to steering and maneuverability. Therefore, a personal watercraft operator should take it upon herself/himself to stay clear of other boaters and especially swimmers.

A beginner should operate the craft at a low speed until the operator feels comfortable and is aware of its limitations and responses to steering.

While riding the craft, keep feet, hands, hair, and clothing away from the pump intake.

Full Throttle Operation

Prolonged operation at full throttle will raise the cylinder head temperature by about 100° above normal. Some units are equipped with a thermo-switch engine temperature sensor. This sensor is connected directly to the CDI unit and grounded to the engine kill switch. If the engine temperature exceeds a maximum value specified by

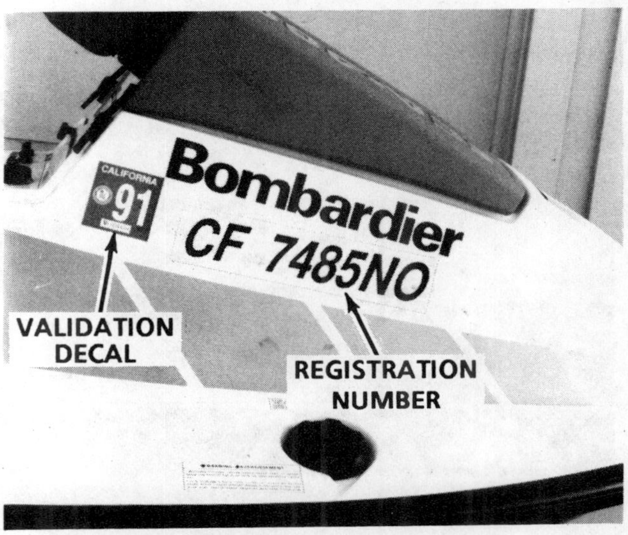

The validation decal, required by most states, and the registration number, required by all states, must be prominently displayed -- port and starboard -- on the forward portion of the hull.

the manufacturer, the ignition will be grounded and the engine shut down to prevent damage through overheating.

Jumping Waves

In the very early days of personal watercraft, jumping could cause extensive internal engine damage. When the craft left the water, unchecked engine rpm rapidly increased to as high as 12,000 rpm. As soon as the craft re-entered the water, engine rpm instantly decreased to about 6,000 rpm. This sudden increase and decrease in engine speed dramatically shortened engine life. In those days, repeated jumps were known to cause bent crankshafts, sheared flywheel keys, and broken connecting rods.

Kawasaki engineers developed an rpm limiter system to prevent high engine rpm and an almost "runaway" condition with the design of a rpm limiter. This limiter restricts engine speed to about 8,000 rpm. Thus, wave jumping, using a ramp, and a host of other almost unbelievable manuevers could be performed without undue stress on the engine. Today, with an rpm limiter system installed, a bent crankshaft or broken connecting rod is almost non-existent.

Alcohol and Substance Use

Operating a personal watercraft while under the influence of either alcohol or drugs is both illegal and life threatening. This fact cannot be stated too strongly. Such substances affect vision and slow reaction times. The danger is quadrupled in a craft on the water compared to on the shore.

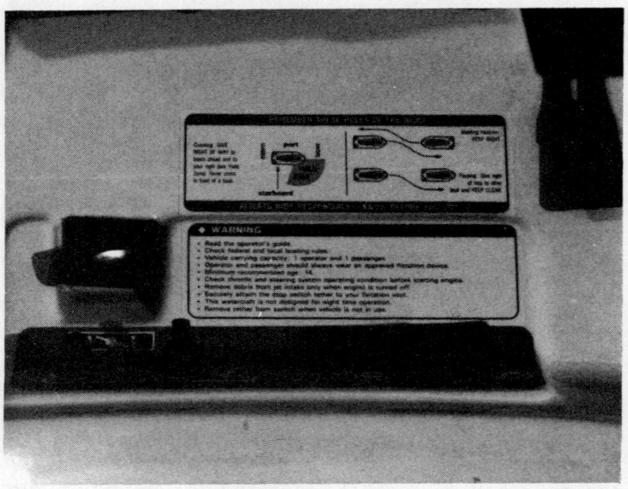

Decals located under the handlebars to remind the operator of the most basic "Rules of the Road".

Age Restrictions

Children under 14 years of age should not operate a watercraft unless accompanied on a two passenger craft by an adult experienced operator. Specific minimum age limits have been imposed by many states.

Speed Restrictions

Where no speed limits are posted, commonsense dictates the craft should be operated so it will pose no danger to other water users or the operator. Many states now have enforced speed limits for certain areas and conditions.

When entering "no wake" or 5 mph zones, such as fishing areas, swimming areas or marinas, the operator must slow to headway speed, the slowest speed at which appropriate steering capability can be maintained.

Individuals responsible for law enforcement on the water are keeping an ever vigilant eye on personal watercraft in many areas. Horseplay such as "wetting down" or "buzzing" other water users may result in a "ticket" for negligent operation.

Operation at Night

Personal watercraft are not equipped with lights and therefore, it is illegal to operate such a craft on the water after sunset.

The law requires any craft operated between sunset and sunrise to display certain colored lights mounted in specific locations, depending on the size and class of the craft. These lights will indicate the position and intended direction of the craft to operators of other watercraft in the area.

Personal watercraft are extremely difficult to see at night because of their small size. Even with lights, because of their almost unbelievable maneuverability, confusing signals would be sent to other traffic.

Shallow Water Operation

Under normal conditions, personal watercraft may be operated in very shallow waters, and it is possible to maneuver the craft directly onto a slopping beach.

Caution must be exercised when moving the craft off a beach to avoid sucking in mud, sand, pebbles, or rocks through the intake. Such ingestion could cause damage to the impeller, the vanes in the pump bowl, or clog engine cooling passages.

WARNING

Expelled sand and small pebbles may cause INJURY to persons in the area.

A grated section set into the hull bottom helps prevent the ingestion of small rocks and pebbles.

Carry about a foot of stiff wire aboard for use in probing into the impeller area to dislodge debris where fingers cannot reach.

Engine Compartment Ventilation

All motorboats (personal watercraft included) built after April 25, 1940 and before August 1, 1980, powered by a gasoline engine or by fuels having a flashpoint of 110° F or less **MUST** have the following, which is quoted from a recent Coast Guard publication:

> At least two ventilation ducts fitted with cowls or their equivalent for the purpose of properly and efficiently ventilating the bilges of every engine and fuel tank compartment. Each duct must be at least two inches in internal diameter.
>
> There shall be at least one exhaust duct installed so as to extend from the lower portion of the bilge to cowls in the open air, and at least one intake duct installed so as to extend to a point at least midway to the bilge or at least below the level of the carburetor air intake.

Flame Arrestors

A flame arrestor, as the name suggests, safeguards against flame caused by engine backfire.

A gasoline engine installed in a motorboat or motor vessel after April 25, 1940, except outboard motors, must have a Coast Guard Approved flame arrestor fitted to the carburetor. This requirement applies to personal watercraft because the engine is enclosed.

Fuel System

All parts of the fuel system should be selected and installed to provide maximum service and protection against leakage.

The capacity of the fuel filter must be large enough to handle the demands of the engine as specified by the engine manufacturer.

All fittings and outlets must come out the top of the tank.

In order to obtain maximum circulation of air around fuel tanks, the tank should not come in contact with the hull except through the necessary supports. The supporting surfaces and hold-downs must fasten the tank firmly and they should be insulated from the tank surfaces. This insulation material should be non-abrasive and non-absorbent material.

Taking On Fuel

The fuel tank should be kept almost full to allow for expansion, but still prevent water from entering the system through condensation caused by temperature changes. Water droplets forming are one of the greatest enemies of the fuel system. By keeping the tank almost full, the air space in the tank is kept to an absolute minimum and there is less room for moisture to form. It is a good practice not to store fuel in the tank over an extended period, say for six months. Today, fuels contain ingredients that change into gums when stored for any length of time. These gums and varnish products will cause carburetor problems and poor spark plug performance. An additive (Sta-Bil) is available and can be used to prevent gums and varnish from forming.

FLAME ARRESTOR

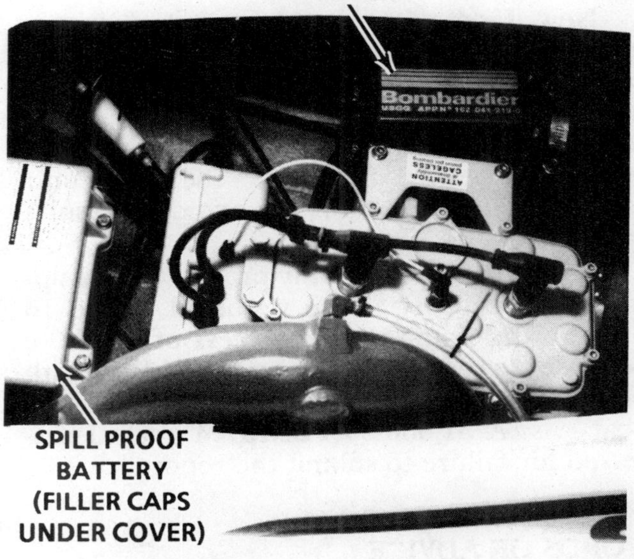

SPILL PROOF BATTERY (FILLER CAPS UNDER COVER)

Carburetors on personal watercraft are required to be fitted with a flame arrestor because the engine compartment is an enclosed area. Other safety features include spill-proof battery filler caps and one-way check valves on the fuel and oil tanks, to prevent spillage under all conditions.

Excessive Noise

Every manufacturer designs a stock exhaust system to keep noise to a minimum. A poorly maintained or modified exhaust system will produce excessive and disturbing noise. This is a form of air polution and is illegal in most states.

Automotive Replacement Parts

When replacing fuel, electrical, and other parts, check to be sure they are marine type and Coast Guard Approved. Automotive parts are not made to the high standards of marine parts. The carburetors must **NOT** leak fuel; electrical equipment **MUST** be able to operate in a gas fume enclosed area without exploding; etc. Automotive parts could cause a fire endangering the operator, passengers, and the craft. The part may look the same and even have a similar number, but if it is not **MARINE** it is not safe to use and will not pass Coast Guard inspection.

Overloading

Each personal watercraft is designed to carry a specific number of persons. Attempting to board more passangers than the craft is designed to carry, will upset weight distribution causing a marked decrease in performance and possible loss of control.

2-5 BOATING ACCIDENT REPORTS

New federal and state regulations require an accident report to be filed with the nearest state boating authority within 48 hours if a person is lost, disappears, or is injured to the degree of needing medical treatment beyond first aid. The time limit is reduced to 24 hours if the accident results in a loss of life.

Accidents involving only property or equipment damage **MUST** be reported within 10 days, if the damage is in excess of $500.00. Some states require reporting of accidents with property damage less than $500.00 or a total boat loss. A $1,000.00 **PENALTY** may be assessed for failure to submit the report.

WORDS OF ADVICE

Take time to make a copy of the report to keep for your records or for the insurance company. Once the report is filed, the Coast Guard will not give out a copy, even to the person who filed the report.

The report must give details of the accident and include:

1- The date, time, and exact location of the occurrence.

2- The name of each person who died, was lost, or injured.

3- The number and name of the vessel.

4- The names and addresses of the owner and operator.

If the operator cannot file the report for any reason, each person on board **MUST** notify the authorities, or determine that the report has been filed.

In nautical terms, the front of the craft is the **bow;** the rear is the **stern;** the right side, when facing forward, is the **starboard** side; and the left side is the **port** side. One easy way to remember this basic fundamental is to consider the words "port" and "left" both have four letters and go together.

All directional references in this manual use this terminology. Therefore, the direction from which an item is viewed is of no consequence, because **starboard** and **port** **NEVER** change no matter where the individual is located, or standing, even on his or her head.

2-6 SECURITY

As mentioned in the opening paragraphs of this chapter, personal watercrafting is one of the fastest growing leisure activites in the world. Like most personal possessions, personal watercraft have value and valuables are subject to be lost to unscrupulous characters -- thieves. The best advice for a personal watercraft owner is "lock it or lose it".

Do not leave the craft unattended on the beach or dock. **ALWAYS** remove the kill switch tether from the switch and of course, the ignition key, if the unit is equipped with these features. Certain aftermarket security devices are available for installation to prevent theft of personal watercraft. These include: A "secret" ignition grounding switch, which grounds one side of the ignition coil, and a hidden fuel shut-off valve.

Most owners transport their craft on trailers or in the bed of a pickup truck. If loaded on a trailer, chain and lock the craft to the trailer. Invest in a hitch lock so the trailer cannot be detached from the towing vehicle. If the craft is loaded on a pickup truck, chain and lock it to the bed, or to a large and heavy object.

3 TUNING

3-1 INTRODUCTION

The efficiency, reliability, fuel economy and enjoyment available from the engine performance are all directly dependent on having the engine properly tuned. The importance of performing service work in the sequence detailed in this chapter cannot be overemphasized. Before making any adjustments, check the specifications in the Appendix. **NEVER** rely on memory when making critical adjustments.

Before beginning to tune any engine, check to be sure the engine has satisfactory compression. An engine with worn or broken piston rings, burned pistons, or scored cylinder walls, cannot be made to perform properly no matter how much time and expense is spent on the tune-up. Poor compression must be corrected or the tune-up will not give the desired results.

A practical maintenance program that is followed throughout the year, is one of the best methods of ensuring the engine will give satisfactory performance at any time.

The extent of the engine tune-up is usually dependent on the time lapse since the last service. A complete tune-up of the entire engine would entail almost all of the work outlined in this manual. A logical sequence of steps will be presented in general terms. If additional information or detailed service work is required, the chapter containing the instructions will be referenced.

Each year higher compression ratios are built into modern marine engines and the electrical systems become more complex, especially with electronic (capacitor discharge) units. Therefore, the need for reliable, authoritative, and detailed instructions becomes more critical. The information in this chapter and the referenced chapters fulfill that requirement.

3-2 TUNE-UP SEQUENCE

During a major tune-up, a definite sequence of service work should be followed to return the engine to the maximum performance desired. This type work should not be confused with attempting to locate

A clean, properly tuned, and adequately maintained power unit can multiply by many fold the enjoyment derived from owning and operating a personal watercraft.

problem areas of "why" the engine is not performing satisfactorily. Such work is classified as "troubleshooting". In many cases, these two areas will overlap, because many times a minor or major tune-up will correct the malfunction and return the system to normal operation.

The following list is a suggested sequence of tasks to perform during the tune-up service work. The tasks are merely listed here. In most cases, procedures are given in subsequent sections of this chapter. For more detailed instructions, see the referenced chapter, for the unit being serviced.

1- Perform a compression check of each cylinder. See Chapter 7 -- **IGNITION.**

2- Inspect the spark plugs to determine their condition. Test for adequate spark at the plug. See Chapter 7 -- **IGNITION.**

3- Start the engine in a body of water or connect a "Flushing Kit" with a garden hose and check the water flow through the engine. See Chapter 8 -- **ENGINE.**

4- Check the carburetor adjustments and the need for a carburetor overhaul. See Chapter 6 -- **FUEL**

5- Check fuel pump for adequate performance and delivery. See Chapter 6 -- **FUEL.**

6- Make a general inspection of the ignition system. See Chapter. 7 -- **IGNITION.**

7- Test the cranking motor and the solenoid. See Chapter 9 -- **ELECTRICAL**

8- Check the internal wiring.

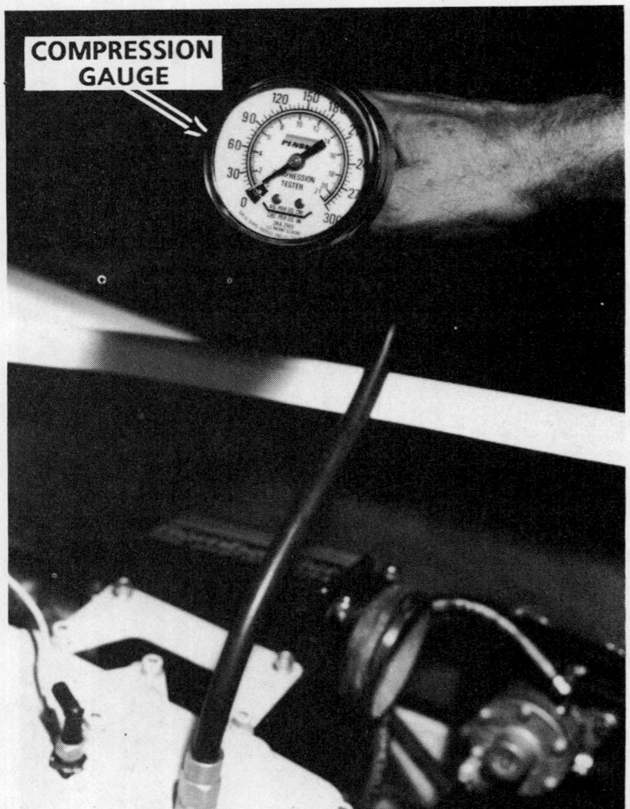

Performing a compression test on a two cylinder engine. Both spark plug leads have been grounded to prevent placing an unnecessary extra load on the ignition coil.

3-3 COMPRESSION CHECK

A compression check is extremely important, because an engine with low or uneven compression between cylinders **CANNOT** be tuned to operate satisfactorily.

Therefore, it is essential that any compression problem be corrected before proceeding with the tune-up procedure.

If the engine shows any indication of overheating, such as discolored or scorched paint, inspect the piston skirts visually thru the exhaust ports, using a mirror and flashlight, for possible scoring. It is possible for a cylinder with satisfactory compression to be scored slightly.

Checking Compression

Remove the spark plug wires. **ALWAYS** grasp the molded cap and pull it loose with a twisting motion to prevent damage to the connection. Remove both spark plugs and remember from which cylinder they came for evaluation later. Ground both spark plug leads to the engine to render the ignition system inoperative while performing the compression check.

Insert a compression gauge into the forward spark plug opening. Crank the engine with the cranking motor, thru at least four complete revolutions of the crankshaft, with the throttle at the wide-open position, to obtain the highest possible reading. Record the highest reading. Repeat the test and record the compression for the other cylinder. The manufacturer specifies the compression pressure for a new engine should be 142.2 psi (981 kpa). A variation between cylinders is far more important than the actual readings.

A variation of more than 15 psi between the two cylinders indicates the lower compression cylinder is defective and requires attention. The problem may be worn, broken, or sticking piston rings, scored pistons, or worn cylinders.

Use of an engine cleaner will help to free stuck rings and to dissolve accumulated carbon. Follow the directions on the can of cleaner.

3-4 SPARK PLUG INSPECTION

Inspect both spark plugs for badly worn electrodes, glazed, broken, blistered, or lead fouled insulators. Replace both the plugs, if either shows signs of excessive wear.

Make an evaluation of the cylinder performance by comparing the spark plug condition with those shown in Chapter 7 -- **IGNITION**. Check each spark plug to be sure they are both of the same manufacturer and have the same heat range rating.

Inspect the threads in the spark plug opening in the head and clean the threads before installing the plug.

When purchasing new spark plugs, **ALWAYS** ask the dealer if there has been a spark plug change for the engine being serviced. Set the spark plug gap to 0.024" (0.060mm) for all series engines covered in this manual.

Crank the engine through several revolutions of the crankshaft to blow out any material which might have become dislodged during cleaning.

Install the spark plugs and tighten them to a torque value of 14 ft lb (20Nm). **ALWAYS** use a new gasket and wipe the seats in the head clean. The gasket must be fully compressed on clean seats to complete the heat transfer process and to provide a gas tight seal in the cylinder. If the torque value is too high, the heat will dissipate too rapidly. Conversely, if the torque value is too low, the dissipation of heat will be too slow.

3-5 ELECTRICAL POWER SUPPLY

BATTERY SERVICE

Remove the battery from the engine compartment for service. Inspect and service the battery, cables and connections. Check for signs of corrosion. Inspect the battery case for cracks or bulges, dirt, acid, and electrolyte leakage.

Check to be sure the breather pipe is attached to the battery and not pinched by any item in the engine compartment.

Clean the top of the battery. The top of a 12-volt battery should be kept especially clean of acid film and dirt, because of the high voltage between the battery terminals and to prevent contact between the acid and the plastic battery case. For best results, first wash the battery with diluted ammonia or baking soda solution to neutralize any acid present. Flush the solution off the battery with clean water. Keep the vent plugs tight to prevent the neutralizing solution or water from entering the cells.

Filling the Battery

Remove each filler cap using a pair of pliers. Fill each cell to the proper level with distilled water. The level of electrolyte should be between the upper and lower level marks during operation. When filling the battery, fill to the upper line and allow the battery to stand for twenty minutes. Check the level again and replenish as necessary.

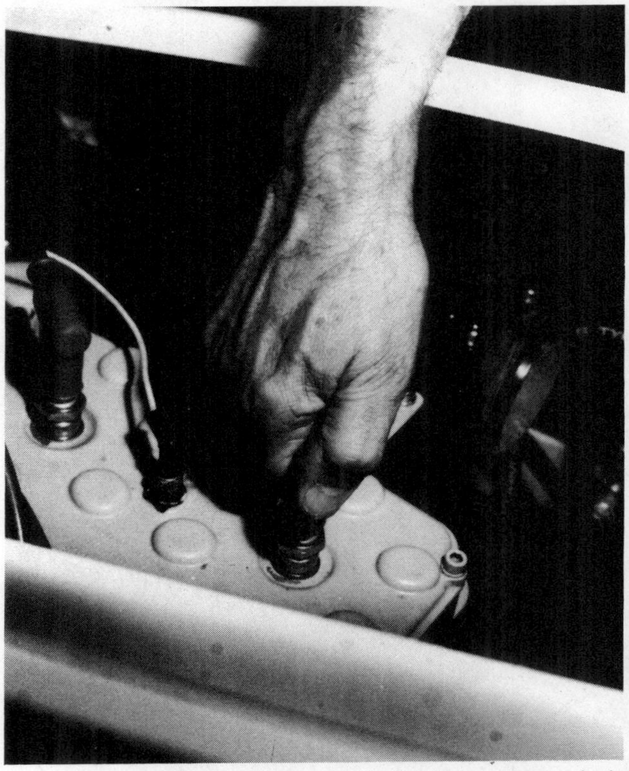

After a compression check or inspection of the spark plugs, install the high tension leads firmly over the spark plugs.

Use a temperature corrected hydrometer to test the specific gravity of the electrolyte. At 20°C (70°F) the hydrometer reading should be 1.26 on the scale. If it is necessary to charge the battery, leave the filler caps lightly resting on the cell openings to allow the gases to escape. The charging current should not exceed 1.9 amps for 10 hours.

After charging, push each filler cap into place and wipe the top of the battery clean before installation into the watercraft.

Clean the battery posts and battery cable ends with a wire brush to ensure good, clean connections. Clean the top surface of the battery and identify the "POS" or ("+") and "NEG" or ("-") embossed symbols. Correctly connect the battery cables to the battery observing polarity. If the cables are connected backwards, the ignition system **WILL BE DESTROYED** the first time the engine is started.

Check to be sure the battery is fastened securely in position. The battery box and cover screws **MUST** be tightened securely to prevent any movement of the battery in the holder under the most violent maneuvers of the watercraft.

*Due to the violent maneuvers experienced by a personal watercraft -- more so than any other type craft -- the battery **MUST** be extremely well-secured with **NO** movement of any kind possible.*

If the battery posts or cable terminals are corroded, the cables should be cleaned separately with a baking soda solution and a wire brush. Apply a thin coating of Multipurpose Lubricant to the posts and cable clamps before making the connections. The lubricant will help prevent corrosion.

Jumper Cables

If a booster battery is used to start the engine, the jumper cables must be connected correctly and in the proper sequence to prevent damage to either battery, or to the rectifier diodes in the ignition system.

ALWAYS connect a cable from the positive terminal of the "dead" battery to the positive terminal of the good battery **FIRST**. Next, connect one end of the other cable to the negative terminal of the good battery. Finally, connect the other end of the cable to the far side of the **ENGINE** for a good ground. By making the ground connection on the engine, any spark created will not be near the battery. An arc near the battery could result in an explosion, destroying the battery and causing serious personal injury.

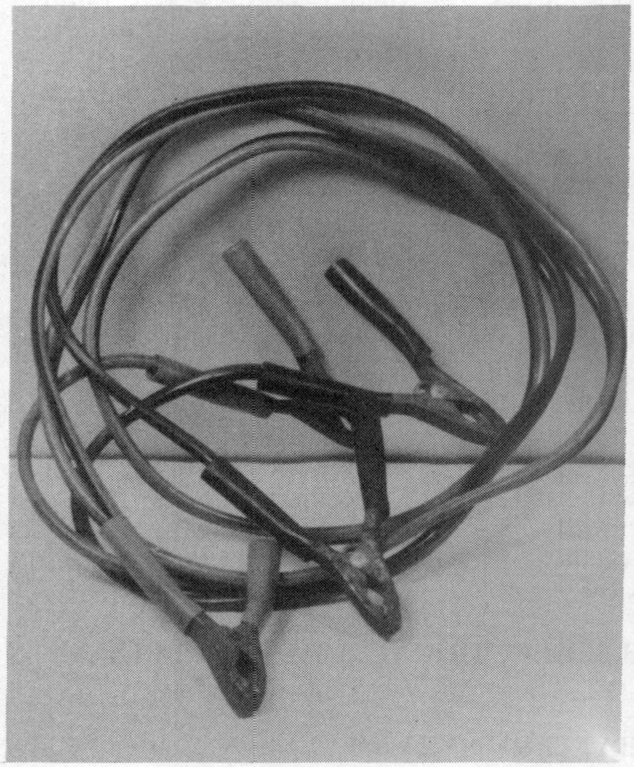

*Common set of jumper cables for use with a second battery to crank and start the engine. **EXTREME** care should be exercised when using a second battery, as explained in the text.*

If a trickle charger is used on a "dead" battery installed in the craft, one battery lead **MUST** be disconnected prior to connecting the charger, to prevent destroying the diodes in the rectifier.

NEVER use a trickle charger as a booster to start the engine, because the diodes in the rectifier will be **DAMAGED.**

3-6 FUEL SYSTEM

Fuel and Fuel Tanks

Take time to check the fuel tank and all fuel lines, fittings, couplings, valves, flexible tank fill, and vent. Turn on the fuel supply valve. If gas was not drained at the end of the previous season, make a careful inspection for gum formation. When gasoline is allowed to stand for long periods of time, particularly in the presence of copper, gummy deposits form. This gum can clog the filters, lines, and passageways in the carburetor.

Commercial additives, such as Sta-Bil, may be used to keep the fuel in the tank fresh. Under favorable conditions, such additives will prevent the fuel from "souring" for up to twelve months.

If the condition of the fuel is in doubt, drain, clean, and fill the tank with fresh fuel, with the proper fuel/oil mixture recommended by the manufacturer.

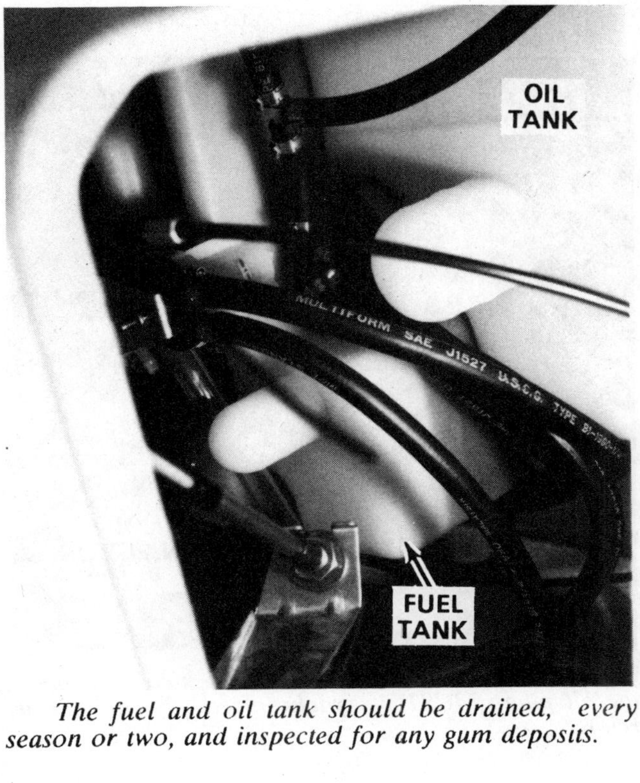

The fuel and oil tank should be drained, every season or two, and inspected for any gum deposits.

CARBURETOR ADJUSTMENTS

Five fuel adjustments may be made on the Mikuni carburetor, for both single and twin carburetor installations, as follows:

a- Main Jet Selection.
b- Throttle Cable Adjustment -- **ALWAYS** followed by oil injection synchronization.
c- Low Speed Mixture Adjustment.
d- Idle Speed Adjustment -- **ALWAYS** followed by oil injection pump synchronization.
e- Choke Cable Adjustment.

Each of the listed adjustments are explained in detail in the following paragraphs.

Main Jet Selection

Several main jet sizes are available to compensate for variations in altitude and temperature. The out-the-door stock main jet is intended for operation at sea level. The jet size is stamped on one face. The following main jet sizes are available: 70, 75, 80, 85, 90, 95, and 100. When operating the craft at higher altitudes, try changing the main jet in increments of 0.005" or 0.010", until an acceptable level of performance is achieved.

Operating the engine on an excessively lean mixture may cause overheating, preignition, and a decrease in performance. Any time the main jet size is altered, keep a close eye on the condition of the spark plug tip. If the tip begins to show signs of overheating, change to a colder heat range. Check with the local dealer for the correct number spark plug. Some spark plug manufacturers use a higher number for colder plug, and others use a lower number for colder plug.

The main jet is located externally. Therefore, the carburetor need not be removed to change a main jet. Refer to Page 6-15 for location of the main jet. The jet is threaded into the fuel chamber, on the same side as the fuel supply hose. When removing the main jet, look for the **O**-ring, which will remain in the jet bore. When installing a new main jet, install the **O**-ring onto the jet first, and then install the jet. **DO NOT** overtighten the jet. Tighten it just enough to prevent fuel from escaping.

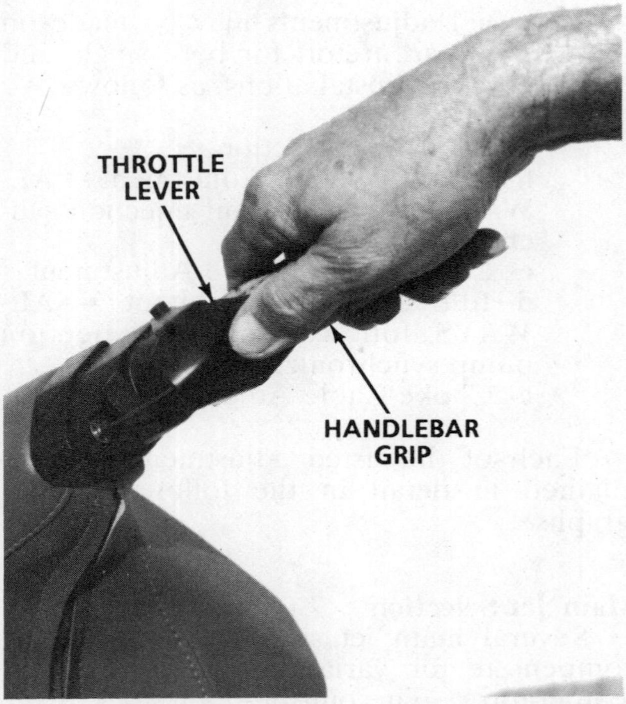

The throttle valve is wide open when the throttle lever is fully applied -- actually making contact with the handlebar grip.

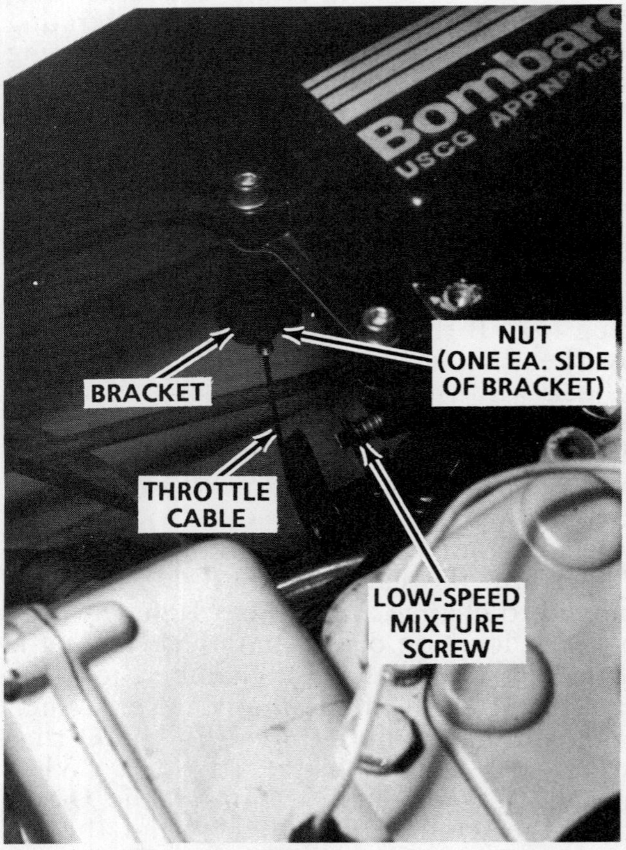

Location of the low-speed mixture screw and throttle cable adjustment nuts.

If working on a twin carburetor installation -- both jets **MUST** be the same size.

Throttle Cable Adjustment

When no throttle is applied, the idle speed screw must contact the stopper on the carburetor. When the throttle lever is fully applied -- actually making contact with the handlebar grip -- the stop lever on the throttle linkage must contact the other side of the stopper on the carburetor. If these two conditions are not met, an adjustment on the throttle cable is necessary at the cable support bracket. Adjust the throttle cable length by loosening or tightening the two nuts -- one on each side of the support bracket.

After a throttle cable adjustment is made, **ALWAYS** perform a check to ensure the oil injection pump is synchronized with the carburetor. If the oil injection pump is not properly synchronized with the carburetor, serious damage may be caused to the engine due to lack of adequate lubrication.

Low Speed Mixture Screw

The low speed mixture screw must **NEVER** be used to alter engine rpm. Such action could result in severe engine damage. This screw is used to smooth engine operation at a set idle speed. The idle speed is changed by adjustment of the idle speed screw, covered in the following paragraph. If working on a twin carburetor in-

*After a throttle cable adjustment has been made, the oil pump **MUST** be synchronized with the carburetor.*

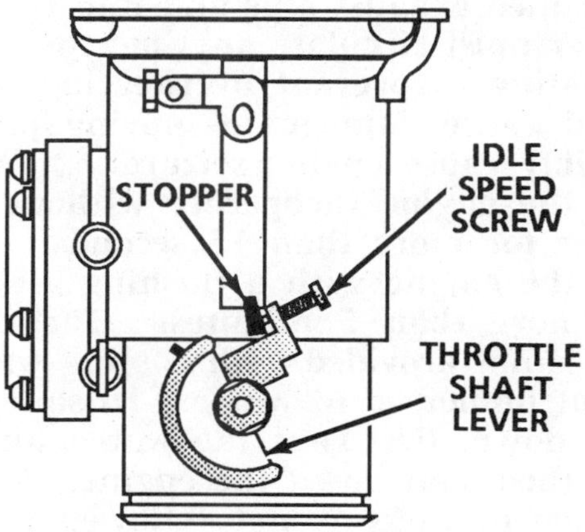

NO THROTTLE APPLIED

Line drawing to indicate the throttle cable adjustment with the throttle closed -- released.

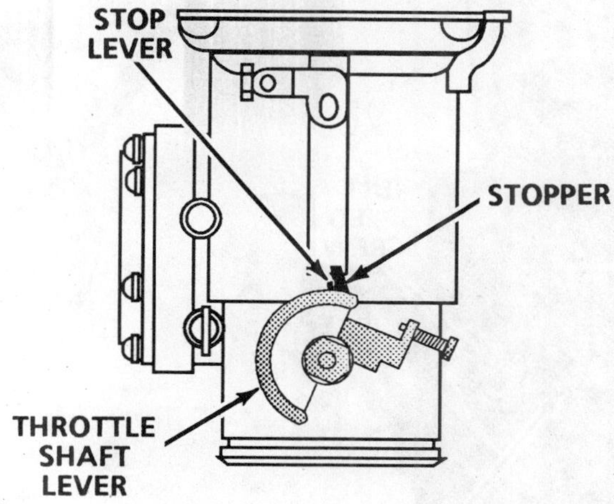

FULL THROTTLE (WOT)

Line drawing to indicate the throttle cable adjustment with the throttle open -- at the WOT position.

stallation, perform the required work on each carburetor in turn.

Now, with the engine shut down, tighten the low speed mixture screw, identified in the accompanying illustration, until the screw is **LIGHTLY** seated, and then back it out 1-1/2 to 2 full turns as a preliminary setting.

Start the engine and allow it to warm to operating temperature.

CAUTION

The following procedure MUST be followed whenever the engine is started with a hose attached for cooling the engine. Start the engine, and then IMMEDIATELY turn the water on. The engine MUST be started first to prevent water from backing up into the tuned exhaust pipe and into the engine. Water must circulate anytime the engine is operating to prevent overheating -- which could cause damage to moving parts and possibly cause engine seizure. NEVER allow the engine to operate without cooling water for more than 15 seconds. NEVER run the engine with a flushing attachment for more than 5 minutes. The driveline seal is not provided with cooling water if it is not immersed in water. To shut the engine down, first shut the water supply off, and then shut down the engine. If this sequence of tasks is not followed as stated, the engine could be severely damaged.

REMEMBER

Start the engine.
Turn the water on.
Turn off the water.
Shut down the engine.

To lean the mixture -- rotate the low speed mixture screw **CLOCKWISE.**

To enrichen the mixture -- rotate the low speed mixture screw **COUNTERCLOCKWISE.**

Rotate the low speed mixture screw in the direction which will result in the smoothest engine operation.

Idle Speed Adjustment

The idle speed adjustment screw is located on the throttle shaft linkage. If working on a twin carburetor installation, only the aft carburetor is equipped with an idle speed screw. A link rod connects the two throttle shafts together. Therefore, when an adjustment is made at the idle speed screw on the aft carburetor, the adjustment holds good for both carburetors.

When no throttle is applied, the tip of the screw rests against a stopper cast into the carburetor. To set the idle speed, obtain an induction type tachometer.

To increase engine rpm -- rotate the idle speed screw **CLOCKWISE.**

To decrease engine rpm -- rotate the idle speed screw **COUNTERCLOCKWISE.**

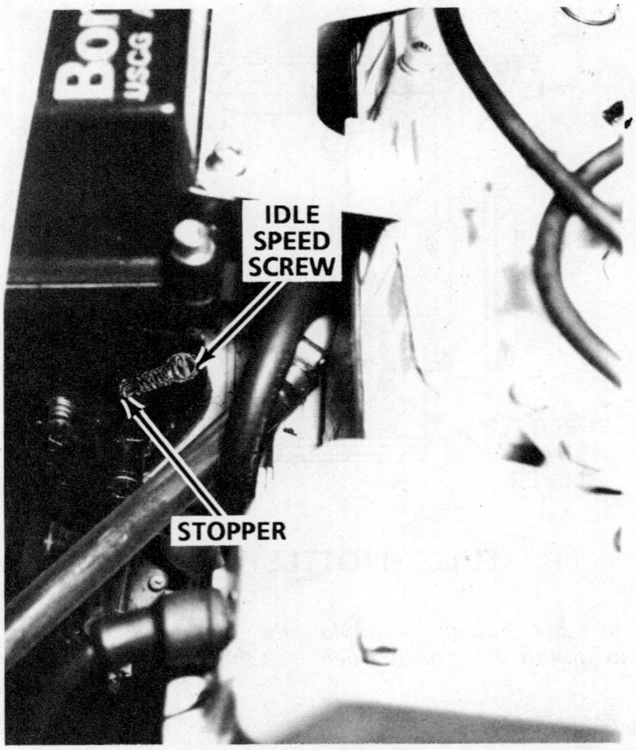

Location of the idle speed screw used to make the idle speed adjustment on a Mikuni carburetor.

Start the engine and allow it to warm to operating temperature.

CAUTION
The following procedure MUST be followed whenever the engine is started with a hose attached for cooling the engine. Start the engine, and then IMMEDIATELY turn the water on. The engine MUST be started first to prevent water from backing up into the tuned exhaust pipe and into the engine. Water must circulate anytime the engine is operating to prevent overheating -- which could cause damage to moving parts and possibly cause engine seizure. NEVER allow the engine to operate without cooling water for more than 15 seconds. NEVER run the engine with a flushing attachment for more than 5 minutes. The driveline seal is not provided with cooling water if it is not immersed in water. To shut the engine down, first shut the water supply off, and then shut down the engine. If this sequence of tasks is not followed as stated, the engine could be severely damaged.

Rotate the idle speed screw until the engine idles at 1500 rpm -- if using a flush attachment, or 1200 rpm, if the craft is in the water.

After a throttle cable adjustment is made, **ALWAYS** perform a check to ensure the oil injection pump is synchronized with the carburetor. If the oil injection pump is not properly synchronized with the carburetor, serious damage may be caused to the engine due to lack of adequate lubrication. See Section 6-41 for oil pump synchronization.

Choke Cable Adjustment
Push the choke knob inward. Observe the choke pivot arm on the side of the carburetor. The arm should be in the position shown in the accompanying illustration,

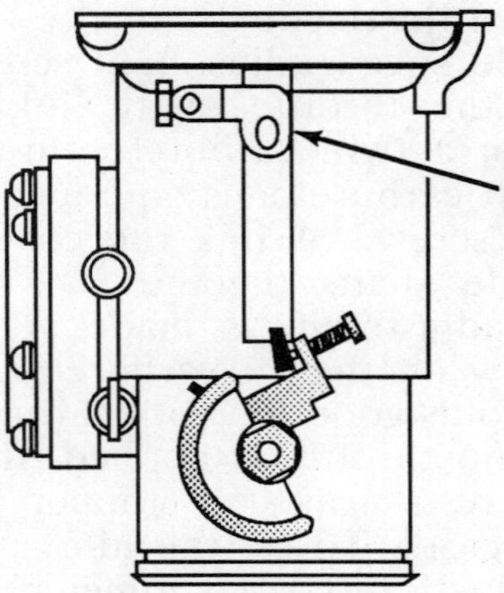

CHOKE IN OPEN POSITION

*Line drawing to indicate the choke cable adjustment with the choke in the fully **OPEN** position.*

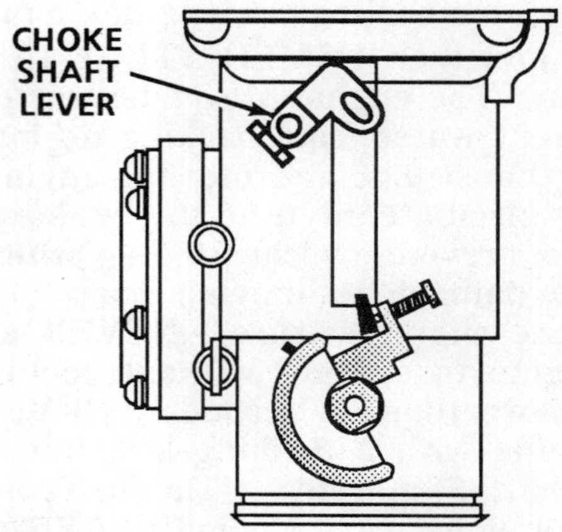

CHOKE IN CLOSED POSITION

*Line drawing to indicate the choke cable adjustment, with the choke in the **CLOSED** position.*

with minimum slack in the choke cable. If the choke needs adjustment: Loosen the small screw securing the choke cable end to the barrel on the choke lever. Push the choke knob inward until it rests against the control panel. Set the choke pivot arm in the fully open position, and then tighten the screw securing the cable to the barrel.

Choke Cable -- Excessive Slack

If there is too much slack in the cable, loosen the top nut around the cable at the cable support bracket and tighten the bottom nut until only a small amount of slack remains.

If the cable is tight -- with no slack -- loosen the bottom nut around the cable at the cable support bracket and tighten the top nut until there is just a small amount of slack in the cable.

If the cable has stretched, or has become kinked or frayed, or tends to stick, the entire cable should be replaced.

3-7 FUEL PUMPS

Many times, a defective fuel pump diaphragm is mistakenly diagnosed as a problem in the ignition system. The most common problem is a tiny pin hole in the diaphragm. Such a small hole will permit gas to enter the crankcase and wet foul both spark plugs at idle speed. During high speed operation, gas quantity is limited, the plugs are not fouled and will, therefore, fire in a satisfactory manner.

Location of the remote fuel pump on all models prior to 1991.

Since 1991, the fuel pump is an integral part of the carburetor.

If the fuel pump fails to perform properly, an insufficient fuel supply will be delivered to the carburetor. This lack of fuel will cause the engine to run lean, lose rpm, or cause piston scoring.

NEVER use liquid Neoprene on fuel line fittings. Use Permatex when making fuel line connections. Permatex is avail-

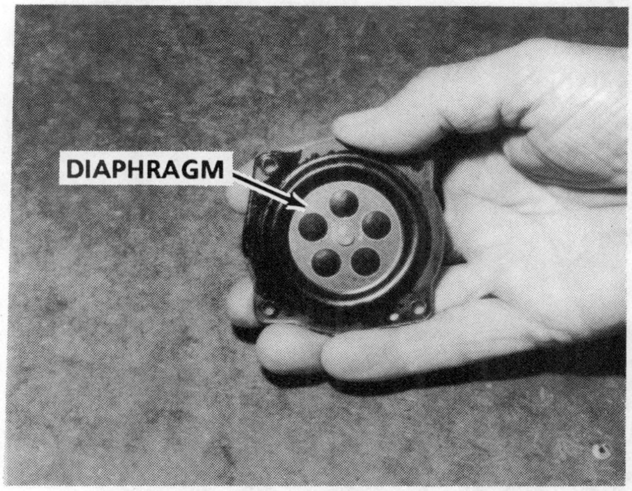

Close view of a diaphragm from a Mikuni carburetor with integral fuel pump. The diaphragm must be carefully inspected for the tiniest hole or tear. Such damage will "wet foul" the spark plugs at idle speed.

able at almost all marine and hardware stores.

To service the fuel pump, see Chapter 6 -- **FUEL.**

3-8 CRANKING MOTOR AND SOLENOID

Cranking Motor Test

Check to be sure the battery is fully charged. Would you believe, many cranking motors are needlessly disassembled, when the battery is actually the culprit.

Lubricate the pinion gear and screw shaft with No. 10 oil.

Connect one lead of a voltmeter to the positive terminal of the cranking motor. Connect the other meter lead to a good ground on the engine. Check the battery voltage under load by depressing the ignition switch and observing the voltmeter reading.

If the reading is 9-1/2 volts or greater, and the cranking motor fails to operate, repair or replace the unit. See Chapter 9 -- **ELECTRICAL.**

Preliminary Tasks Before Testing Cranking Motor Solenoid

Remove the battery cover.

Disconnect the three Red leads from the upper left terminal of the cranking motor solenoid.

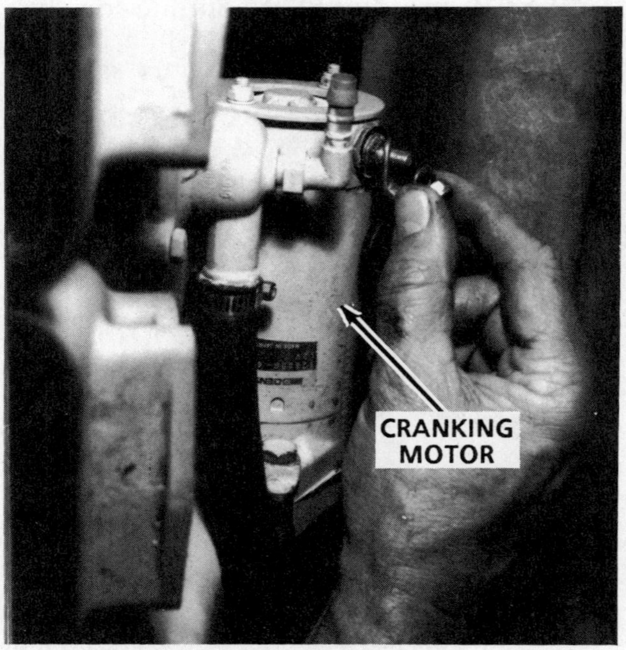

*Cranking motor electrical leads being disconnected prior to making tests on the motor. See Chapter 9 -- **ELECTRICAL,** to service the cranking motor.*

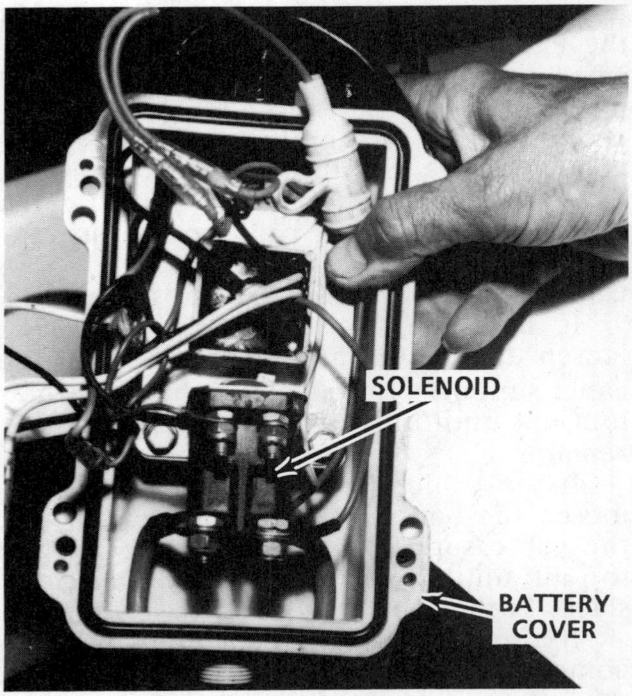

The cranking motor solenoid is located inside the battery cover.

Disconnect the single Red large lead from the upper right terminal.

Disconnect the Yellow/Red lead from the lower left terminal.

Disconnect the two Brown leads from the lower right terminal.

Cranking Solenoid Testing

1- Select the Rx1 ohm scale on the ohmmeter. Make contact with the two meter leads across the two upper solenoid terminals. The meter should read an infinite resistance. If the meter shows resistance of less than infinity, the solenoid is defective and **MUST** be replaced.

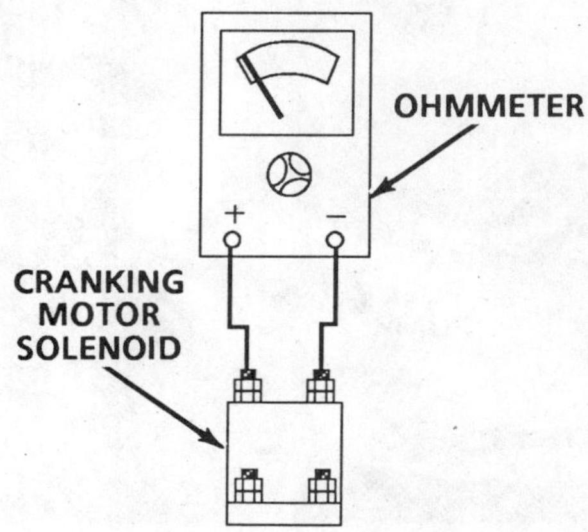

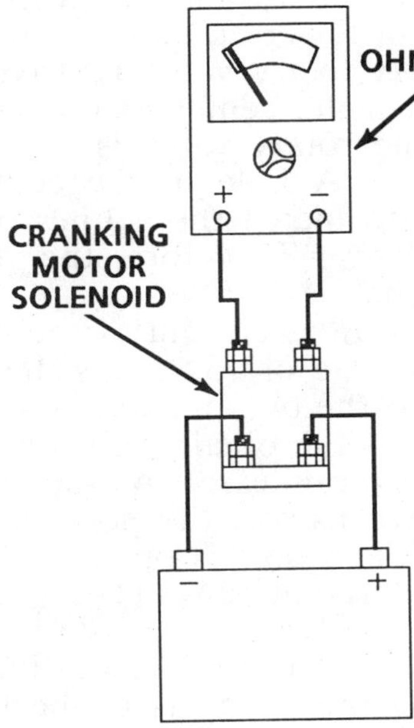

2

2- Keep the ohmmeter on the same scale. Obtain a 12-volt battery and a pair of jumper leads. Using the set of jumper leads, connect the battery across the two lower terminals. Make contact with the two meter leads across the two upper solenoid terminals. If the solenoid emits a "click" and the ohmmeter registers zero ohms, the solenoid is good. If no "click" is heard, or if the meter registers a high or infinite resistance, the solenoid is defective and **MUST** be replaced.

The "bottom line" on a jet pump is the condition and efficiency of the impeller. Adequate care, maintenance and timely replacement will result in maximum fun and safety for the owner and guests using the craft.

If the solenoid failed the test, it must be replaced. Two bolts secure the solenoid to the battery cover. When installing the replacement solenoid, apply Loctite 242 to the threads of the bolts and install the new solenoid in place. Tighten both bolts securely.

Connect the two Brown leads to the lower right terminal.

Connect the Yellow/Red lead to the lower left ternimal.

Connect the three smaller Red leads to the upper left terminal.

Place the cover over the battery. Install and tighten the four securing screws.

3-9 JET PUMP

At press time, only the Bombardier Sea-Doo GT and a very few other personal watercraft were equipped with a reverse gate. If the unit being serviced has a reverse gate, the gate **MUST** be properly adjusted to obtain maximum performance from the craft. When properly adjusted, the gate will permit the pump to deliver its full potential of thrust with minimum drag.

Gate Position

The shift rod lever adjustment should be checked from time to time to ensure the gate is firmly against the pump housing, when the unit is in the **FORWARD** position (wide open).

With the gate against the housing, any rattle noises will be avoided as the craft moves through the water. Proper positioning of the gate in forward gear will prevent wave action from accidentally shifting the gate into reverse as the craft is operated through violent maneuvers.

A simple "tuning" task with the jet pump is to make a thorough check of the linkage and gate movement. Such a check will ensure maximum thrust and use of the horsepower developed by the engine.

Impeller

Excessively rounded jet impeller edges will reduce the efficiency of the jet drive.

The term cavitation "burn" is a common expression used throughout the world among people working with pumps, impeller blades, and forceful water movement.

"Burns" on the impeller blades are caused by cavitation air bubbles exploding with considerable force against the impeller blades. The edges of the blades may develop small "dime size" areas resembling a porous sponge, as the aluminum material is actually "eaten" by the condition just described.

3-10 ROTARY VALVE DESCRIPTION AND OPERATION

The most unique feature of the engine used in all Bombardier Sea-Doo watercraft -- the Rotax 587 -- is the rotary valve. Rotary valve two-stroke engines are widely used in GP road racing motorcycle engines.

The purpose of the rotary valve is to mechanically time the oil/air/fuel mixture to be injected at the precise moment into the crankcase. The timing on this engine is so critical a special section is included in Chapter 8 -- **ENGINE**. Incorrect timing of the rotary valve could lead to difficult starting and backfiring during engine operation.

The oil/air/fuel mixture timing must not be confused with ignition/spark timing. Ignition/spark timing is electronically controlled by the ignition module and cannot be adjusted.

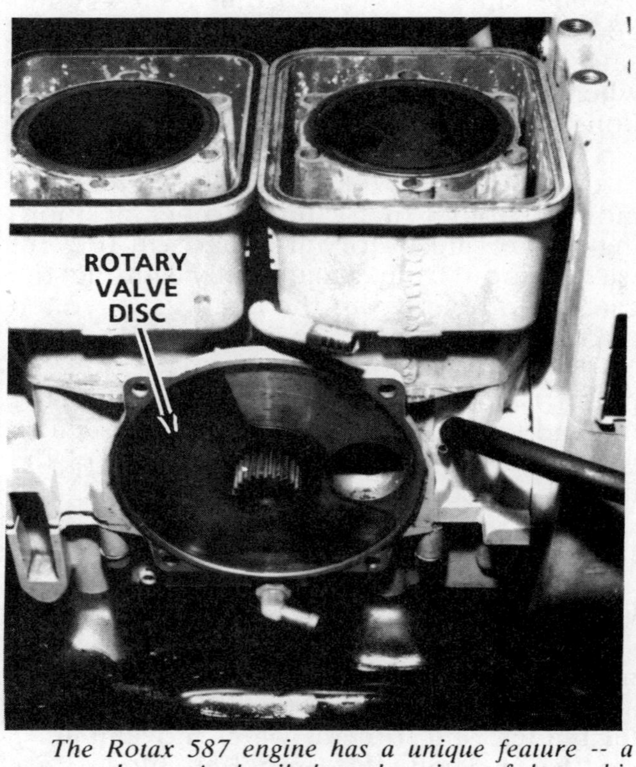

The Rotax 587 engine has a unique feature -- a rotary valve. A detailed explanation of how this valve functions, is given in the text.

The intake ports are machined into the starboard side of lower crankcase, one port per cylinder. A rotary valve is driven off the crankshaft in the center of the two intake ports. The rotary valve is actually a thin metal disc. A hole in the center of the disc is cut to match the splines on the rotary valve shaft. Therefore, the disc is driven by the shaft.

A large window is cut out of the disc -- a pie shaped wedge of 132°. As this window rotates past the two intake ports -- one after the other -- the oil/air/fuel mixture is injected into the crankcase. As soon as the trailing edge sweeps past the port, the port is closed off. Each port is opened at 132° BTDC and is closed at 52° ATDC.

If the rotary valve is not installed correctly, the mixture will be injected into the crankcase when the piston is in the wrong position -- at the wrong part of the stroke. The resulting ignition of the mixture will not reach maximum effectiveness and will produce a backfire.

Correct positioning of the rotary valve is quite simple. One edge of the window **MUST** align with a notch cast into the rotary valve housing. It is extremely unlikely the rotary valve -- once installed properly -- will shift during engine operation. The rotary valve housing is an integral part of the upper crankcase. Unfortunately, if a replacement of the upper crankcase is to be made, the rotary valve housing will not have a notch. In this case, the rotary valve **MUST** be aligned with the rotary valve shaft using a degree wheel on the replacement crankcase. This alignment is relatively easy and is covered in Chapter 8 -- **EN-**

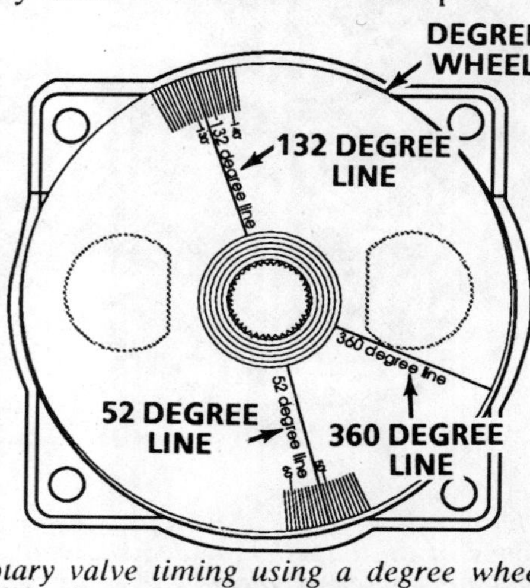

Rotary valve timing using a degree wheel, as explained in the text. A degree wheel suitable for cutting out and using is included in the Appendix.

GINE. A cutout paper degree wheel is also provided on a separate page in the Appendix and may be cutout for use.

ROTARY VALVE DISC MODIFICATION

The stock factory "out-the-door" disc is machined to provide maximum power for a broad range of rpm operation. The window is a set size to control port timing. The experienced performance seeker, will increase the valve closing delay in very **SMALL** increments, not more than two degrees (2o) at a time until the desired results are obtained. Indiscriminate snipping at the valve disc, especially on the port closing side of the window, will cause an appreciable delay in closing the port after the piston has started to move downward from TDC. Such action would cause low speed blow back.

Retarding the intake port closing, will move the engine power peak higher and at the same time will reduce the power at the lower end of the rpm range.

The intake port opening point has some influence on engine power output, but it is not significant compared to port closing timing. Early intake port opening provides a minimal increase in power. Polishing the intake ports and smoothing out any rough edges on the rotary disc will help engine performance.

3-11 MODIFYING FOR HIGHER PERFORMANCE

Normally Seloc manuals only cover stock engine and jet pump components. However, personal watercraft enthusiasts are notorious for modifying their equipment to gain higher performance. Therefore, a large aftermarket industry has mushroomed in just a few short years to accommodate this need for speed. A few well chosen words on modifying for performance seem to be in order.

The following general guidelines will provide improved two stroke performance. The tasks listed below, in order of effectiveness, or possibly a combination of modifications and components, may give the desired results.

a- To increase bottom end power:
- Increase cylinder compression.
- Advance the timing.
- Use a smaller diameter carburetor.
- Use a long chamber exhaust.
- Use a low pitch impeller.
- Use a larger nozzle bore.

b- To increase top end power:
- Install a larger diameter carburetor.
- Use a short exhaust chamber.
- Go to higher porting specifications.
- Use a higher pitched impeller.
- Use a ride plate generating lift.
- Try nitrous oxide injection.

c- To increase acceleration:
- Increase cylinder compression.
- Use a lighter weight driveshaft.
- Install an aluminum flywheel.
- Check for optimum air flow.
- Try nitrous oxide injection.

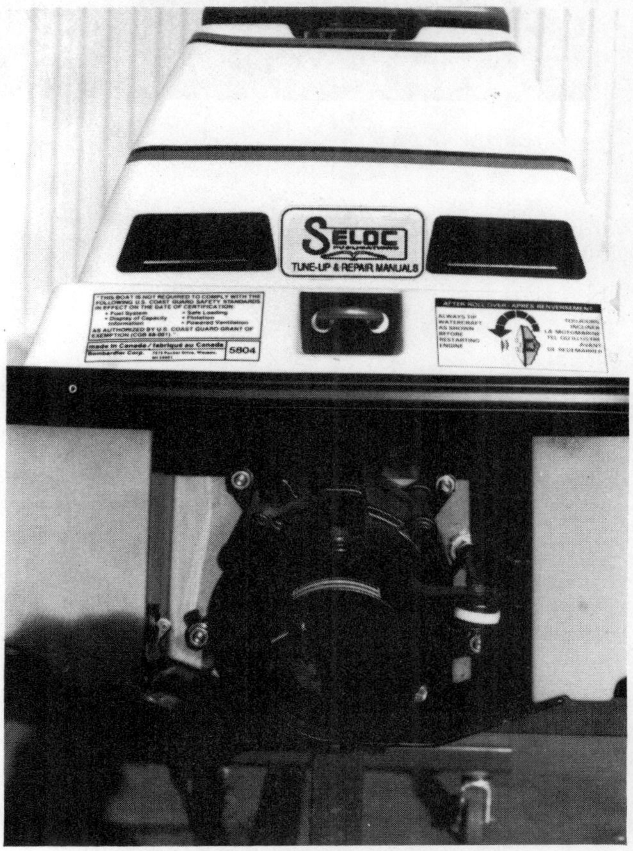

A clean, properly tuned, and adequately maintained power unit has a long list of rewards for the owner.

"I've got mine free, how are you doin'".

4
MAINTENANCE

4-1 INTRODUCTION

The watercraft being prepared for service probably represents a sizeable investment. In order to protect this investment and for the owner to receive the maximum amount of enjoyment from the craft it must be cared for properly while being used and when it is out of the water. Always store the watercraft with the bow slightly higher than the stern and be sure to remove the inner hull drain plug/s. If any type of cover is used to protect the craft -- plastic, canvas, whatever -- be sure to allow for some movement of air through the hull. Proper ventilation will assure evaporation of any condensation formed due to changes in temperature and humidity.

GOOD WORDS

The authors estimate 75% of engine repair work can be directly or indirectly attributed to lack of proper care for the engine. This is especially true of care during the off-season period. There is no way on this green earth for a mechanical engine, particularly a marine engine, to be left sitting idle for an extended period of time, say for six months, and then be ready for instant satisfactory service.

Imagine, if you will, leaving your automobile for six months, and then expecting to turn the key, have it roar to life, and be able to drive off in the same manner as a daily occurrence.

It is critical for a marine engine to be run at least once a month, preferably, in the water, but if this is not possible, then a flush attachment **MUST** be connected whether directly to the engine block or in the hose line from the jet pump to the engine.

Move the craft to a body of water or connect a garden hose to the engine cooling water supply fitting or flush fitting on the cylinder head.

CAUTION

The following procedure MUST be followed whenever the engine is started with a hose attached for cooling the engine. Start the engine, and then IMMEDIATELY turn the water on. The engine MUST be started first to prevent water from backing up into the tuned exhaust pipe and into the engine. Water must circulate anytime the engine is operating to prevent overheating -- which could cause damage to moving parts and possibly cause engine seizure. NEVER allow the engine to operate without cooling water for more than 15 seconds. NEVER run the engine with a flushing attachment for more than 5 minutes. The driveline seal is not provided with cooling water if it is not immersed in water. To shut the engine down, first shut the water supply off,

There is no substitute and few valid excuses for not developing good habits leading to regular maintenance practices. The results of a good program are evident in top performance, maximum enjoyment, and "pride of ownership", all at minimum cost.

During the storage period, and especially after a prolonged period of storage, the steering should be operated from hard-over to hard-over to verify the system is in proper working order.

and then shut down the engine. If this sequence of tasks is not followed as stated, the engine could be severely damaged.

REMEMBER
Start the engine.
Turn the water on.
Turn off the water.
Shut down the engine.

NEVER, AGAIN NEVER, operate the engine at high speed with a flush device attached. The engine, operating at high speed with such a device attached, might very likely **RUNAWAY** from lack of load on the impeller, causing extensive damage.

After the engine has been operated, the shift mechanism, if the unit is so equipped,

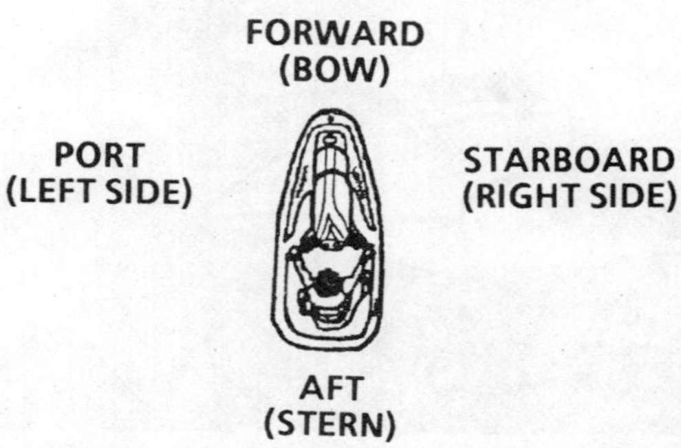

Common terminology used throughout the world and extensively in this manual for reference designation on water craft of ALL sizes. "Port", "Starboard", "Forward", and "Aft", never change, even if standing on your head.

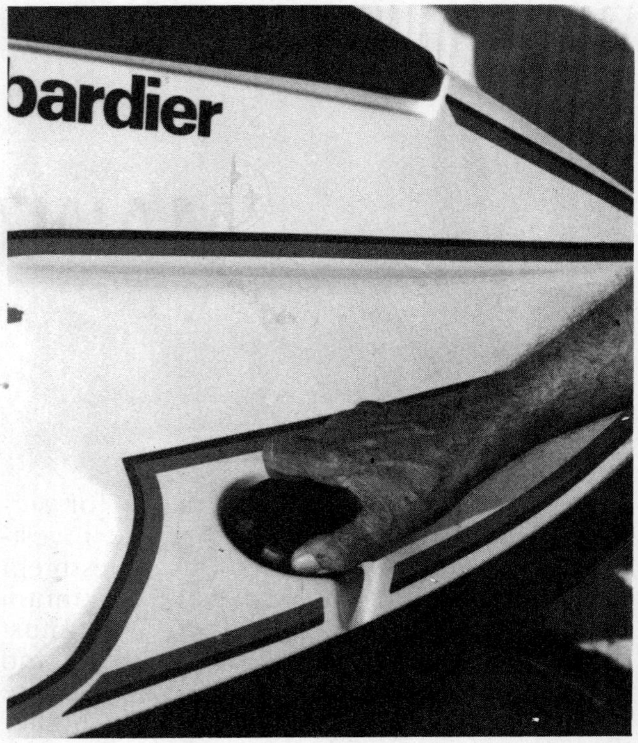

After a fun day on the water and before moving the craft out of the water, ***SLOWLY*** *open the fuel tank cap to release built-up pressure in the tank.*

should be operated through the full range several times. The steering should also be operated from hard-over to hard-over several times.

Only through a regular maintenance program can the owner expect to receive long life and satisfactory performance at minimum cost.

The material presented in this chapter is divided into four general areas.

1- General information every owner should know.

2- Maintenance tasks to be performed periodically to keep the craft operating at minimum cost.

3- Care necessary to maintain the appearance of the craft and give the owner a "Pride of Ownership" look.

4- Winter storage practices to minimize deterioration during the off-season when the craft is not in use.

In nautical terms, the front of the craft is the **BOW**; the rear is the **STERN**; the right side, when facing forward, is the **STARBOARD** side; and the left side is the **PORT** side. All directional references in this manual use this terminology. Therefore, the direction from which an item is viewed is of no consequence, because **STARBOARD** and **PORT** never change no matter where the individual is located or his/her position -- even if standing on one's head.

AFTER USE TASKS

At the end of a day's use, make a habit of opening the engine compartment to allow fumes to escape and sponging out all water in the bottom of the hull before trailering the craft. Leave the engine cover off when the craft is stored overnight to allow any moisture in the compartment to evaporate. Slowly open the fuel tank cap to release built up pressure. Pressure in the tank will increase considerably during hot weather. Releasing the pressure reduces the chance of an explosion and insures smooth fuel flow when the engine is next started.

If the craft will remain unused for a week, spray storage oil inside the engine and a light film of lubricant, such as WD-40, Super Lube, or Justice Bros. 80, on the engine exterior. Keeping the engine exterior lightly oiled at all times will result in water sliding off the engine and draining away rather than accumulating in casting webs -- resulting in corrosion.

Cooling

The engine cooling passages must be flushed with clear water on a regular basis. This accomplished by simply connecting a "Flushing Kit" per the manufacturer's instructions **OR** by disconnecting the cooling hose from the pump to the engine and connecting a garden hose to the disconnected hose. Such an easy maintenance practice should become a habit and will prevent accumulation of debris which could restrict engine coolant water flow and cause overheating.

Controlling Corrosion

Since man first started out on the water, corrosion has been his enemy. The first form was merely rot in the wood and then it was rust, followed by other forms of destructive corrosion in the more modern materials.

The higher the concentration of salt in the water, the worse the corrosion which will take place.

One defense against corrosion is to use similar metals in the manufacture of the parts exposed to the water.

A second defense against corrosion is to insulate dissimilar metals. This can be accomplished by using an exterior coating of Sea Skin or by insulating them with plastic or rubber gaskets.

Pump Impeller

The pump impeller is a precisely machined and dynamically balanced aluminum spiral. Observe the drilled recesses at exact locations to achieve the delicate balancing.

Excessive vibration of the jet pump may be attributed to an out-of-balance condition caused by the impeller striking rocks, gravel, or by cavitation "burn".

The term cavitation "burn" is a common expression used throughout the world among people working with pumps, impeller blades, and forceful water movement.

"Burns" on the impeller blades are caused by cavitation air bubbles exploding with considerable force against the impeller blades. The edges of the blades may develop small "dime size" areas resem-

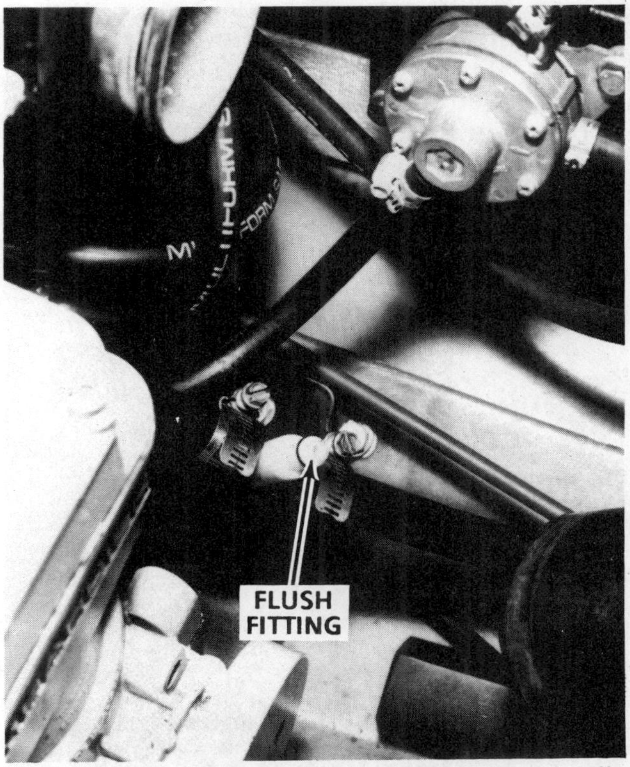

Location of the flush fitting on the Bombardier Sea-Doo engine. A garden hose connected to this fitting is used to flush the engine -- a very important task, especially following operation in salt water.

The state license identification number must be prominently displayed on the forward portion of the hull -- port and starboard.

bling a porous sponge, as the aluminum is actually "eaten" by the condition just described.

Excessive rounding of the impeller edges will reduce efficiency and performance. Therefore, the impeller should be inspected at regular intervals. If rounding is detected, the impeller should be removed from the housing, placed on a work bench, and the edges rounded as close to a 1/32" radius as possible, using a file.

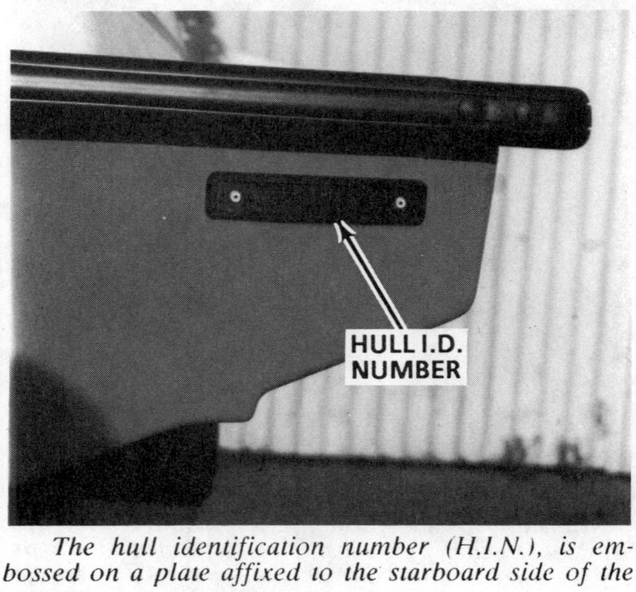

The hull identification number (H.I.N.), is embossed on a plate affixed to the starboard side of the transom.

Refer to Chapter 10 -- **Jet Pump** for detailed procedures on removal and installation of the impeller.

4-2 SERIAL NUMBERS

Most manufacturers use three different identification numbers stamped, embossed, or decaled somewhere on the craft.

The engine serial number is the manufacturer's key to engine changes. This number identifies the year of manufacture, the qualified horsepower rating, and the parts book identification. If any correspondence or parts are required, the engine serial number **MUST** be used or proper identification is not possible.

U.S. Federal regulations require all personal watercraft to have a Hull Identification Number (HIN) permanently attached to the craft in two separate locations: one on the transom on the starboard side, above the waterline, and the other in an unexposed location, which is usually left to the discretion of the manufacturer. This "HIN" number can be most helpful when attempting to retrieve a stolen craft. The "HIN" number is required in most states in order to properly register craft and ensure the owner pays all taxes due.

Some manufacturers also use a Primary I.D. Number stamped on a plate attached to the inside of the engine compartment.

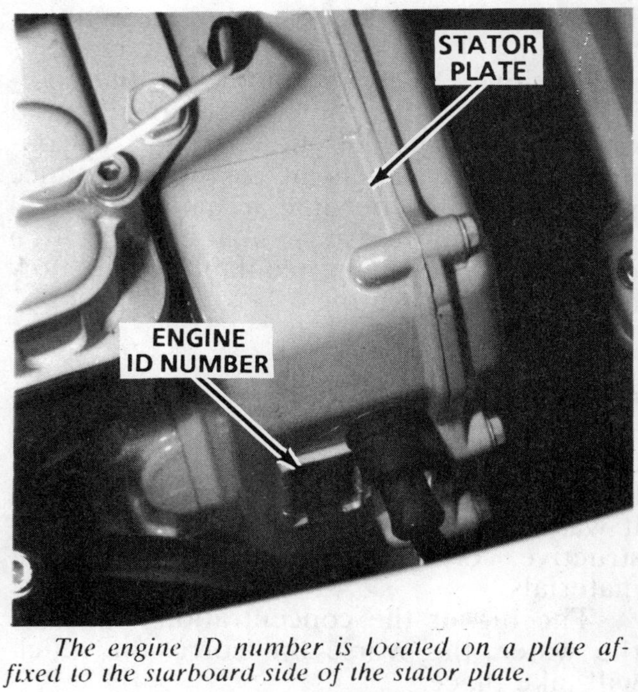

The engine ID number is located on a plate affixed to the starboard side of the stator plate.

ONE MORE WORD

As a theft prevention measure, all metal plates with any type of identification number are especially made "wafer" thin. Any attempt to remove these plates will result in cracks across the serial number.

4-3 LUBRICATION -- COMPLETE UNIT

With every type of mechanical invention with moving parts, lubrication plays a prominent role in operation, enjoyment, and longevity of the unit at minimum cost.

If a personal watercraft is operated in salt water, the frequency of applying lubricant to fittings is usually cut in half as compared with the unit being used in fresh water. The few minutes involved in moving around the craft applying lubricant and at the same time making a visual inspection of its general condition will pay in rich rewards with years of continued service.

The first personal watercraft arrived on the scene in the United States in the early 1970's. It is not uncommon to see outboard units well over 20 years of age moving a boat through the water as if the unit had recently been purchased from the current line of models. An inquiry with the proud owner will undoubtedly reveal his main credit for its performance to be regular periodic maintenance. There is no reason this same longevity of performance cannot be true for personal watercraft.

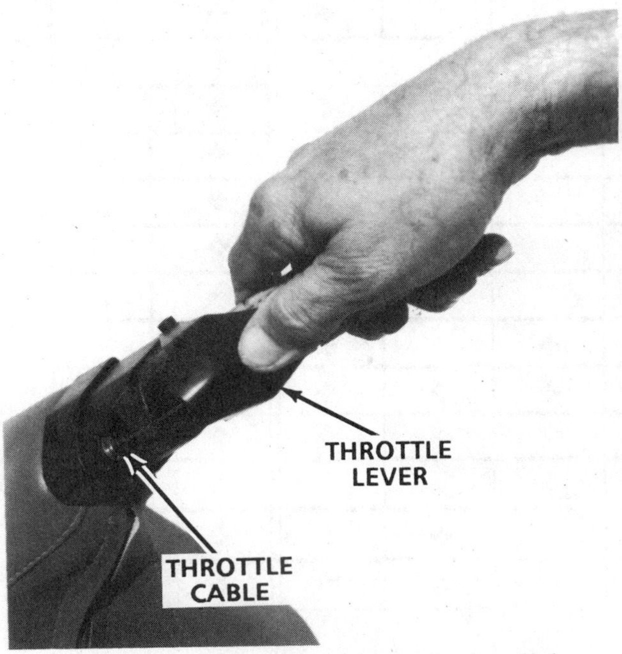

Lubricate the throttle cable at the handlebar every 100 hours of engine operation.

The accompanying chart can be used as a guide to the frequency of maintenance while the craft is being used during the season. Unless otherwise stated use a good brand name multi-purpose water resistant lubricant. Detailed lubrication instructions are given below for locations required special attention.

In addition to the normal lubrication listed in the lubrication chart, the prudent owner will inspect and make checks on a regular basis, as listed in the accompanying chart.

Throttle Cable Lubrication

Hold the throttle lever in the full open position against the grip. Use a small screwdriver and pry out the small seal from the groove of the throttle stop and a little way along the cable. Spray rust inhibitor along the exposed length of cable. Fill the groove with multi-purpose water resistant lubricant, and then pull the seal down along the cable and back into place inside the groove. Wipe away any excess lubricant.

Steering Cable Lubrication

Pry the barrel of the steering cable free of the ball joint on the steering post arm. Coat the ball joint with lubricant, and then snap the barrel back onto the ball joint.

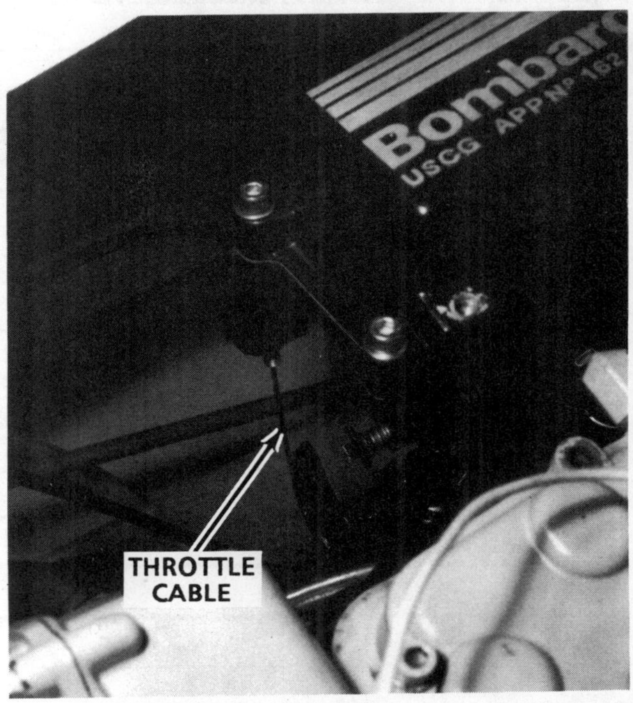

Lubricate the carburetor end of the throttle cable at the same time the handlebar end is lubricated.

Lubrication Points:

	Initially every 10hrs	Initially every 50hrs	Initially every 100hrs	Then every 100hrs	Then every 200hrs
Carburetor throttle shaft			❖	❖	
Throttle cable - engine end			❖	❖	
Throttle cable - handlebar stop			❖	❖	
Choke cable - engine end			❖	❖	
Choke knob shaft			❖	❖	
Steering nozzle pivot shaft			❖	❖	
Steering post arm ball joint			❖	❖	
Reverse gate pivot pins			❖	❖	
Jet pump end cap reservoir			❖	❖	
Seat latch			❖	❖	

Maintenance Tasks:

	Initially every 10hrs	Initially every 50hrs	Initially every 100hrs	Then every 100hrs	Then every 200hrs
Flame arrestor element	❖	❖	❖	❖	
Clean and gap spark plugs	❖	❖	❖	❖	
Adjust carburetor idle	❖		❖	❖	
Adjust carburetor mixture	❖		❖	❖	
Drain and inspect fuel tank					❖
Clean and inspect fuel filter	❖	❖	❖	❖	
Inspect fuel line			❖	❖	
Inspect check valve			❖	❖	
Adjust throttle control			❖	❖	
Adjust choke system			❖	❖	
Adjust steering system			❖	❖	
Inspect impeller		❖	❖	❖	
Check impeller to pump case clearance			❖	❖	
Inspect rubber coupling					❖
Inspect electrical connections	❖	❖	❖	❖	
Check battery electrolyte level	❖	❖	❖	❖	
Clean bilge strainer			❖	❖	
Inspect rubber water seal	❖		❖	❖	
Retorque head bolts		❖	❖	❖	
Inspect driveshaft boot	❖	❖	❖	❖	

Fuel/Oil Mixture

A new or rebuilt engine must be properly "broken-in" if it is to develop maximum performance. An engine which is incorrectly "broken-in" will never be able to achieve the maximum rated horsepower. During the initial breaking-in period, the fuel/oil ratio should be very rich. Extra two-stroke oil should be added to the fuel tank at a rate of 2 oz per gallon, for the first ten hours of operation after the unit is brought out of storage. This added oil will **ENSURE** adequate lubrication of moving parts which have been drained of oil during the storage period. Operation of an engine without the proper amount of oil will quickly lead to a significant loss of horsepower.

After the first ten hours of operation, the oil injection system will deliver an adequate amount of oil for lubrication.

Experienced two-stroke mechanics run their own units on a higher mixture of oil on the premise it is cheaper to replace a fouled spark plug than to have a complete engine overhaul with rebored cylinders, oversize pistons and rings. These same mechanics who own oil injected units add an extra ounce of oil per each gallon of fuel. The addition of this extra oil may prevent engine seizure in the event of an oil pump failure or clog in an oil delivery line.

Use only outboard **MARINE** oil in the mixture, never automotive oils. Four-stroke automotive engine oil is not formulated to burn completely -- only to lubri-

The steering cable ball joint at the outlet nozzle requires periodic lubrication.

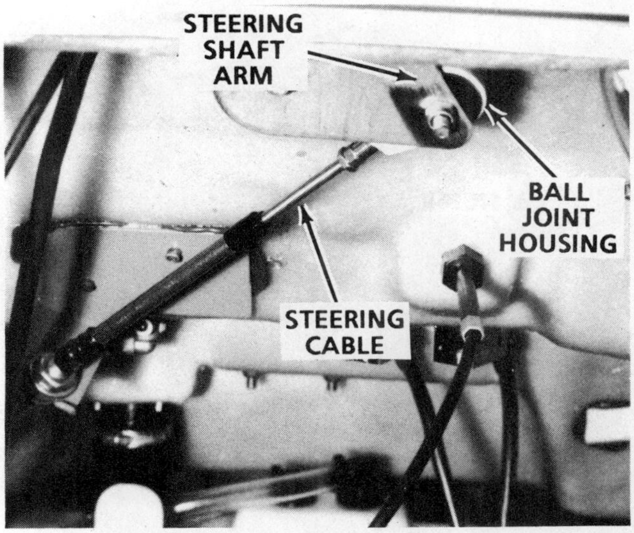

The steering cable ball joint at the steering shaft arm also requires periodic lubrication.

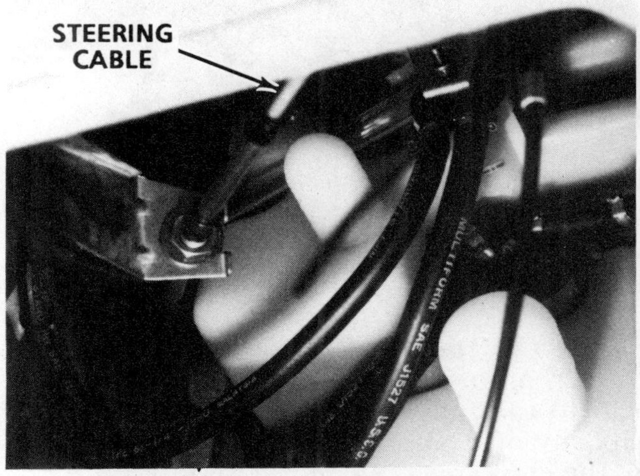

All exposed cable should be lubricated on a regular basis.

Checking the oil level in the oil tank.

The in-line oil filter is located between the oil tank and the oil injection pump.

cate. Therefore, automotive oil, if used in a two-stroke engine will leave an undesirable residue.

Engine Oil Tank and Filter

A remote oil injection tank is mounted next to the fuel tank. The tank has an oil capacity of about 95 fl oz (2.8L). Oil level in the tank may be checked by using an externally mounted dipstick -- shown in the accompanying illustration on the previous page.

An in-line oil filter is installed in the line between the oil tank and the oil injection pump. The manufacturer recommends the filter be inspected on a monthly basis during the use season. The level of oil in the oil tank should be checked each day the craft is operated.

Jet Pump Lubrication

The forward end of the pump shaft is supported by the flywheel. A lubrication fitting on the flywheel hub serves to lubricate the forward end of the driveshaft. Use a standard pump and force bearing lubricant into the fitting just until the boot begins to expand.

The aft end of the pump shaft is supported by a bearing in the impeller housing. This bearing is lubricated through two holes in the casting from a reservoir in the end cap at the aft end of the shaft. The lubricant level in the reservoir should also be checked every three months during the season of operation.

SPECIAL WORDS

The pump may remain in place on the craft, but the nozzle assembly must be removed to gain access to the end cap reservoir. See Chapter 10 -- **JET PUMP** for detailed procedures to remove and install the nozzle.

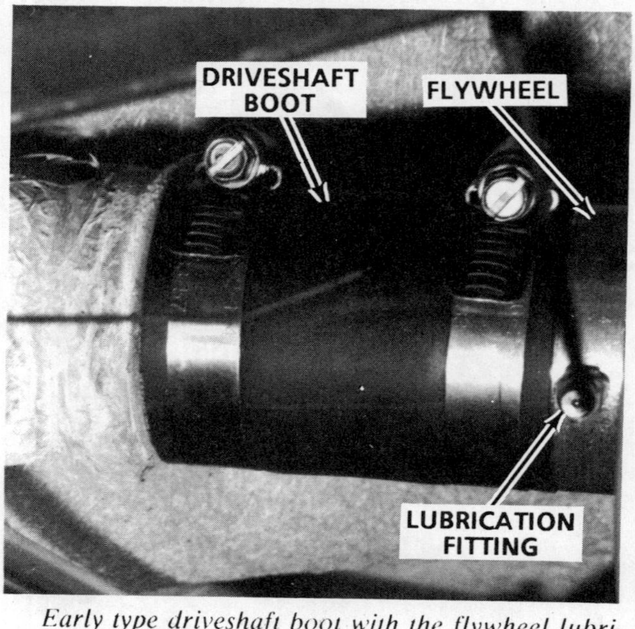

Early type driveshaft boot with the flywheel lubrication fitting identified.

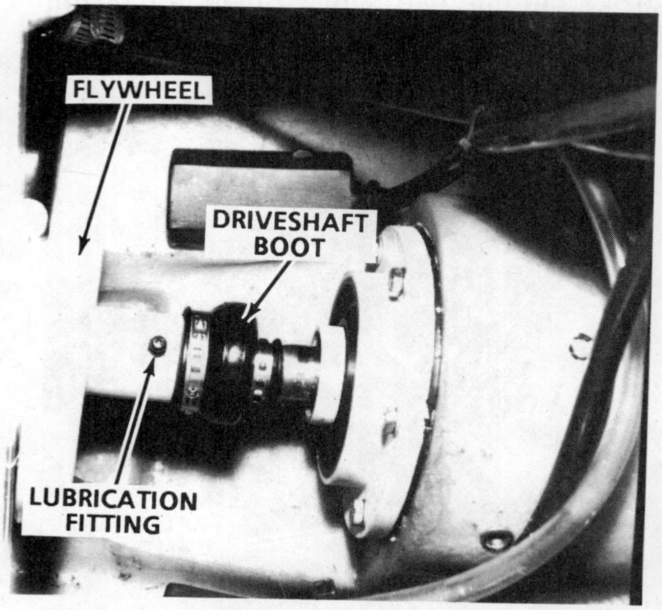

Late type driveshaft boot with flywheel lubrication fitting identified.

To fill the reservoir, after the nozzle is clear, remove both plugs. Use only Sea-Doo Jet Pump Synthetic Oil, P/N 293-600-011. **NEVER** mix this synthetic oil with a mineral oil. Pour the oil into the top filler hole until it begins to escape from the lower hole.

Apply a light coating of Loctite 515, P/N 293-800-007 to threads of both plugs. Install and tighten the plugs securely. Install the nozzle.

4-4 INSPECTION AND SERVICE

Fuel Tank Check Valve and Oil Tank Check Valve

Before disturbing any hose be sure to have a new hose tie wrap or a small hose clamp to replace the one removed. Original equipment tie wraps must be cut through to remove them from a hose.

Disconnect the ventilation hose from the fitting on the fuel tank pickup. Remove the two tie wraps, one on each end of the check valve.

Blow through the fitting on one face of the valve. Now, blow through the fitting on the other face of the valve. The air should pass through the valve **ONLY** in

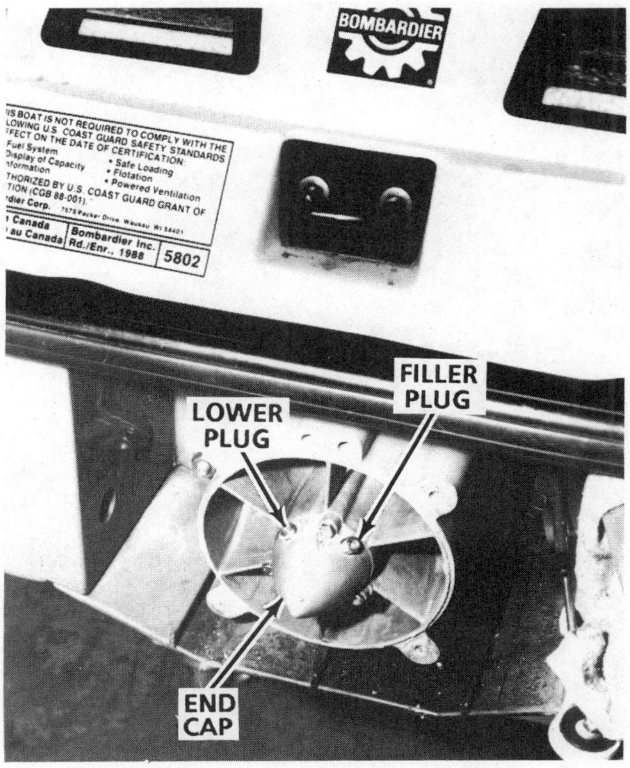

The impeller shaft is lubricated from a reservoir in the end cap at the aft end of the shaft.

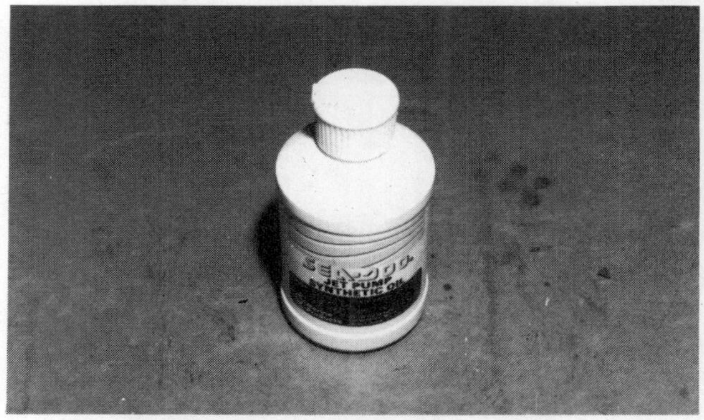

Only Sea-Doo Synthetic Oil should be used to lubricate the jet pump.

one direction. A directional arrow should be embossed on the valve indicating the direction of fuel flow -- and the direction air passes. If there is no directional arrow embossed on the valve, use a magic marker, or other device, and mark an arrow to indicate the direction of air and fuel flow. If the check valve will not allow air to pass in one direction, the fuel will remain in the tank and not be drawn out by the fuel pump.

If the valve allows air to pass in the wrong direction, it will also allow fuel to flow out of the tank through the valve and spill into the hull when the craft is subjected to violent maneuvers. Therefore, replace the valve if it fails to function properly -- allows air to pass in only one direction.

Perform the same check for the oil tank check valve. Both the oil and fuel valves are identical.

A faulty fuel tank check valve will starve the engine of fuel. If the condition becomes really serious, the engine will faulter and die.

A faulty oil check valve will reduce critical lubrication to the engine. Such a condition will quickly cause extensive and expensive damage.

Fuel Filler Cap

Remove the fuel filler cap and inspect the condition of the rubber seal. Replace the seal if it is worn or cracked.

Check the tightness of the large hose clamp around the fuel tube to the tank.

Drain Plug

Remove the drain plug from the transom. Inspect the condition of the O-ring. Replace if necessary. Clean any dirt or

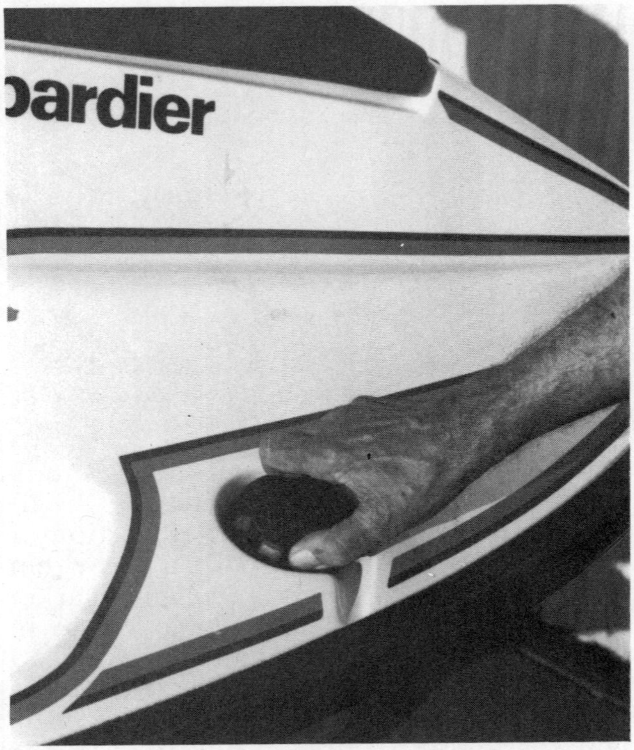

Removing the fuel filler cap to inspect the condition of the rubber seal.

sand from the plug threads. Install the plug and tighten it securely.

4-5 IMPELLER-TO-IMPELLER HOUSING WEAR RING CLEARANCE

SPECIAL WORDS

The following procedure may be performed with the watercraft elevated, as on a set of saw horses, or with the craft resting on its side. Either of these positions will provide access to the rock grate and the impeller. The accompanying illustration shows the craft tilted up for clearance.

If the impeller-to-wear ring clearance is to be determined with the water craft raised on saw horses, the battery may be left secured in the craft. **HOWEVER**, if the watercraft is to be positioned on its side, first disconnect the electrical leads at both battery terminals, and then remove the battery.

Once the craft is secured in a favorable position, remove the four bolts securing the rock grate to the jet pump. If the seal between the hull and the pump housing is broken, or the foam packing is damaged, reseal the housing and intake area with silicone sealant to prevent impeller cavitation.

Insert a feeler gauge in the center between each impeller blade and the wear ring. Determine the clearance.

If the clearance is less than 0.040" (1.00mm) for each blade, the condition is considered satisfactory. No action is necessary.

If the clearance is more than the service limit, inspect the condition of the wear

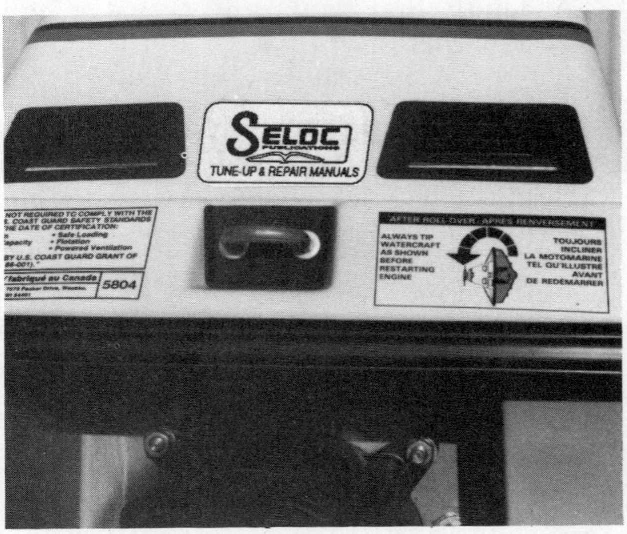

*When preparing to inspect the rock grate, observe the warning decal on the stern regarding which direction the craft must be tilted -- **ALWAYS** to port.*

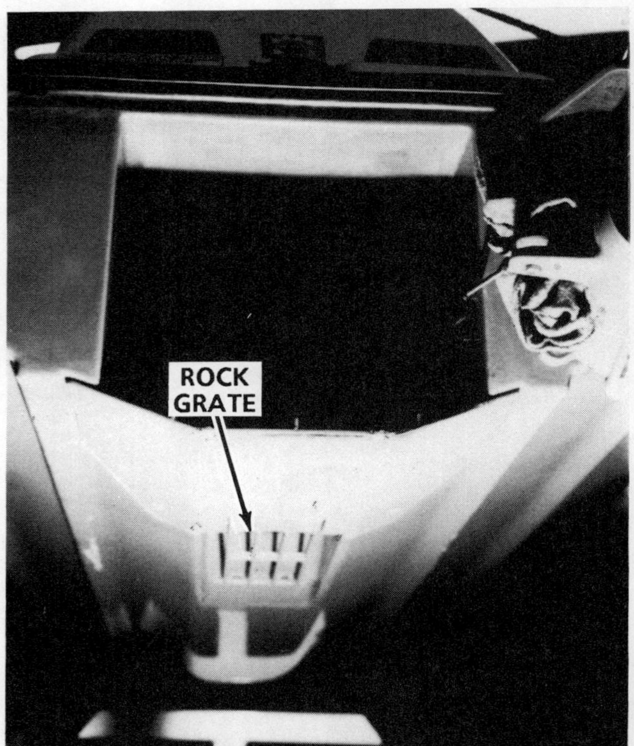

This craft is set on a stand with the bow raised to gain access to the rock grate.

ring. If the wear ring has deep scratches, replace the wear ring. If the wear ring appears in good condition, the problem must be with the impeller. Visually inspect the impeller for nicks, scratches, pitting or a "mushroomed" edge.

If the clearance is not kept to a minimum, the jet pump will lose efficiency by allowing excess water to pass between the blades and the housing. This water has no thrust and will actually slow the stream of water under pressure from the impeller.

The jet pump must be removed from the craft to replace the wear ring or to service the impeller. See Chapter 10 -- **JET PUMP** for detailed illustrated instructions.

Install the rock grate and tighten the attaching bolts securely.

Engine Alignment

The pump shaft is indexed directly into the splined hub on the flywheel. Bombardier Sea-Doo watercraft do not use a rubber disc type coupler between the flywheel and the pump shaft, as found on many other watercraft. Engine alignment is set at the factory and should **NEVER** require adjustment. The rubber engine mounts and shim material pieces should not be disturbed.

If an engine mount should ever need replacing, due to deterioration, **TAKE CARE** to save and use the original pieces of shim material over the mounts to maintain engine alignment.

4-6 PRE-SEASON PREPARATION

Satisfactory performance and maximum enjoyment can be realized if a little time is spent in preparing the unit for service at the beginning of the season. Assuming the unit has been properly stored, as outlined in Section 4-10 of this chapter, a minimum amount of work is required to prepare the unit for action on the water.

The following steps outline an adequate and logical sequence of tasks to be performed before using the craft the first time in a new season.

1- Lubricate the craft according to the manufacturer's recommendations. Refer to the lubrication chart on Page 4-6. Remove, clean, inspect, adjust, and install the spark plugs with new gaskets (if they require gaskets). Make a thorough check of the ignition system. This check should include: all electrical connections; condition of the wiring; battery electrolyte level; and status of the battery charge.

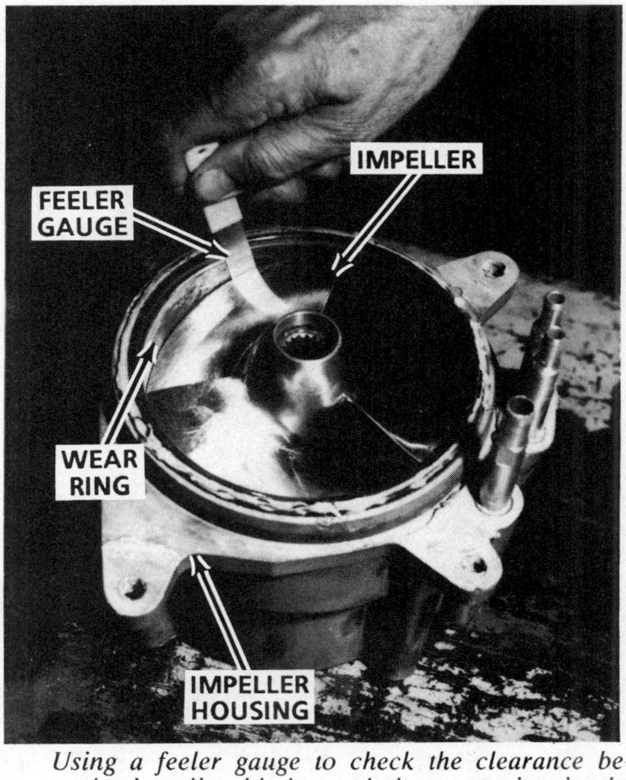

Using a feeler gauge to check the clearance between the impeller blades and the wear ring in the impeller housing, as explained in the text.

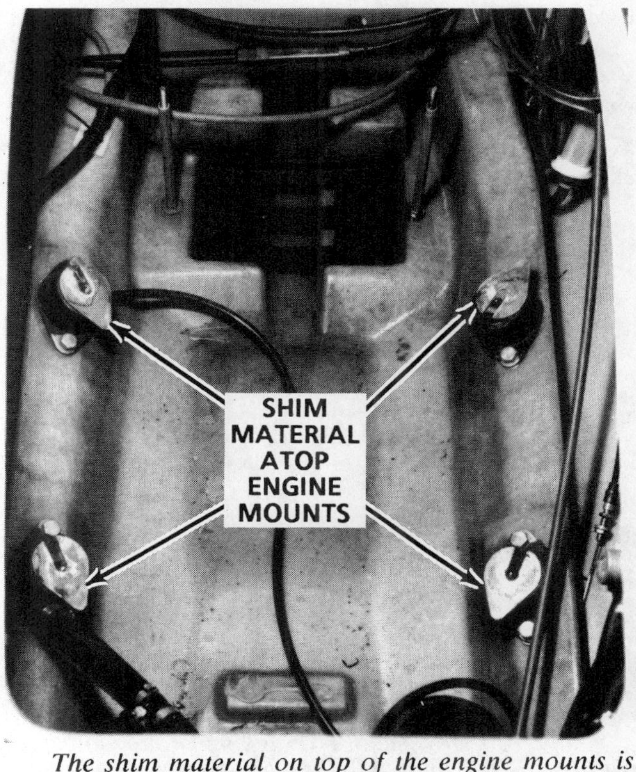

The shim material on top of the engine mounts is determined at the factory and should never need to be changed.

2- Take time to check the gasoline tank and all fuel lines, fittings, couplings, valves, including the flexible tank fill and vent. Turn on the fuel supply valve at the tank. If the fuel was not drained at the end of the previous season, make a careful inspection for gum formation. When gasoline is allowed to stand for long periods of time, gummy deposits form. This gum can clog the filters, lines, and passageways in the carburetor.

GOOD WORDS

All manufacturers recommend the fuel and oil filters be replaced at the start of each season or at least once a year.

3- Close all water drains. Check and replace any defective water hoses. Check to be sure the connections do not leak. Replace any spring-type hose clamps, if they have lost their tension, or if they have distorted the water hose. The new clamps should be the band-type clamps.

4- The engine can be run with the jet pump in water to flush it. If this is not practical, a flush attachment may be used. This device is attached to the engine water supply hose, between the engine and the pump. Connect a garden hose to the flush fitting.

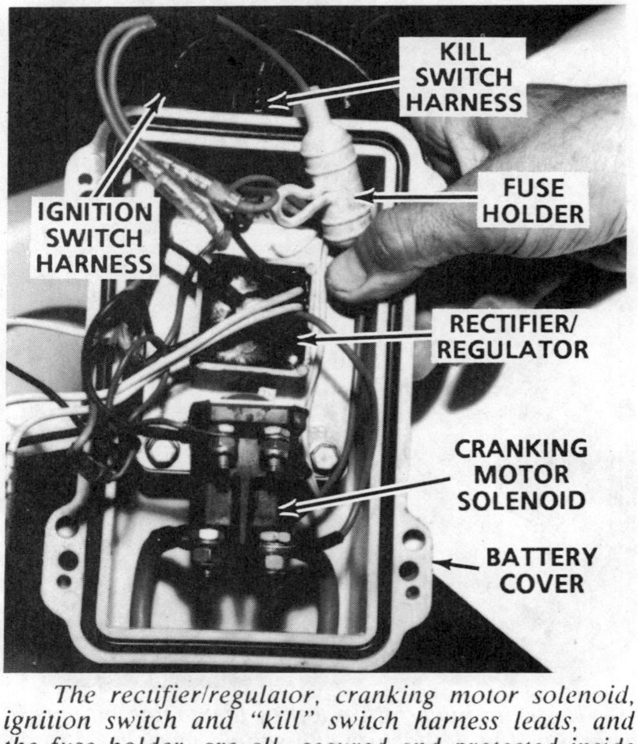

The rectifier/regulator, cranking motor solenoid, ignition switch and "kill" switch harness leads, and the fuse holder, are all secured and protected inside the battery case cover.

CAUTION

The following procedure MUST be followed whenever the engine is started with a hose attached for cooling the engine. Start the engine, and then IMMEDIATELY turn the water on. The engine MUST be started first to prevent water from backing up into the tuned exhaust pipe and into the engine. Water must circulate anytime the engine is operating to prevent overheating -- which could cause damage to moving parts and possibly cause engine seizure. NEVER allow the engine to operate without cooling water for more than 15 seconds. NEVER run the engine with a flushing attachment for more than 5 minutes. The driveline seal is not provided with cooling water if it is not immersed in water. To shut the engine down, first shut the water supply off, and then shut down the engine. If this sequence of tasks is not followed as stated, the engine could be severely damaged.

REMEMBER

Start the engine.
Turn the water on.
Turn off the water.
Shut down the engine.

Check the exhaust outlet for water discharge. Check for leaks.

The flame arrestor should be kept fairly clean and the rubber gasket around the arrestor must remain in good condition.

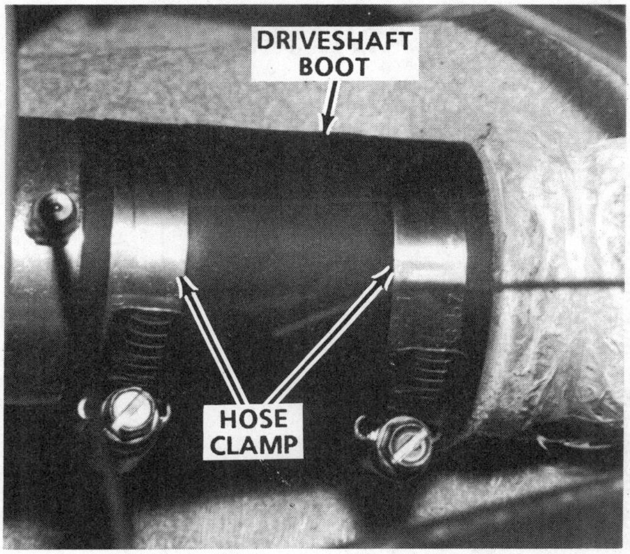

Old style driveshaft boot secured with standard hose clamps.

5- Check the electrolyte level in the battery and the voltage for a full charge. Clean and inspect the battery terminals and cable connections. **TAKE TIME** to check the polarity, if a new battery is being installed. Cover the cable connections with lubricant or special protective compound as a prevention to corrosion formation. Check all electrical wiring and grounding circuits.

6- Check all electrical parts on the engine and lower portions of the hull to be sure they are of the type to prevent ignition of an explosive atmosphere. Rubber caps help keep spark insulators clean and reduce the possibility of arcing. Electric cranking motors and high-tension wiring harnesses should be of a marine type to prevent an explosive mixture from igniting.

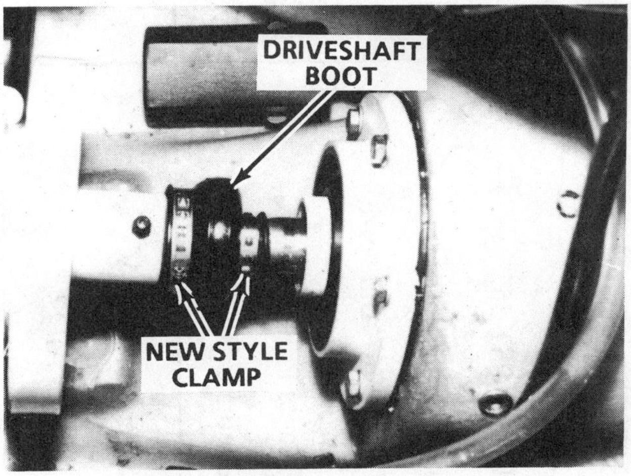

New style driveshaft boot secured with new state-of-the-art clamps.

4-7 SEALANTS, ADHESIVES, LUBRICANTS, AND FUEL STABILIZERS

It is common practice for the larger manufacturers of personal watercraft to market their own line of products for use on their craft. Bombardier chemical engineers have developed such a line available for use with Bombardier personal watercraft and snowmobiles. Throughout this manual, the authors recommend the application of the manufacturer's products as a first choice. All products listed are alternatives of equal value, and may be used with confidence, if the manufacturer's line is not available.

Sealants and Adhesives

Many sealants and adhesives are recommended and they are **NOT** interchangeable. Each is designed to perform under a different set of conditions. Follow the directions on the package for cleaning and preparing surfaces. Sealants and adhesives **MUST** be applied **ONLY** to clean dry sur-

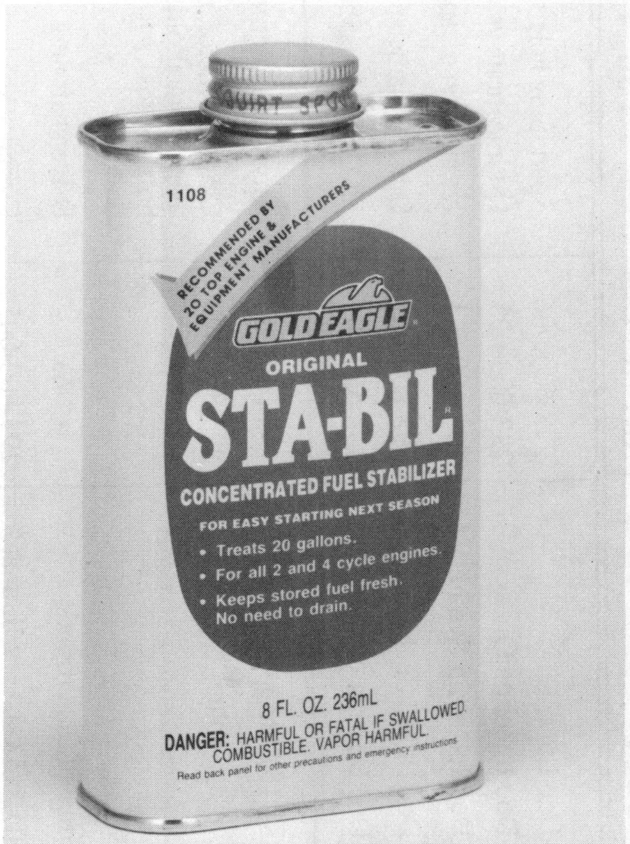

A gasoline stabilizer and conditioner may be used to prevent fuel from "souring" for up to twelve full months.

FACTORY RECOMMENDED LUBRICANTS AND ADHESIVES

PRODUCT	PART NUMBER	USE
Sea-Doo Lube	293-600-006	Spark plug hole threads
Sea-Doo Injection Oil	293-600-005	Oil injection system and rotary valve face.
Sea-Doo Jet pump Synthetic Oil	293-600-011	Jet pump end cap reservoir
Sea-Doo PL-32 Sealant	293-800-001	Outside diameter of all oil seals
Sea-Doo Bearing Grease	293-550-005	Jet Pump thrust bearing and washer
Loctite 157 Stripper	295-000-040	Anywhere Loctite 515 Gasket Eliminator was used.
Loctite 242 (blue)	293-800-002	Cranking motor, rock grate, steering nozzle, engine support, engine mount, exhaust manifold, flame arrestor, rotary valve, and stator plate mounting bolt threads. Also crankshaft taper and all electrical component mounting bolts.
Loctite 414 Superbond	N/A	Drive shaft plug
Loctite 515 Gasket Eliminator	293-800-007	Cylinder head, crankcase, and nozzle-to-impeller housing mating surfaces.
Loctite 517 or 587 Superflex Ultra Blue	293-800-016	Impeller housing mounting holes, ride shoe perimeter, and all oil seal lips.
Loctite 592	N/A	Exhaust outlet water fittings
Loctite PST 567	293-800-013	All other water fittings, and crankcase drain plug threads.
Loctite 648 (green)	290-899-788	Distance sleeve on rotary valve shaft
Loctite 767 Anti-seize	293-550-001	Crankshaft/flywheel threads, drive shaft and impeller shaft threads, and all bearing race exteriors.
Loctite Primer N	293-600-012	Internal aluminum threads
Loctite U Blue	293-800-006	All exposed cable threads
Silicone 732 RTV	N/A	Electrical harness grommets, hull screw threads, and wear ring lip.
Armour All ®	N/A	Vinyl seat

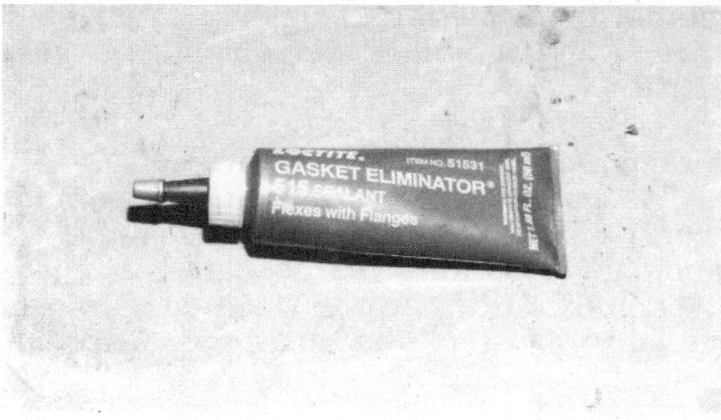

One of the many Loctite products the manufacturer endorses for use in sealing engine parts.

faces. Apply sparingly -- excessive amounts may block oil passageways and cause serious damage.

The accompanying table lists all the required products to lubricate and correctly seal any work or maintenance performed on the craft.

Fuel Stabilizer

"Sta-Bil" Fuel Conditioner and Stabilizer is recommended during engine operation and during the storage period. This fluid absorbs water in the fuel system and protects against corrosion.

If used during operation, this fuel additive will prevent the formation of gum and varnish deposits and greatly extend the period between required carburetor overhauls.

When added to the fuel during storage, the additive will prevent the fuel from "souring" for up to twelve full months.

Bombardier distributes many products under their own label, such as this Multi-Purpose Lubricant.

4-8 FIBERGLASS HULLS

Fiberglass reinforced plastic hulls are tough, durable, and highly resistant to impact. However, like any other material they can be damaged. One of the advantages of this type construction is the relative ease with which it may be repaired. Because of its break characteristics, and the simple techniques used in restoration, these hulls have gained popularity throughout the world.

A fiberglass hull has almost no internal stresses. Therefore, when the hull is broken or stove-in, it retains its true form. It will not dent to take an out-of-shape set. When the hull sustains a severe blow, the impact will be either absorbed by deflection of the laminated panel or the blow will result in a definite, localized break. Repairs are usually confined to the general area of the rupture.

Cleaning, Waxing, and Polishing

Any craft should be washed with clear water after each use to remove surface dirt and any salt deposits from use in salt water. Regular rinsing will extend the time between waxing and polishing. It will also give "pride of ownership", by having a sharp, "squared away" appearance. Elbow grease, a mild detergent, and brush will be required to remove stubborn dirt, oil, and other unsightly deposits.

Avoid harsh abrasives or strong chemical cleaners. A white buffing compound can be used to restore the original gloss to a scratched, dull, or faded area. The finish should be thoroughly cleaned, buffed, and polished at least once each season. Take care when buffing or polishing with a marine cleaner not to overheat the surface being worked, because the area will become "burned".

4-9 SUBMERGED ENGINE SERVICE

A submerged engine is always the result of an unforeseen accident. Once the craft is recovered, special care and service procedures **MUST** be closely followed in order to return the unit to satisfactory performance.

CRITICAL WORDS

Never attempt to start an engine immediately after it has been submerged. If water has entered a cylinder, the piston will not be able to compress the liquid. Any attempt to start the engine will most likely result in a bent connecting rod.

NEVER, again we say **NEVER** allow an engine that has been submerged to stand more than a couple hours before following the procedures outlined in this section and making every effort to get it running. Such delay will result in serious internal damage. If all efforts fail and the engine cannot be started after the following instructions have been performed, the engine should be disassembled, cleaned, assembled, using new gaskets, seals, and **O**-rings, and then started as soon as possible.

Submerged engine treatment is divided into three unique problem areas: submersion in salt water; submerged while engine was running; and a submerged unit in fresh water.

The most critical of these three circumstances is the engine submerged in salt water, with submersion while running, a close second.

Salt Water Submersion

NEVER attempt to start the engine after it has been recovered from salt water submersion. This action will only result in additional parts being damaged and the cost of restoring the engine will be increased considerably. If the engine was submerged in salt water the complete unit **MUST** be disassembled, cleaned, and assembled with new gaskets, **O**-rings, and seals. See Chapter 8 -- **ENGINE.** The corrosive effect of salt water can only be eliminated by the complete job being properly performed.

SUBMERGED WHILE RUNNING

SPECIAL INSTRUCTIONS

If the engine was running when it was submerged, the chances of internal engine damage is greatly increased. Remove the spark plugs to prevent compression in the cylinders. Use a socket wrench on the flywheel nut to rotate the crankshaft. If the attempt fails, the chances of serious internal damage is greatly increased. Serious damage might include, but not be limited to: bent connecting rod/s, bent crankshaft, or damaged cylinder. If the crankshaft cannot be rotated, the engine must be completely disassembled. See Chapter 8 -- **ENGINE.**

Submerged Engine -- Fresh Water

As an aid to performing the restoration work, the following steps are numbered and should be followed in sequence. However, illustrations are not included with the procedural steps because the work involved is general in nature.

1- Recover the craft as quickly as possible.

2- Remove the engine compartment cover and the spark plugs.

3- Remove the carburetor diaphragm cover.

4- Flush the outside of the engine with fresh water to remove silt, mud, sand, weeds, and other debris. **DO NOT** attempt to start the engine if sand has entered any part of the engine. Such action will only result in serious damage to engine components. Sand in the engine means the unit must be disassembled. See Chapter 8 -- **ENGINE.**

CRITICAL WORDS

Never attempt to start an engine immediately after it has been submerged. If water has entered a cylinder, the piston will not be able to compress the liquid. Any attempt to start the engine will most likely result in a bent connecting rod.

5- Remove as much water as possible from the engine. Most of the water can be eliminated by turning the craft over with the spark plug opening facing **DOWN**. With the craft in this position, rotate the flywheel with a socket wrench on the flywheel nut. Rotate the crankshaft through at least 10 complete revolutions. After you are satisfied there is no more water in the cylinders, proceed with the next step.

6- Alcohol will absorb moisture. Therefore, position the craft in an upright position, and then pour alcohol into the carburetor throat. Again rotate the flywheel through several revolutions of the crankshaft.

7- Pour alcohol into the spark plug openings and again rotate the crankshaft.

8- With the craft in the upright position, remove the two crankcase drain plugs -- Allen head bolts -- located under the oil injection pump on the starboard side. Allow all the liquid -- oil and possibly water -- to drain.

9- With the craft still upright, pour approximately one teaspoon of engine oil into each spark plug opening. Rotate the crankshaft to distribute the oil in the cylinders.

10- Install and connect the spark plugs.

11- Install the carburetor diaphragm cover.

Move the craft to a body of water or connect a garden hose to the engine cooling water supply fitting or flush fitting on the cylinder head.

12- Obtain **FRESH** fuel and attempt to start the engine.

CAUTION

The following procedure MUST be followed whenever the engine is started with a hose attached for cooling the engine. Start the engine, and then IMMEDIATELY turn the water on. The engine MUST be started first to prevent water from backing up into the tuned exhaust pipe and into the engine. Water must circulate anytime the engine is operating to prevent overheating -- which could cause damage to moving parts and possibly cause engine seizure. NEVER allow the engine to operate without cooling water for more than 15 seconds. NEVER run the engine with a flushing attachment for more than 5 minutes. The driveline seal is not provided with cooling water if it is not immersed in water. To shut the engine down, first shut the water supply off, and then shut down the engine. If this sequence of tasks is not followed as stated, the engine could be severely damaged.

REMEMBER

Start the engine.
Turn the water on.
Turn off the water.
Shut down the engine.

13- If the engine fails to start, determine the cause, electrical or fuel, correct the problem and again attempt to get it running. **NEVER** allow a recovered submerged engine to remain unstarted for more than a couple hours without following the procedures in this section and attempting to get it running. If attempts to start the engine fail, the unit should be disassembled, cleaned, assembled, using new gaskets, seals, and O-rings, as **SOON** as possible.

4-10 WINTER STORAGE

Taking extra time to store the craft properly at the end of each season, will increase the chances of satisfactory service when brought out for the next season's operation. **REMEMBER**, idleness is the greatest enemy of a marine engine. The unit should be run on a monthly basis. The steering and shifting mechanism should also be worked through complete cycles several times each month. The owner who spends a small amount of time involved in such maintenance will be rewarded by satisfactory performance, and greatly reduced maintenance expense for parts and labor.

Proper storage involves adequate protection of the unit from physical damage, rust, corrosion, and dirt.

If a submerged engine is not attended to without delay, the unit may suffer serious damage and keep the owner off the water much longer than necessary, until expensive repairs are completed.

The following steps provide an adequate maintenance program for storing the unit at the end of each season.

1- Empty all fuel from the carburetor. For many years there has been the widespread belief simply shutting off the fuel at the tank and then running the engine until it stops is the proper procedure before storing the engine for any length of time. Right? **WRONG! ! !**

First, it is **NOT** possible to remove all fuel in the carburetor by operating the engine until it shuts down from lack of fuel. Considerable fuel is trapped in the float chamber and other passages and in the line leading to the carburetor. The **ONLY** guaranteed method of removing **ALL** fuel is to take the time to remove the carburetor, and drain the fuel.

Secondly, if the engine is operated with the fuel supply shut off until it quits from lack of fuel, the fuel and oil mixture inside the block is removed, leaving bearings, pistons, rings, and other parts without any protective lubricant.

Remove the spark plugs and pour about an ounce of two-cycle engine oil into each spark plug opening. Wait a minute or two, and then slowly rotate the flywheel through two complete revolutions of the crankshaft. This action will evenly distribute the oil on the cylinder walls. Install the spark plugs.

2- Drain the fuel tank and the fuel lines. Pour approximately one quart (0.96 liters) of benzol (benzine) into the fuel tank, and then rinse the tank and pickup filter with the benzol. Drain the tank. Store the craft with the engine compartment cover **OPEN** to allow air to circulate around the tank. Leave the tank fuel vent **OPEN** to permit air to circulate through the tank.

3- Clean the in-line fuel filter with benzol. See Chapter 6 -- **FUEL**.

4- Lubricate the throttle and shift linkage. Lubricate the steering pivot shaft with water resistant multi-purpose lubricant.

5- Plug the exhaust outlet with a shop towel or similar item to prevent any type of contaminant from entering the exhaust system.

BATTERY STORAGE

Remove the battery from the craft and keep it charged during the storage period. Clean the battery thoroughly of any dirt or corrosion, and then charge the battery to the full specific gravity reading. After charging, store the battery in a clean cool dry place where it will not be damaged or knocked over.

NEVER store the battery with anything on top of it or cover the battery in such a manner as to prevent air from circulating around the filler caps. All batteries, both new and old, will discharge during periods of storage, more so if they become hot than if they are stored in a place to remain cool.

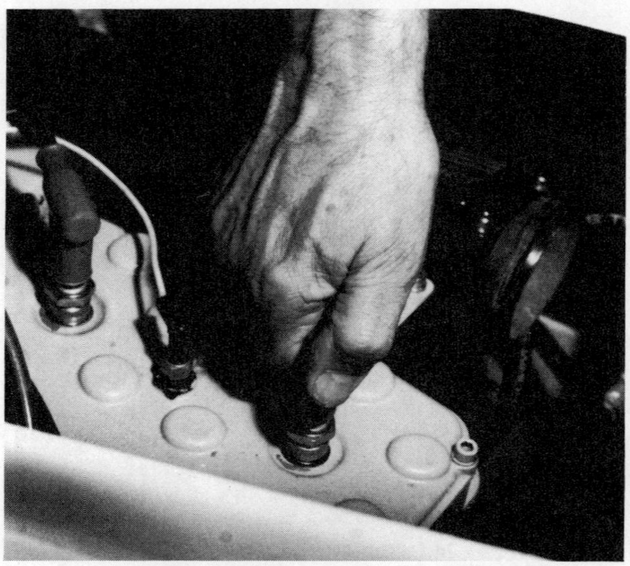

When preparing the craft for off-season storage, remove the spark plugs and insert about an ounce of oil into each opening, as explained in the text.

The battery should always be removed from the craft during extended periods of storage. Keep it dry, cool, and charged.

Therefore, the electrolyte level and the specific gravity should be checked at regular intervals. A drop in the specific gravity reading is cause to charge the battery back to a full reading.

One of the largest personal watercraft battery manufacturers recommends a battery in storage should be checked every two weeks and charged if necessary. The electrolyte can also be replenished at this time to prevent sulfation from occurring inside the battery.

In cold climates, **EXERCISE CARE** in selecting the battery storage area. A fully charged battery will freeze at about 60° below zero.

4-11 PRE-SEASON CHECK

Before attempting to start the craft after winter storage, check the steering and throttle action for smooth operation, without any indication of binding. If the battery was removed for storage, check to be sure it has a full charge.

FILLING THE BATTERY

Remove each filler cap using a pair of pliers. Fill each cell to the proper level with distilled water. The level of electrolyte should be between the upper and lower level marks. When filling the battery, fill to the upper line and allow the battery to stand for about twenty minutes. Check the level again and replenish, as necessary.

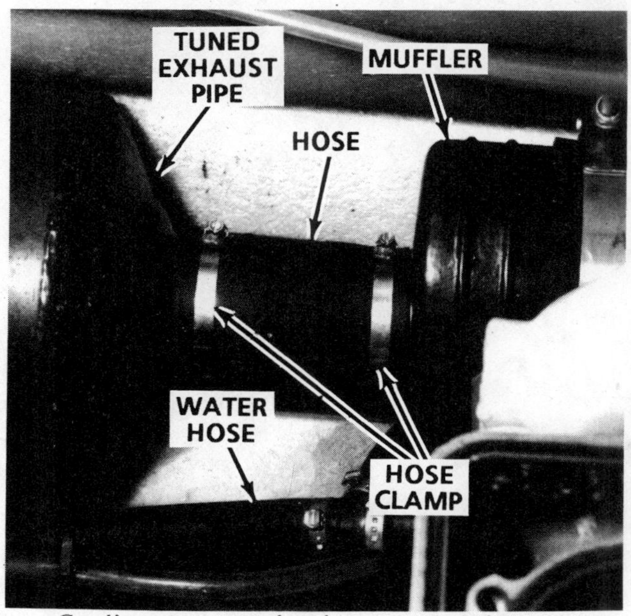

Cooling water and exhaust system hose connections should be checked as part of the pre-season routine.

Use a temperature corrected hydrometer to test the specific gravity of the electrolyte. At 20°C (68°F), the hydrometer reading should be 1.26 on the scale. If it is necessary to charge the battery, leave the filler caps lightly resting on the cell openings to allow the gases to escape. The charging current should not exceed 1.9 amps for 10 hours.

After charging, push each filler cap into place and wipe the top of the battery clean before installation.

Clean the battery posts and battery cable ends with a wire brush to ensure good, clean connections. Clean the top surface of the battery and identify the "**POS**" or "+" and the "**NEG**" or "-" embossed symbols. Correctly connect the battery cables to the battery observing polarity. If the cables are connected backwards, expensive parts in the ignition system will be **DESTROYED** the first time an attempt is made to start the engine.

Check to be sure the breather pipe is in place on the battery and the pipe was not pinched when the battery was installed.

FIRST TIME START

If the proper maintenance was observed at the end of the previous season, both the carburetor and fuel tank should be completely empty. Fill the gas tank with the proper mixture and open the fuel valve, if equipped. Open the choke. **DO NOT** attempt to crank the engine until it starts, **AT THIS TIME**. Considerable cranking time will be spent before the oil/fuel mixture engine bearings and other moving parts. Ex-

*The bilge inlet and hose should be checked at the beginning of each season to **ENSURE** the system will work on demand in an emergency.*

pensive parts may be damaged due to lack of lubrication and the cranking motor will become excessively hot, which is also **BAD NEWS**.

If the craft cannot be moved to a body of water, connect a garden hose to the flush fitting.

CAUTION

The following procedure MUST be followed whenever the engine is started with a hose attached for cooling the engine. Start the engine, and then IMMEDIATELY turn the water on. The engine MUST be started first to prevent water from backing up into the tuned exhaust pipe and into the engine. Water must circulate anytime the engine is operating to prevent overheating -- which could cause damage to moving parts and possibly cause engine seizure. NEVER allow the engine to operate without cooling water for more than 15 seconds. NEVER run the engine with a flushing attachment for more than 5 minutes. The driveline seal is not provided with cooling water if it is not immersed in water. To shut the engine down, first shut the water supply off, and then shut down the engine. If this sequence of tasks is not followed as stated, the engine could be severely damaged.

REMEMBER

Start the engine.
Turn the water on.
Turn off the water.
Shut down the engine.

The correct procedure for starting the engine is to obtain an **AEROSOL** can of WD-40. The product **MUST** be an aerosol canister because the liquid in the container must be atomized. A simple squirt can will not do the job. **DO NOT** use ether or a commercial starting fluid because they will evaporate almost instantly. These products work well with four-stroke engines equipped with an oil pump, but they do not contain a lubricant essential for a two-stroke engine. WD-40 is an extremely flammable oil.

Check to be sure both spark plugs are tight to prevent the mist from the canister from igniting. Spray the WD-40 through the flame arrestor and continue spraying with **NO** throttle advance until the engine starts. As soon as the engine is operating, cease spraying the WD-40. The engine may emit some smoke for a few seconds after it starts. Watch the sediment bowl (if so equipped). As soon as the bowl fills with fuel, the engine should operate on the regular oil/fuel mixture. Work the throttle to keep the engine at idle speed until it reaches operating temperature, then close the choke. Advance the throttle a few times and the engine should then idle smoothly.

Check to be sure the battery electrolyte reaches the upper mark. During engine operation, the electrolyte should remain between the upper and lower level marks.

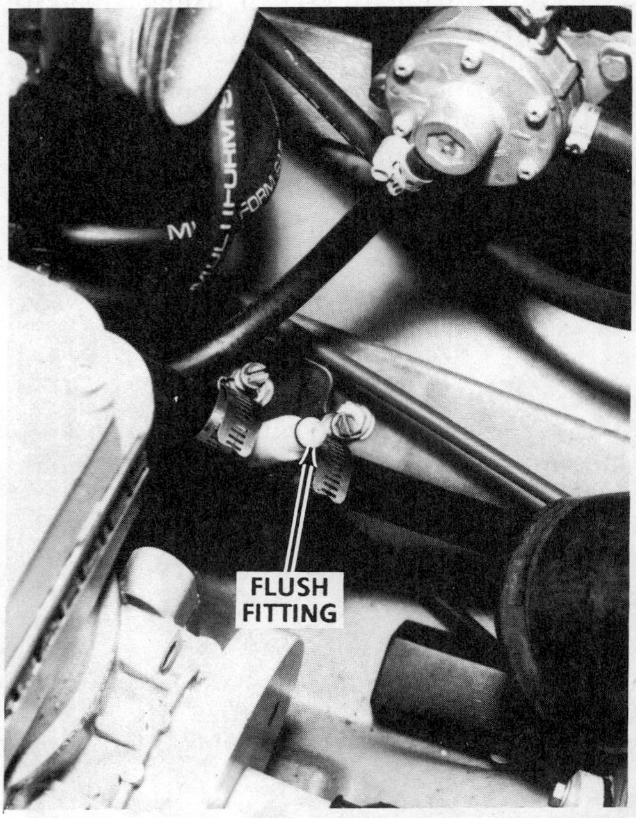

*Location of the engine flush fitting. If the engine is to be operated out of the water, a garden hose **MUST** be attached to this fitting to provide cooling water, as explained with **CAUTIONS** in the text.*

5
TROUBLESHOOTING

5-1 INTRODUCTION

This chapter is divided into five main sections as follows:

5-2 Mechanical engine problems.
5-3 Fuel system problems.
5-4 Ignition system faults.
5-5 Cranking system failures.
5-6 Charging system malfunctions.
5-7 Troubleshooting Charts

If a change in engine rpm is experienced at normal operating speed, such a change is usually a symptom of an engine problem or trouble in the jet pump. An increase or a decrease in rpm may be an indication of the area to be checked. The following generalizations may be helpful in isolating the problem.

A noticeable change in craft performance, as evidenced by one of the following symptoms, may be attributed to one or more of the areas listed:

Lower than Normal Engine Rpm

Mechanical problems in the engine:
- Plugged flame arrestor.
- Fouled spark plug/s, incorrect heat range or gap.
- Shorted or weak ignition coil.
- Contaminated fuel.
- Restricted exhaust system.
- Inadequate fuel pump pressure.
- Loss of compression due to blown head gasket.

Mechanical problems in the jet drive:
- Worn or damaged impeller.
- Main bearing failure.
- Driveshaft spline failure.

Higher than Normal Engine Rpm

Mechanical problem in the engine:
- None.

Mechanical problems in the jet drive:
- Clogged intake grate -- possibly plastic bag or other debris.
- Line, sea weed, etc. entangled in impeller.
- Ingested sand.

If a reduction in speed is experienced, accompanied with no loss or gain of engine rpm, the loss of performance can fairly accurately be attributed to some type of drag on the craft or restriction of the water ejected from the jet pump.

Drag on the hull is usually caused by damage to the craft.

Many times, drag from the jet pump is caused by a partially lowered reverse gate, if so equipped, or from dragging debris caught in the intake grate.

A partially lowered reverse gate may be caused by incorrect cable adjustment or by a weak or broken return spring.

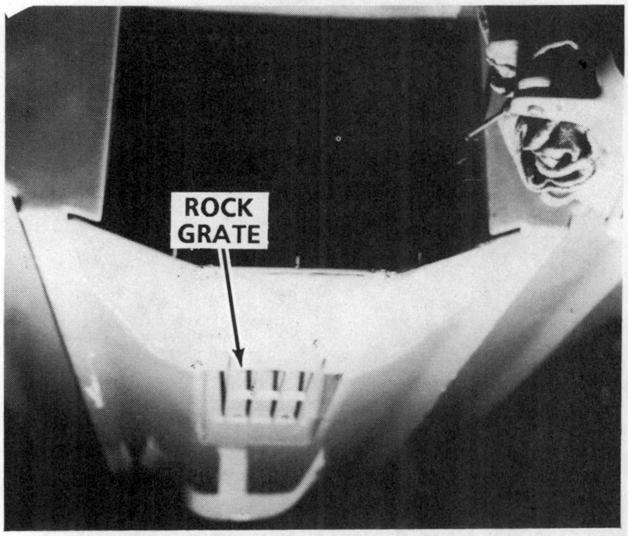

One of the most hazardous conditions for a personal watercraft is restriction of water through the rock grate caused by a plastic bag, plastic wrap, or by some type of sea weed. Even the smallest reduction in water volume will affect jet pump performance and could quickly lead to engine overheating.

Engine Troubleshooting

Troubleshooting must be a well thought-out procedure. Always attempt to proceed with the troubleshooting in an orderly manner. The "shot in the dark" approach will only result in wasted time, incorrect diagnosis, replacement of unnecessary parts, and frustration.

Obviously, if the instructions are to be of maximum benefit as a guide, they must be fully understood and followed in the proper sequence.

When an engine fails to start, the trouble must be localized to one of four general areas: cranking system; ignition system; fuel system; or compression. Each of these areas must be systematically inspected until the trouble is isolated. At that time, detailed tests of the system or area must be made to determine the part causing the problem. The last section of this chapter describes charging system troubleshooting.

Troubleshooting Check

When using this Troubleshooting Check, proceed sequentially through each test until the defect is uncovered. Then skip to the detailed testing procedure and check for that system. For example, if, when using the Troubleshooting Check procedure, the first two systems, cranking and ignition tests OK, but the third test shows there is trouble in the fuel system, then proceed to the detailed test under the Fuel System Troubleshooting Check in Section 5-3.

Cranking System Test

1- Turn the key switch to the **START** position or press the ignition button. The cranking motor should crank the engine at a normal rate of speed.

If the cranking motor cranks the engine slowly or fails to crank it at all, the trouble is either in the battery or in the cranking system. Test the battery first, then if the battery checks out, proceed directly to Troubleshooting the Cranking System in Section 5-7 for detailed testing procedure to isolate the problem.

Ignition System Test

2- Connect a standard spark tester (available at almost any automotive parts store at very nominal cost), between the spark plug and the spark plug lead. Hold the spark plug lead with an insulated holder (to avoid receiving a "shock" when the current passes through). Crank the engine with the cranking motor. If there is **NO** spark, or if the spark is very weak, the trouble is in the ignition system. Proceed directly to the Ignition Troubleshooting in Section 5-4 for

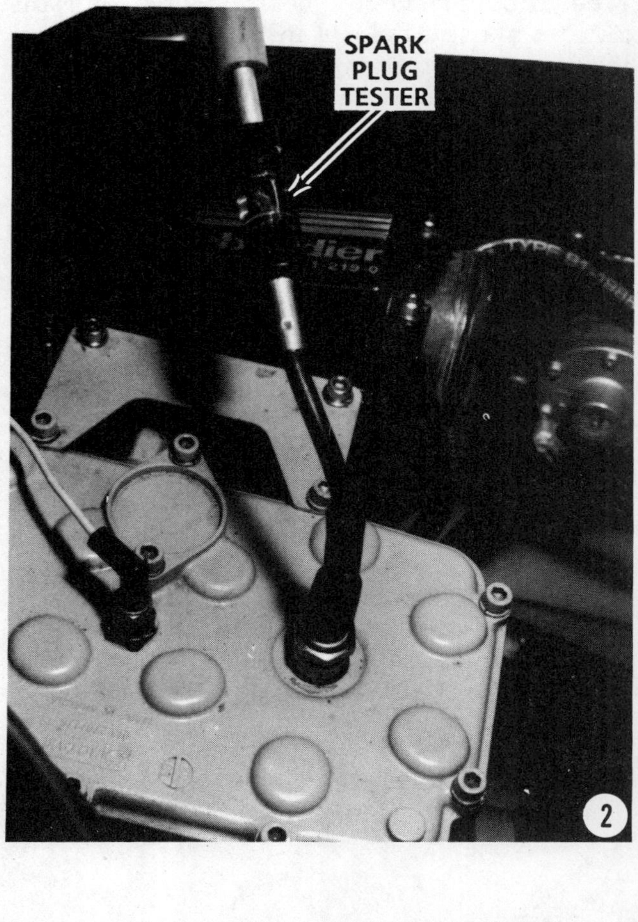

detailed testing procedures to uncover the problem.

3- Remove the flame arrestor. Look down into the throat of the carburetor and at the same time open and close the throttle several times. Observe the throttle valve to verify it moves from the fully closed to the fully open position. Perform the same check with the choke valve.

Attempt to determine the age of the fuel in the fuel tank. Many fuels tend to "sour" in three to four months, especially if an additive such as Sta-Bil has not been added. Other fuels take longer to "sour". In no case should an attempt be made to start the engine if the fuel in the tank is more than 12 months old.

Compression Test

4- Good compression is the key to proper engine performance. An engine with worn piston rings or a blown gasket cannot be made to perform satisfactorily until the mechanical defects contributing to low compression are corrected. Generally, a compression gauge is used to determine the cranking pressure within each cylinder.

To make a compression test, remove both spark plugs and remember or identify from which cylinder they were taken. Evaluation of the spark plug firing end can be most useful in determining how the cylinder is functioning. After the spark plugs have been removed, install a compression gauge into one cylinder and crank the engine.

The throttle valve and choke **MUST** be in the **WIDE OPEN** position in order to obtain maximum readings. Crank the engine through several revolutions and observe the highest reading on the compression gauge.

The cranking compression for a new engine is 142 psi. The significance in a compression test is the variation in pressure readings between the cylinders. As long as this variation is between 20-30 psi, cylinder compression is considered as "satisfactory". If a greater variation exists, then the lower reading cylinder should be checked by making a cylinder leak test. A simple leak test can be made by first inserting a teaspoonful of oil into the spark plug opening of the lower reading cylinder, and then cranking the engine a few times to distribute the oil.

Check the compression again to see if inserting the oil caused a change. The oil helps make a temporary seal around the rings and increases the compression. If the reading increased, the compression loss was probably due to worn rings. If the reading did not change appreciably, the loss may be due to a blown head gasket.

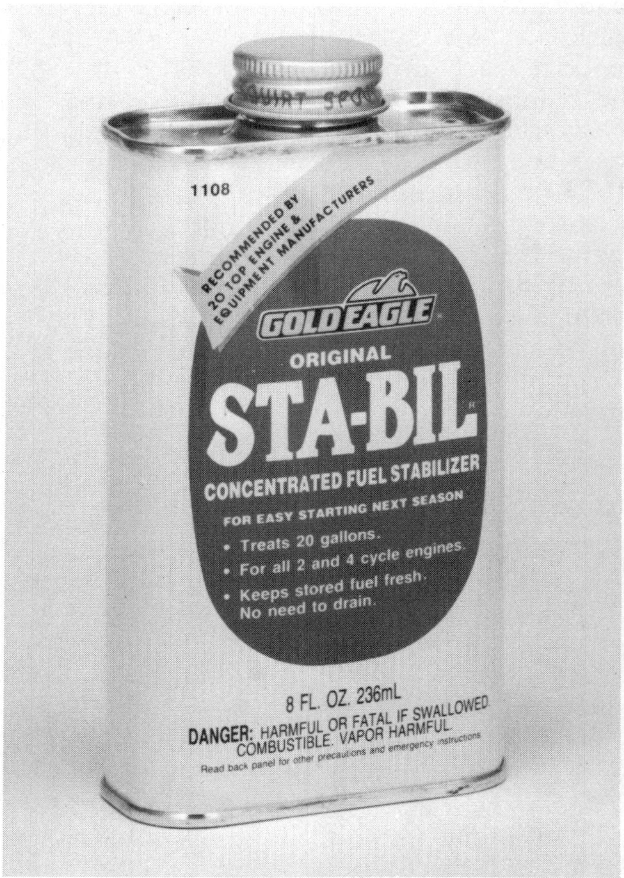

A gasoline stabilizer and conditioner may be used to prevent fuel from "souring" for up to twelve full months.

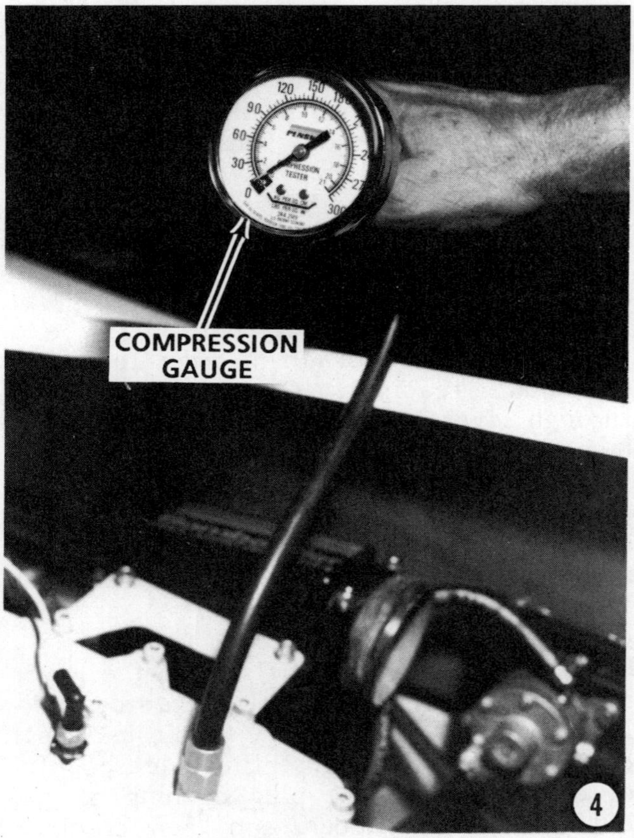

5-2 MECHANICAL ENGINE PROBLEMS

Jumping Waves

Even though "jumping waves" with a personal watercraft may be considered **GREAT** sport, consider the extensive internal engine damage which will surely occur if this type action is consistently practiced.

During wave jumping or using a wet ramp, when the craft clears the water, the engine RPM approaches the limit set by the RPM limiter in an unloaded condition.

Once the craft returns to the water surface, the sudden load on the impeller is tranferred along the driveline. Speed of the rotating pump shaft and the crankshaft is suddenly reduced while the flywheel continues to spin rapidly due to its mass. This condition, especially when repeated over a period of time, often shears the Woodruff key and the flywheel becomes repositioned on the crankshaft.

A very small misalignment will result in a change in ignition timing. The craft operator will notice a drastic loss of power. A worse condition would be the flywheel shifting to the degree engine start would not be possible.

This sudden increase and decrease in engine speed dramatically shortens engine life. Bent crankshafts and broken connecting rods can usually be directly attributed to this type of "fun play".

Easy Check for Twisted Crankshaft

A twisted crankshaft may be detected by first removing the spark plugs and setting up a dial indicator in one of the spark plug openings. If a dial indicator is not available, a pencil, straw, or other small diameter straight object may be used.

With the indicator, or other device, in place, determine the TDC (top dead center) of the No. 1 piston. Now, make two matching marks -- one on the flywheel and the other adjacent on the stator. Next, rotate the flywheel **EXACTLY** 180° -- preferably using a degree wheel on the flywheel. Make a second mark on the stator adjacent to the mark on the flywheel.

NOW, check for TDC of the other piston. The dial indicator should indicate the same reading as for the first piston **AND** the two marks on the stator should be **EXACTLY** opposite each other through the center of the crankshaft. If the indicator is off or if the marks on the stator are not aligned, as described -- the crankshaft is twisted and **MUST** be replaced.

Engine Noises

Engine noises can be generally classified as knocks, slaps, clicks, or squeaks. These noises are usually caused by loose bearings, sloppy pistons, or other moving parts of the engine.

A main bearing knock is usually identified by a dull thud which is noticeable under engine load. Attempting to move the craft

Removal of the crankshaft is not necessary to check for a "twisted" condition, as explained in the text.

under power with it tied to the dock will bring out a main bearing knock. If you pull the spark plug wire from one plug at a time and the noise disappears when one of the wires is removed, then the noise is probably coming from that cylinder. The cause may be either the rod bearing, piston pin, or the piston, which has quieted down because the load has been removed from the part. If a rod bearing has failed, the noise will be loudest when the engine is decelerating. Piston pin noise and piston pin slap are generally louder when a cold engine is first started.

Carbon build up in the combustion chamber can cause interference with a piston.

Many times the source of an unusual sound may be isolated by using a stethoscope or other listening device. One such device is a long shank screwdriver. Allow your hand to extend over the end of the handle, press an ear to your hand, and then probe with the other end of the screwdriver by touching it firmly against the block at each cylinder or noise area. Take care and use good judgment when using this method of attempting to detect the cause of a problem because the noise will travel through other metallic parts of the engine and could lead to a false interpretation of what you are hearing.

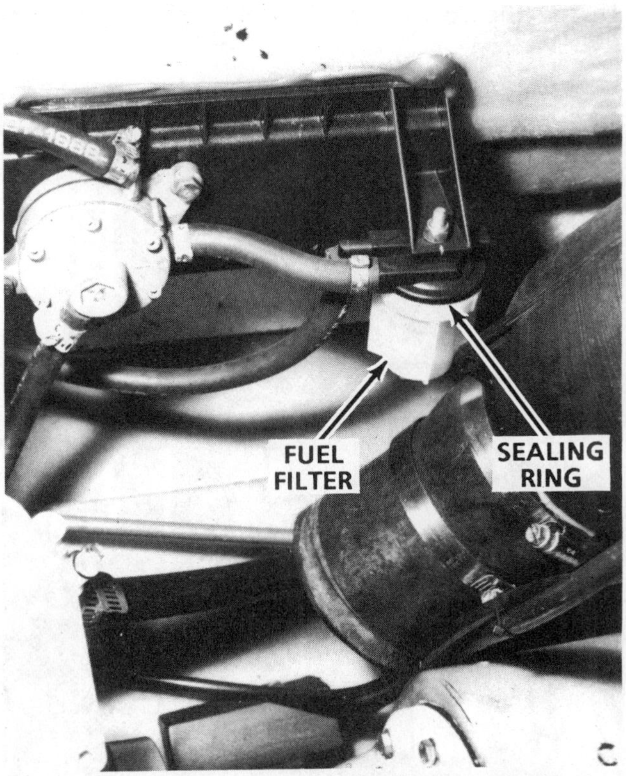

A leaking sealing ring at the fuel filter will draw air into the system. The leak may not be evident until the engine is shut down -- fuel will then leak out past the damaged sealing ring.

5-3 FUEL SYSTEM PROBLEMS

The following paragraphs provide an orderly sequence of tests to pinpoint problems in the system. It is very rare for the carburetor by itself to cause failure of the engine to start.

Many times fuel system troubles are caused by a plugged fuel filter, a defective fuel pump, or by a leak in the fuel line from the fuel tank to the fuel pump or by a vacuum leak in the line between the pump and the crankcase.

Fuel will begin to sour in three to four months and will cause engine starting problems. A fuel additive such as Sta-Bil may be used to prevent gum from forming during storage or prolonged idle periods.

When the engine is hot, the fuel system can cause starting problems. After a hot engine is shut down, the temperature inside the fuel bowl may rise to 200°F and cause the fuel to actually boil. All carburetors are vented to allow this pressure to escape to the atmosphere. However, some of the fuel may percolate over the high-speed nozzle and overflow into the intake manifold.

In order for this raw fuel to vaporize enough to burn, considerable air must be added to lean out the mixture. Therefore,

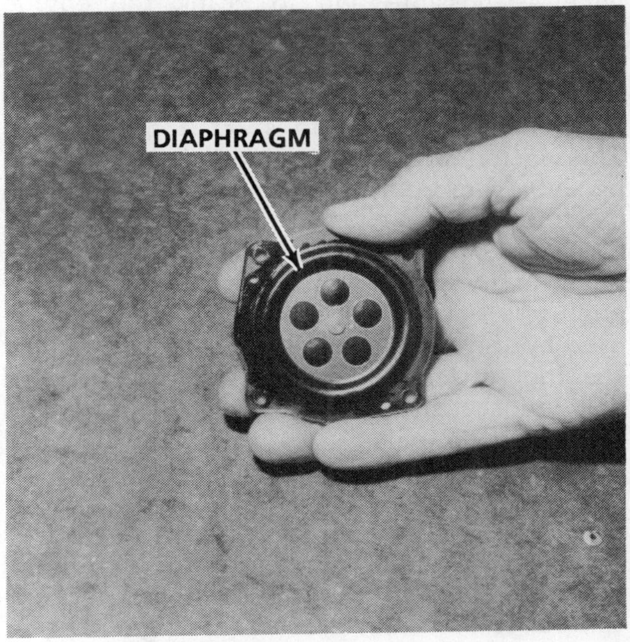

The tiniest "pin" hole in the carburetor diaphragm will seriously affect carburetor performance.

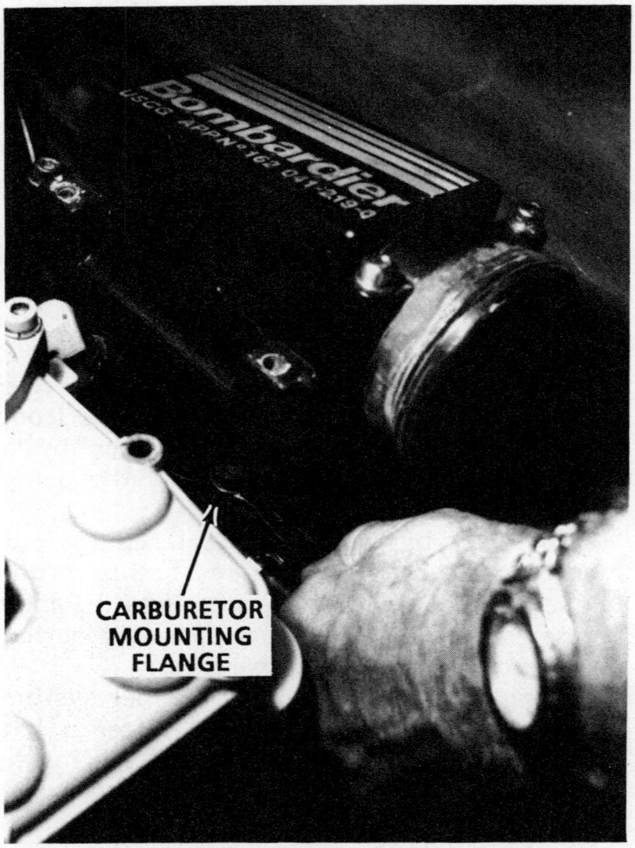

A small leak between the carburetor and the intake manifold, caused by a defective carburetor mounting gasket, will result in rough engine performance at idle speed.

the only remedy is to open the throttle as wide as possible and to crank the engine until enough air is drawn in to provide the proper mixture for the engine to start. **NEVER** move the throttle lever back-and-forth in an attempt to start a hot engine. This action will only compound the problem by adding more fuel to an already too-rich mixture.

A leak between the fuel tank and the pump many times will not appear when the engine is operating because the suction created by the pump sucking fuel will not allow the fuel to leak. Once the engine is turned off and the suction no longer exists, fuel may begin to leak.

Engine Surge

If the engine operates as if the load on the craft was being constantly increased and decreased, even though the operator attempts to maintain a constant engine speed, the problem can most likely be attributed to the fuel pump.

Rough Engine Idle

Problems preventing an engine from running smoothly include: a fouled spark plug; an air leak in the intake manifold; and uneven compression between the cylinders.

Of course any problem in the carburetor affecting the air/fuel mixture will also prevent the engine from operating smoothly at idle speed. These problems usually include: leaking needle valve and seat; a dirty flame arrestor; defective choke; and improper adjustments for idle mixture or idle speed.

5-4 IGNITION SYSTEM FAULTS

WARNING

Check to be sure the engine compartment is well ventilated and free of any gasoline vapors before starting any of the following tests. A spark will be generated creating a potential fire hazard if fuel vapors are present.

1- Check to be sure the battery has a full charge. If not, correct the condition by charging the battery or making a substitution.

2- Check all the terminal connections inside the electrical box. Check all primary

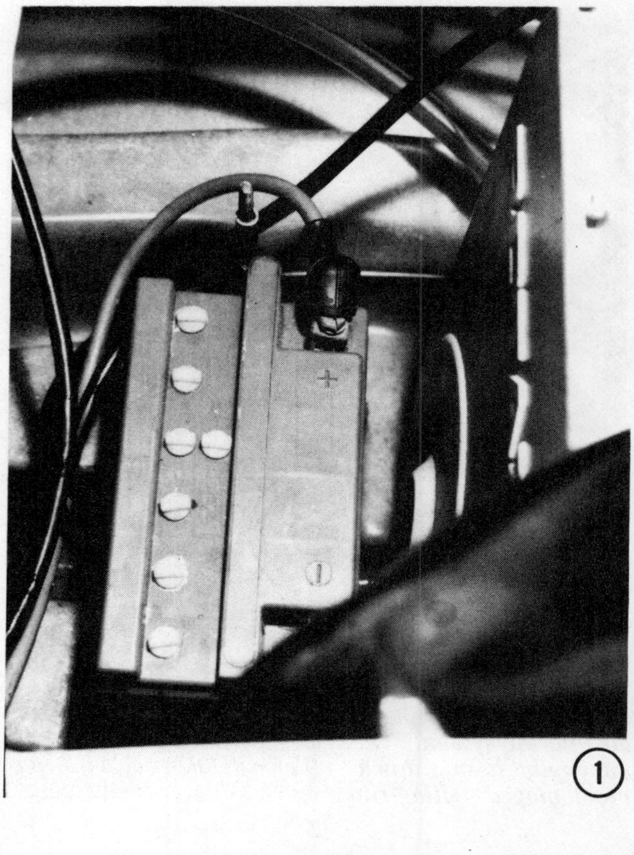

①

and secondary ignition wiring, the engine "kill" switch and the ignition switch.

3- Use a spark tester and check for spark. If a spark tester is not available, hold the plug wire about 1/4" (6.4mm) from the engine. Crank the engine through a few revolutions using the cranking motor and check for spark. A strong spark over a wide gap must be observed when testing in this manner, because under compression a strong spark is necessary in order to ignite the air/fuel mixture in the cylinder. This means it is possible to think a strong spark is present, when in reality the spark will be too weak when the plug is installed. If there is no spark, or if the spark is weak, the trouble is most likely under the flywheel in the stator assembly.

4- Perform the resistance test procedures given in the Ignition chapter to identify the faulty component.

Intermittent or Multiple Problems

Many ignition problems occur only during engine operation, when the component is subject to vibration and/or an increase in temperature. Many of these defective components return to normal once the engine is shut down and allowed to cool.

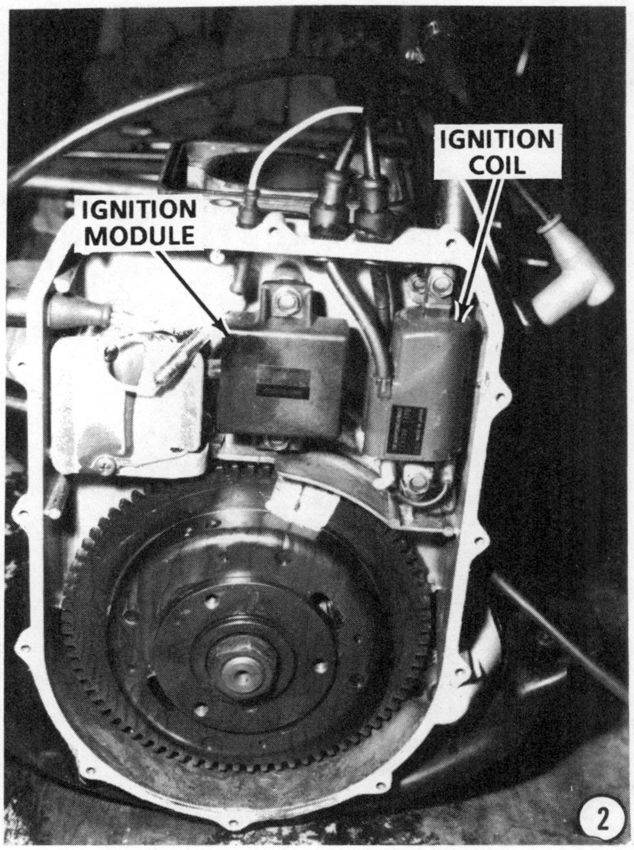

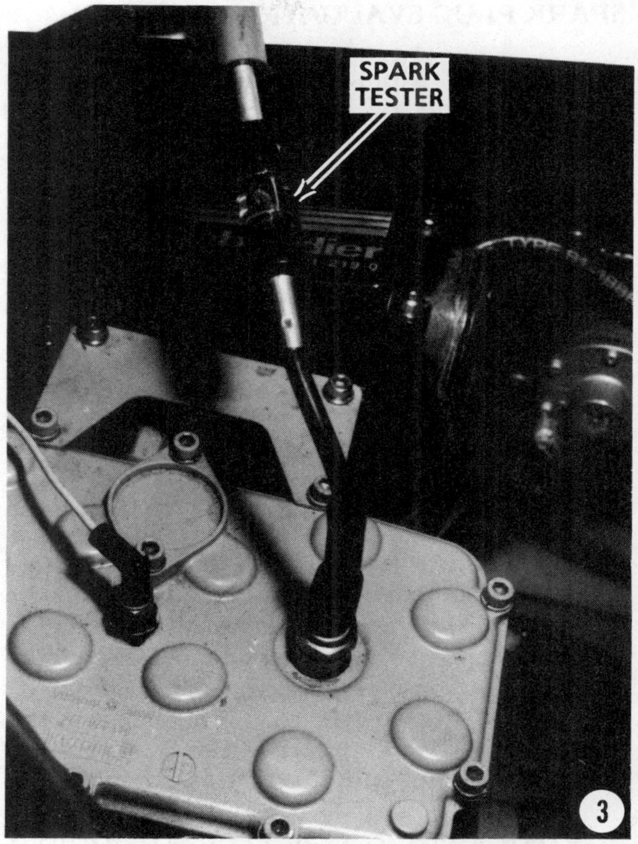

There is also the possibility of more than one defective component in the ignition system. One component failure may have caused a surge of power and "fried" another component.

Make every attempt to avoid the **BUY** and **TRY** method of troubleshooting. In **MOST** cases electrical components are not returnable, once the item leaves the store.

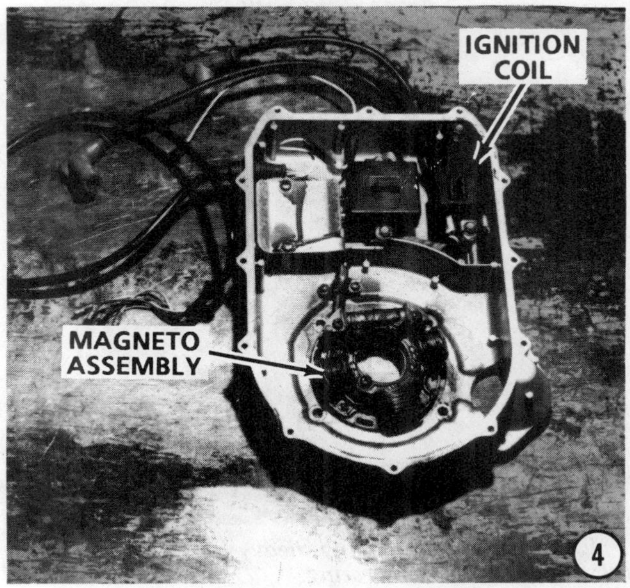

SPARK PLUG EVALUATION

Removal: Remove the spark plug wires by pulling and twisting on only the molded cap. **NEVER** pull on the wire or the connection inside the cap may become separated or the boot damaged. Remove the spark plugs and identify from which cylinder they were taken. **TAKE CARE** not to tilt the socket as you remove the plug or the insulator may be cracked.

Examine: Carefully examine the firing end to determine the firing condition in each cylinder.

Correct Color: A proper firing plug should be dry and powdery. Hard deposits inside the shell indicate the engine is starting to use some oil, but not enough to cause concern. The most important evidence is the light gray color of the porcelain, which is an indication this plug has been running at the correct temperature. This means the plug is one with the correct heat range and also that the air/fuel mixture is correct.

Overheating: A dead white or gray insulator, which is generally blistered, is an indication of overheating and pre-ignition. The electrode gap wear rate will be more than normal and in the case of pre-ignition, will actually cause the electrodes to melt. Overheating and pre-ignition are usually caused by overadvanced timing, detonation from using too-low an octane rating fuel, an excessively lean air/fuel mixture, or problems in the cooling system.

Rich Mixture: A black, sooty condition on both the spark plug shell and the porcelain is caused by an excessively rich air/fuel mixture, both at low and high speeds. The rich mixture lowers the combustion temperature so the spark plug does not run hot enough to burn off the deposits.

Deposits formed only on the shell is an indication the low-speed air/fuel mixture is too rich. At high speeds with the correct mixture, the temperature in the combustion chamber is high enough to burn off the deposits on the insulator.

Too Cool: A dark insulator, with very few deposits, indicates the plug is running too cool. This condition can be caused by low compression or by using a spark plug of an incorrect heat range. If this condition shows on only one plug it is most usually caused by low compression in that cylinder. If both plugs have this appearance, then it is probably due to the plugs having a too-low heat range.

Fouled: A fouled spark plug may be caused by the wet oily deposits on the insulator shorting the high-tension current to ground inside the shell. The condition may also be caused by ignition problems which prevent a high-tension pulse to be delivered to the spark plug.

Carbon Deposits: Heavy carbon-like deposits are an indication of excessive oil consumption. This condition may be the result of worn piston rings.

Example of overheating, heavy load. Use a spark plug with a lower heat rating.

Spark plug fouled by oil. The engine may be in need of an overhaul.

Electrode Wear: Electrode wear results in a wide gap and if the electrode becomes carbonized it will form a high-resistance path for the spark to jump across. Such a condition will cause the engine to misfire during acceleration. If both plugs are in this condition, it can cause an increase in fuel consumption and very poor performance at high-speed operation. The solution is to replace the spark plugs with a rating in the proper heat range and gapped to specification.

Red rust-colored deposits on the entire firing end of a spark plug can be caused by water in the cylinder combustion chamber. This can be the first evidence of water entering the cylinders through the exhaust manifold. This condition **MUST** be corrected at the first opportunity.

5-5 CRANKING SYSTEM FAILURES

Regardless of how or where the solenoid is mounted, the basic circuits of the starting system on all makes of cranking motors are the same and similar tests apply. In the following testing and troubleshooting procedures, the differences are noted.

WARNING

ALWAYS TAKE TIME TO VENT THE BILGE WHEN MAKING ANY OF THE TESTS AS A PREVENTION AGAINST IGNITING ANY FUMES ACCUMULATED IN THAT AREA.

Faulty Symptoms

If the cranking motor spins, but fails to crank the engine, the cause is usually a corroded or gummy Bendix drive. The drive should be removed, cleaned, and given an inspection.

If the cranking motor cranks the engine too slowly, the following are possible causes and the corrective actions that may be taken:

a- Battery charge is low. Charge the battery to full capacity.

b- High resistance connections at the battery, solenoid, or motor. Clean and tighten all connections.

c- Undersize battery cables. Replace cables with sufficient size.

All engine cranking problems fall into one of three problem areas:

1- The cranking motor fails to turn.

2- The cranking motor spins rapidly, but does not crank the engine.

3- The cranking motor cranks the engine, but too slowly.

The following paragraphs provide a logical sequence of tests designed to isolate a problem in the cranking system.

Before wasting too much time troubleshooting the cranking motor circuit, the following checks should be made. Many times, the problem will be corrected.

Powdery deposits have melted and shorted-out this spark plug.

Red, brown, or yellow deposits, are by-products of combustion from fuel and lubricating oil.

a- Battery fully charged.
b- Main fuse is "good" (not blown).
c- All electrical connections clean and tight.
d- Wiring in good condition -- insulation not worn or frayed.

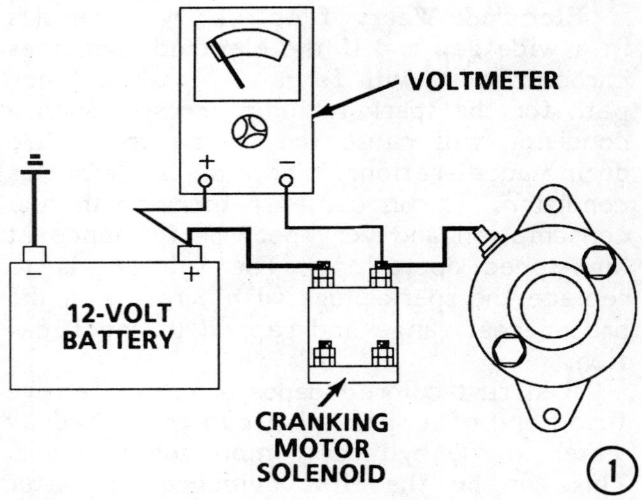

The following troubleshooting procedures are presented in a logical sequence.

Do not operate the cranking motor for more than 15 seconds. Prolonged cranking motor operation will cause overheating and damage the motor.

After each test, allow the cranking motor to cool for a minute or so.

Never depress the **START** button to activate the cranking motor while the engine is operating. Such action will damage the pinion and/or flywheel gears.

Cranking Circuit Tests

1- Remove both spark plug leads from the plugs and ground the ends to the engine to prevent accidental engine start. Keep the leads grounded for the following tests.

Obtain a voltmeter and select the Vx1 DC scale. Observe the starter solenoid. Both large terminals have Red leads connected. One terminal has the battery cable connected and the other terminal has the lead from the cranking motor connected. Make contact with the negative Black meter lead to the terminal on the starter solenoid with the battery cable connected.

Make contact with the positive Red meter lead to the positive battery terminal. The voltmeter should register less then 0.25V DC. If the reading exceeds 0.25V DC, clean the connections at both ends of the positive battery cable and repeat the test. If the reading does not change, replace the cable.

2- Keep the meter on the same scale. Make contact with the negative Black voltmeter lead to the negative terminal on the battery. Make contact with the positive Red voltmeter lead to a good engine ground. The voltmeter should register less then 0.25V DC. If the reading exceeds 0.25V DC, clean the connections at both ends of the negative battery cable and repeat the test. If the reading does not change, replace the cable.

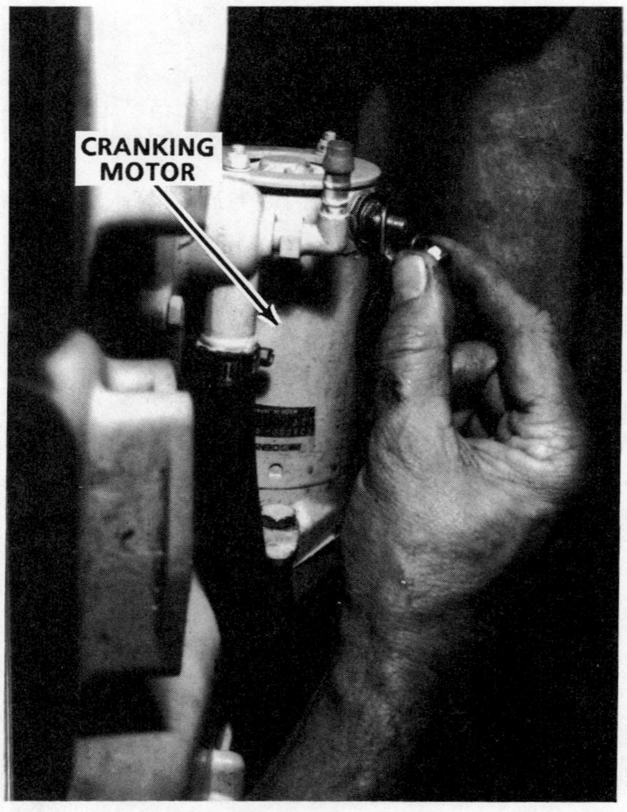

*Disconnecting electrical leads in preparation to performing bench tests on the cranking motor, as described in detail -- Chapter 9 -- **ELECTRICAL.***

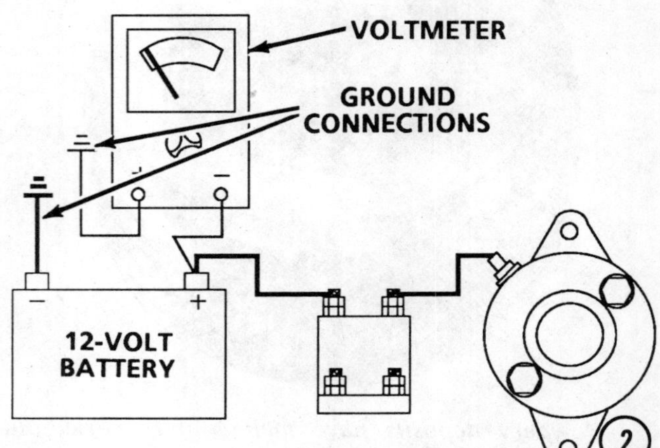

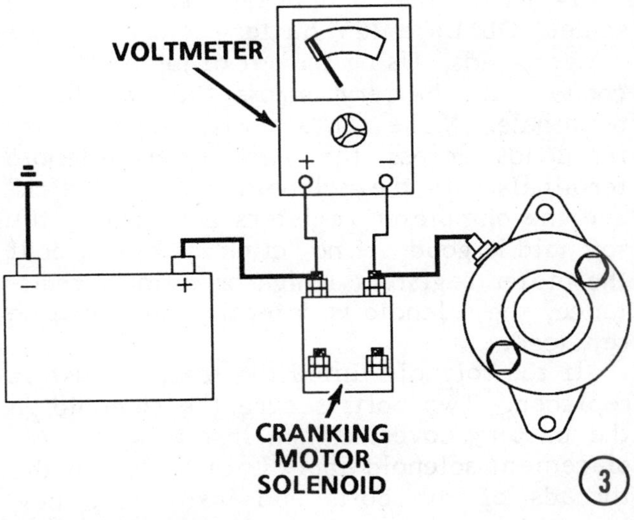

CRITICAL WORDS

During the following tests -- Step 3 and Step 4, both meter leads must be placed on the solenoid terminals **WHILE** the engine is being cranked. **ALSO,** both meter leads must be removed from the solenoid terminals **WHILE** the engine is being cranked and **BEFORE** the engine has stopped cranking. If these precautions are not followed, the voltmeter may be damaged.

3- Keep the meter on the same scale. Obtain the services of an assistant.

Signal the assistant to start cranking the engine. While the engine is being cranked, make contact with the negative Black voltmeter lead to the solenoid terminal which has the lead from the cranking motor connected, and then quickly make contact with the positive Red voltmeter lead to the solenoid terminal which has the battery cable connected.

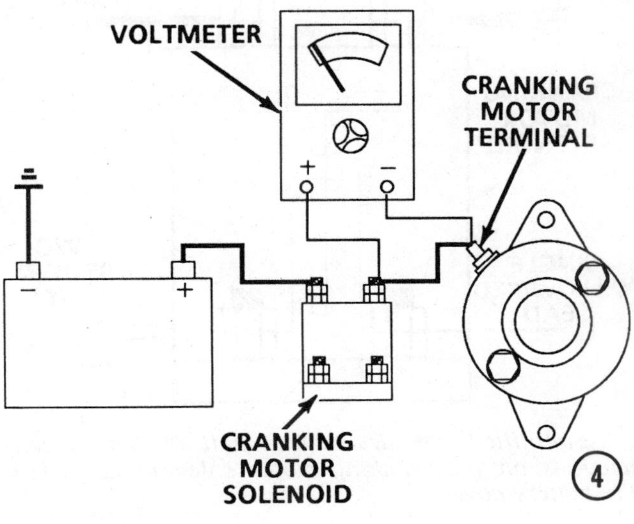

If the meter reading exceeds 0.25V DC, the starter solenoid must be replaced. Remove the meter leads and signal the assistant to cease cranking.

4- Keep the meter on the same scale and the assistant at the **START** button. Observe the same precautions as before in Step 3. While the engine is being cranked, make contact with the negative Black voltmeter lead to terminal on the cranking motor, and then quickly make contact with the positive Red voltmeter lead to the solenoid terminal which has the Red cranking motor connected.

If the meter reading exceeds 0.25V DC, the connections at both ends of the solenoid-to-cranking motor lead should be cleaned and the test repeated. If the reading does not change, the lead must be replaced.

CRANKING MOTOR SOLENOID TESTING

Description

The cranking motor solenoid is actually a switch between the battery and the engine. The switch cannot be serviced. Therefore, if testing indicates the switch is faulty, it **MUST** be replaced. The solenoid is housed inside the battery cover. The solenoid may remain inside the cover during testing, but the cover must be removed, in order to make the necessary connections.

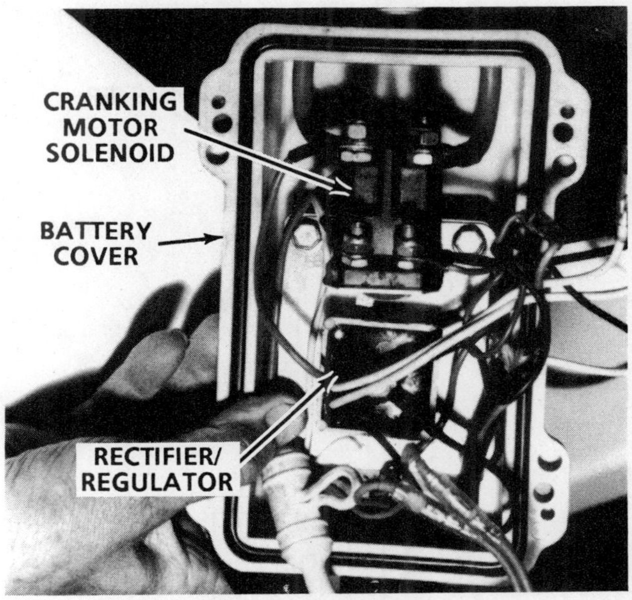

Cover removed from the battery box exposing the rectifier/regulator and the cranking motor solenoid.

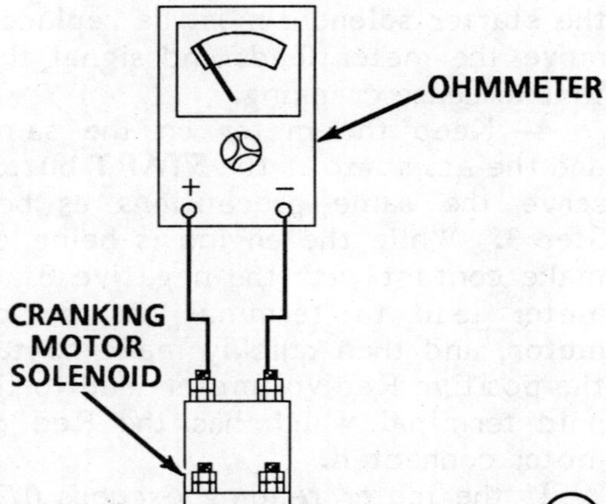

1

Preliminary Tasks

Remove the battery cover.

Disconnect the three Red leads from the upper left terminal.

Disconnect the single Red large lead from the upper right terminal.

Disconnect the Yellow/Red lead from the lower left terminal.

Disconnect the two Brown leads from the lower right terminal.

Cranking Solenoid Testing

1- Select the Rx1 ohm scale on the ohmmeter. Make contact with the two meter leads across the two upper solenoid terminals. The meter should read an infinite resistance. If the meter shows a resistance of less than infinity, the solenoid is defective and must be replaced.

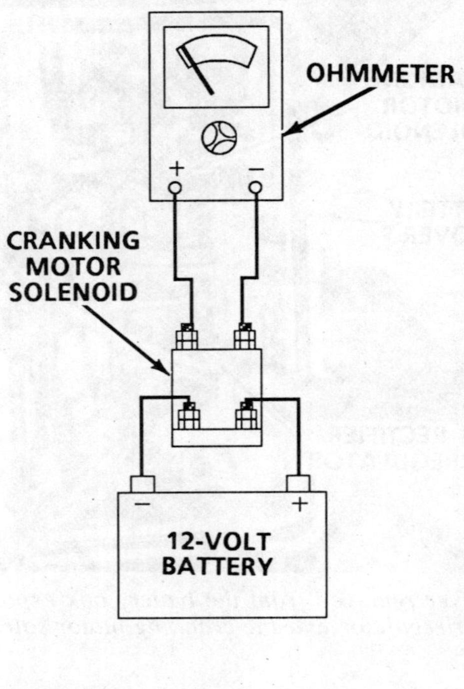

2

2- Keep the ohmmeter on the same scale. Obtain a 12V battery and a pair of jumper leads. Using the set of jumper leads, connect the battery across the two lower terminals. Make contact with the two meter leads across the two upper solenoid terminals. If the solenoid emits a "click" and the ohmmeter registers zero ohms, the solenoid is good. If no "click" is heard, or if the meter registers a high or infinite resistance, the solenoid is defective and must be replaced.

If the solenoid failed the test, it must be replaced. Two bolts secure the solenoid to the battery cover. When installing the replacement solenoid apply Loctite 242 to the threads of the bolts and install the new solenoid in place. Tighten both bolts securely.

Connect the two Brown leads to the lower right terminal.

Connect the Yellow/Red lead to the lower left terminal.

Connect the single large Red lead to the upper right terminal.

Connect the three smaller Red leads to the upper left terminal.

Place the cover over the battery. Install and tighten the four securing screws.

5-6 CHARGING SYSTEM MALFUNCTIONS

The battery charging coil, rectifier/regulator, battery, and the necessary wiring to connect it all together comprise the charging system.

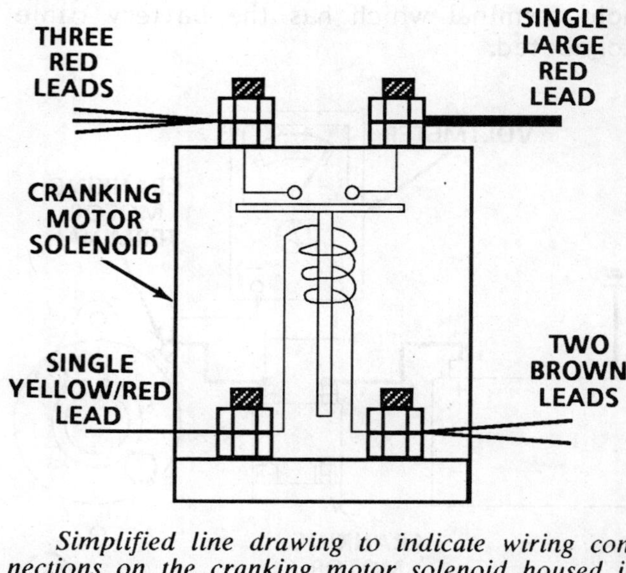

Simplified line drawing to indicate wiring connections on the cranking motor solenoid housed in the battery cover.

The generating coil and battery charging coils are mounted on the magneto assembly. The magneto assembly is secured to the stator plate.

Before the charging system is blamed for battery problems, consider other areas which may be the cause:

1- Operating at low speed for short periods.

2- Voltage losses due to high resistance.

3- Corroded battery cables, connectors, and terminals.

4- Low electrolyte level in the battery cells.

5- Prolonged disuse of the battery causing a self-discharged condition.

Service the battery, refer to Section 5-3. If, after charging, the battery fails to give the required specific gravity reading, then the battery will never hold a charge and must be replaced. This problem may have originated with the battery, a shorted cell perhaps, or may be attributed to the failure of the rectifier/regulator or charging coil. Refer to the Ignition chapter for resistance tests on these two components.

5-7 TROUBLESHOOTING CHARTS

The following six charts are presented as an aid to troubleshooting the engine and jet pump. Refer to the heading in the uppermost box which most closely identifies the problem, and then work downward from there. The probable causes are listed in columns under a specific system, for example: Fuel system, or Electrical system. Once a possible cause is identified, refer to the specific chapter for procedures on testing and service.

When using the charts -- proceed step-by-step through each of the tests or checks until the defect or cause of the problem is uncovered.

FIRST CHART -- Page 5-14 -- sets up the condition of the engine failing to start.

SECOND CHART -- Page 5-15 -- covers situation of engine starting, then stopping, without cause.

THIRD CHART -- Page 5-16 -- deals with an engine misfiring.

FOURTH CHART -- Page 5-17 -- most involved chart of the set -- covers loss of engine power.

FIFTH CHART -- Page 5-18 -- checks out strange unexplained engine noises.

SIXTH CHART -- Page 5-19 -- covers problems involving the jet pump.

ONE FINAL WORD

Before going out on the water, **TAKE TIME** to check and verify the drain plug is installed. Countless number of excursions have had a very sad beginning because the craft was eased into the water only to have the hull begin filling with water.

Keep your gas tank full, the fuel pump pumping, the spark plugs sparking, the charging circuit charging, and the pistons, well -- keep them working too.

Joan & Clarence Coles
and the crew
Seloc Publications

CHART ONE

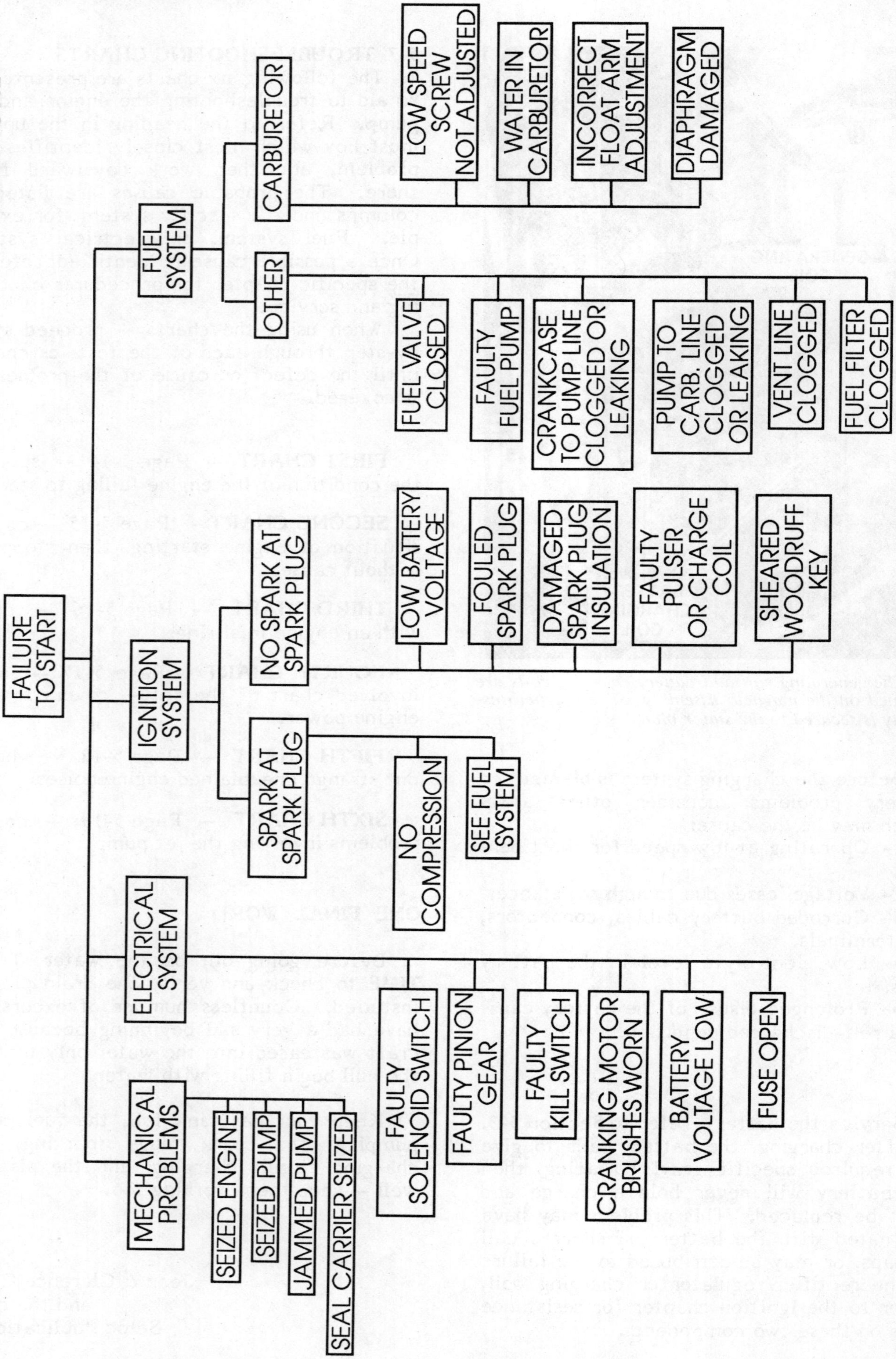
FAILURE TO START
FUEL SYSTEM
CARBURETOR
LOW SPEED SCREW NOT ADJUSTED
WATER IN CARBURETOR
INCORRECT FLOAT ARM ADJUSTMENT
DIAPHRAGM DAMAGED
OTHER
FUEL VALVE CLOSED
FAULTY FUEL PUMP
CRANKCASE TO PUMP LINE CLOGGED OR LEAKING
PUMP TO CARB. LINE CLOGGED OR LEAKING
VENT LINE CLOGGED
FUEL FILTER CLOGGED
IGNITION SYSTEM
NO SPARK AT SPARK PLUG
LOW BATTERY VOLTAGE
FOULED SPARK PLUG
DAMAGED SPARK PLUG INSULATION
FAULTY PULSER OR CHARGE COIL
SHEARED WOODRUFF KEY
SPARK AT SPARK PLUG
NO COMPRESSION
SEE FUEL SYSTEM
ELECTRICAL SYSTEM
FAULTY SOLENOID SWITCH
FAULTY PINION GEAR
FAULTY KILL SWITCH
CRANKING MOTOR BRUSHES WORN
BATTERY VOLTAGE LOW
FUSE OPEN
MECHANICAL PROBLEMS
SEIZED ENGINE
SEIZED PUMP
JAMMED PUMP
SEAL CARRIER SEIZED

CHART TWO

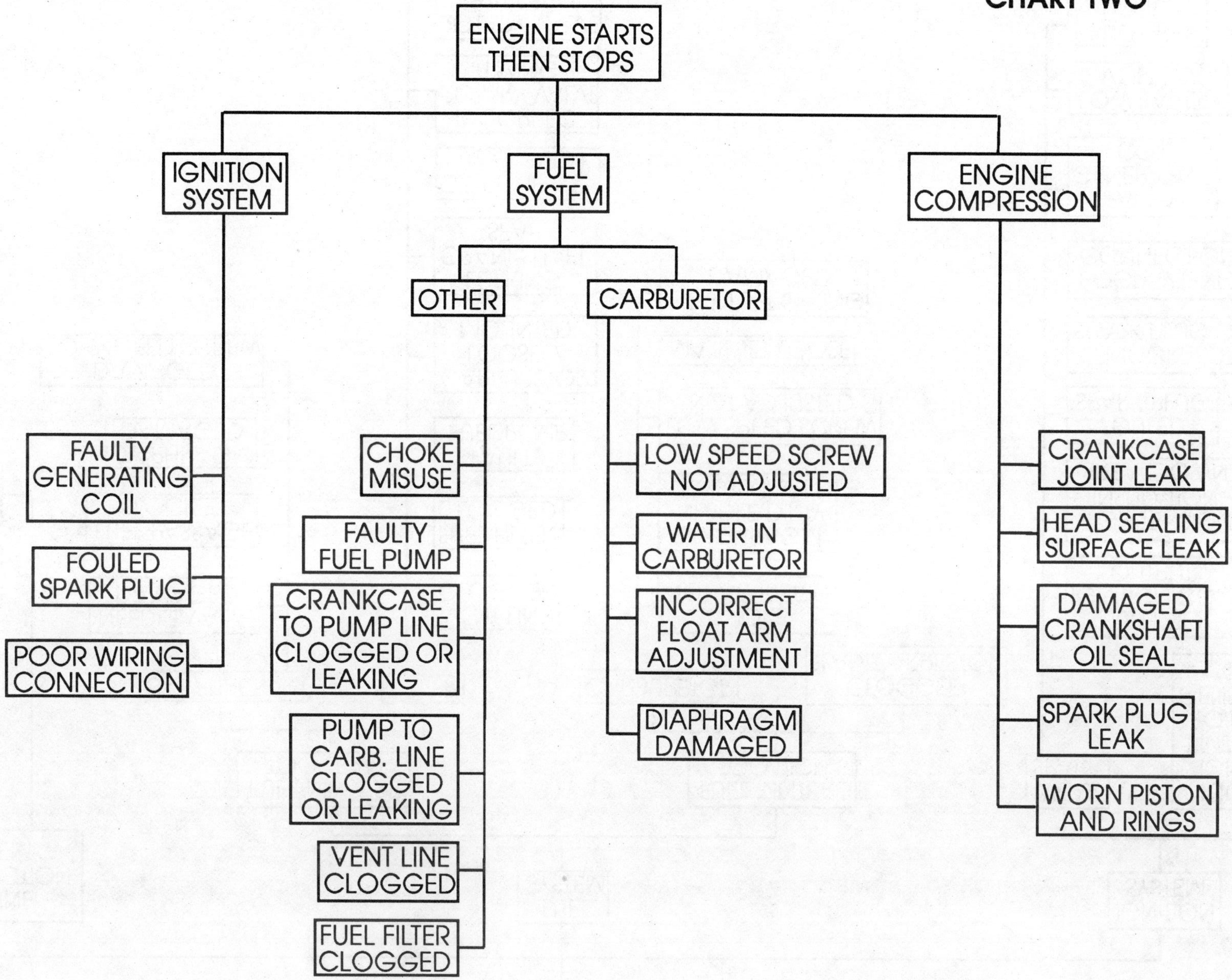
ENGINE STARTS THEN STOPS
IGNITION SYSTEM
FUEL SYSTEM
ENGINE COMPRESSION
OTHER
CARBURETOR
FAULTY GENERATING COIL
FOULED SPARK PLUG
POOR WIRING CONNECTION
CHOKE MISUSE
FAULTY FUEL PUMP
CRANKCASE TO PUMP LINE CLOGGED OR LEAKING
PUMP TO CARB. LINE CLOGGED OR LEAKING
VENT LINE CLOGGED
FUEL FILTER CLOGGED
LOW SPEED SCREW NOT ADJUSTED
WATER IN CARBURETOR
INCORRECT FLOAT ARM ADJUSTMENT
DIAPHRAGM DAMAGED
CRANKCASE JOINT LEAK
HEAD SEALING SURFACE LEAK
DAMAGED CRANKSHAFT OIL SEAL
SPARK PLUG LEAK
WORN PISTON AND RINGS

CHART THREE

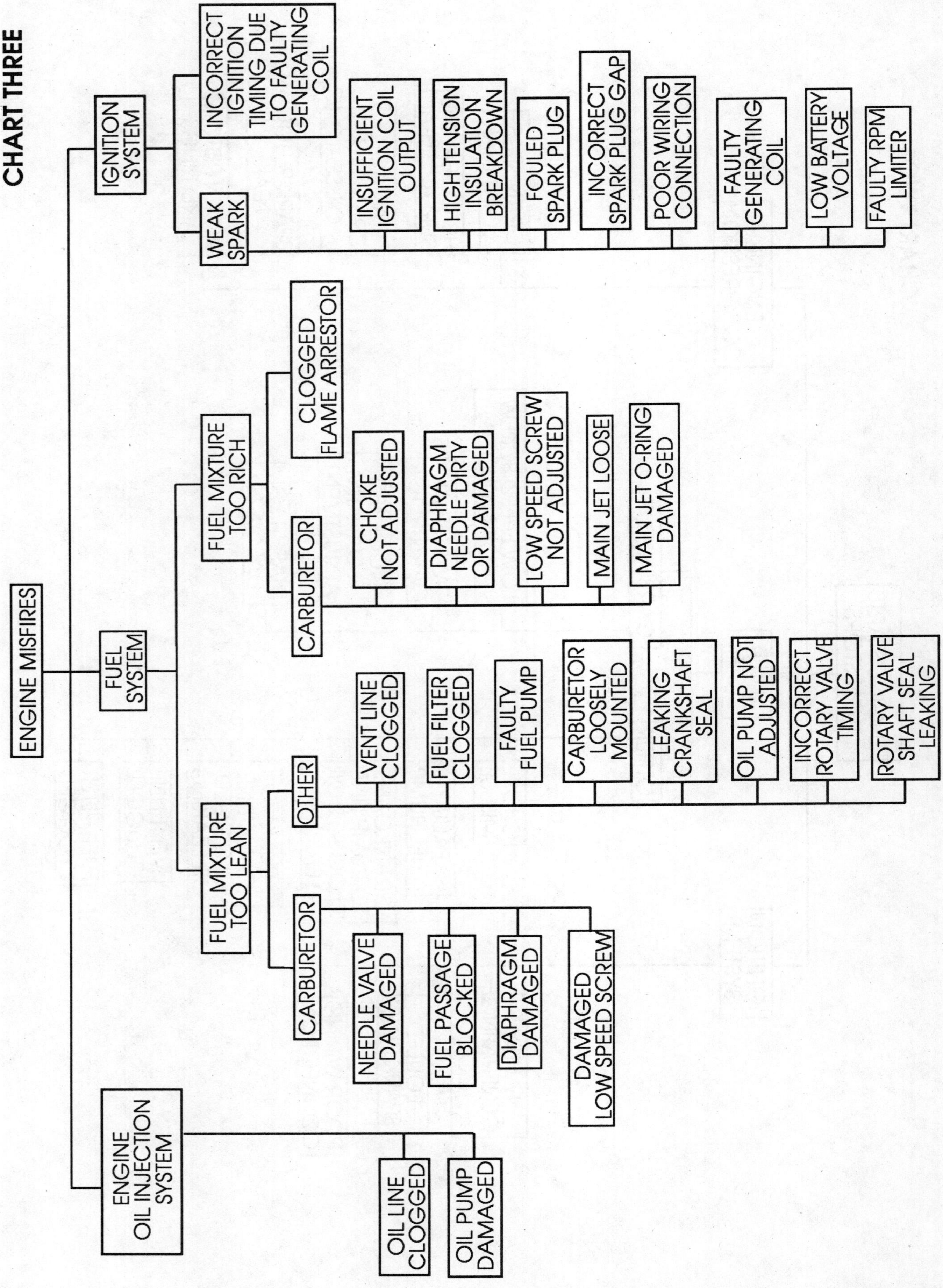
ENGINE MISFIRES
IGNITION SYSTEM
INCORRECT IGNITION TIMING DUE TO FAULTY GENERATING COIL
WEAK SPARK
INSUFFICIENT IGNITION COIL OUTPUT
HIGH TENSION INSULATION BREAKDOWN
FOULED SPARK PLUG
INCORRECT SPARK PLUG GAP
POOR WIRING CONNECTION
FAULTY GENERATING COIL
LOW BATTERY VOLTAGE
FAULTY RPM LIMITER
FUEL SYSTEM
FUEL MIXTURE TOO RICH
CLOGGED FLAME ARRESTOR
CARBURETOR
CHOKE NOT ADJUSTED
DIAPHRAGM NEEDLE DIRTY OR DAMAGED
LOW SPEED SCREW NOT ADJUSTED
MAIN JET LOOSE
MAIN JET O-RING DAMAGED
FUEL MIXTURE TOO LEAN
OTHER
VENT LINE CLOGGED
FUEL FILTER CLOGGED
FAULTY FUEL PUMP
CARBURETOR LOOSELY MOUNTED
LEAKING CRANKSHAFT SEAL
OIL PUMP NOT ADJUSTED
INCORRECT ROTARY VALVE TIMING
ROTARY VALVE SHAFT SEAL LEAKING
CARBURETOR
NEEDLE VALVE DAMAGED
FUEL PASSAGE BLOCKED
DIAPHRAGM DAMAGED
DAMAGED LOW SPEED SCREW
ENGINE OIL INJECTION SYSTEM
OIL LINE CLOGGED
OIL PUMP DAMAGED

CHART FOUR

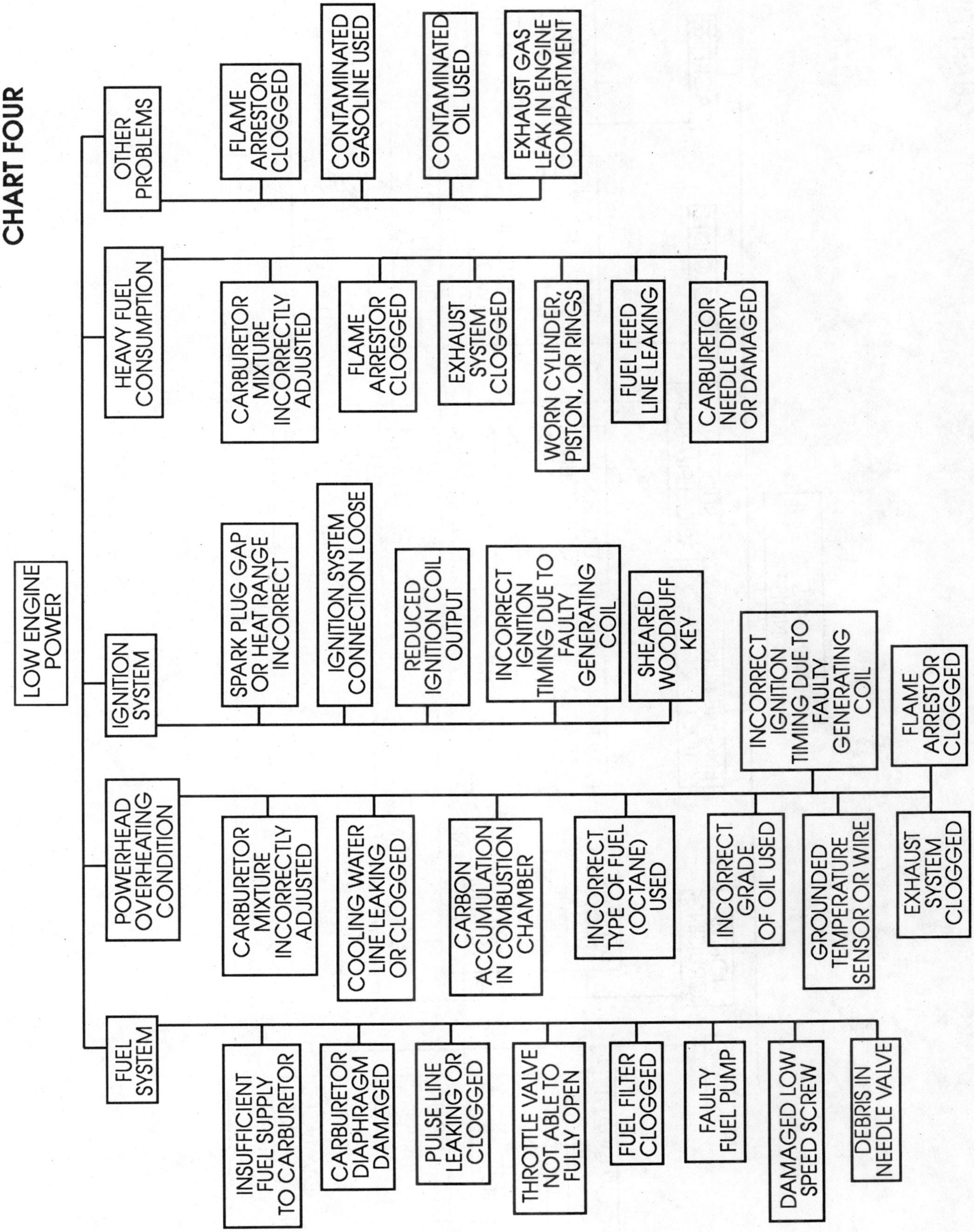
LOW ENGINE POWER
OTHER PROBLEMS
FLAME ARRESTOR CLOGGED
CONTAMINATED GASOLINE USED
CONTAMINATED OIL USED
EXHAUST GAS LEAK IN ENGINE COMPARTMENT
HEAVY FUEL CONSUMPTION
CARBURETOR MIXTURE INCORRECTLY ADJUSTED
FLAME ARRESTOR CLOGGED
EXHAUST SYSTEM CLOGGED
WORN CYLINDER, PISTON, OR RINGS
FUEL FEED LINE LEAKING
CARBURETOR NEEDLE DIRTY OR DAMAGED
IGNITION SYSTEM
SPARK PLUG GAP OR HEAT RANGE INCORRECT
IGNITION SYSTEM CONNECTION LOOSE
REDUCED IGNITION COIL OUTPUT
INCORRECT IGNITION TIMING DUE TO FAULTY GENERATING COIL
SHEARED WOODRUFF KEY
POWERHEAD OVERHEATING CONDITION
CARBURETOR MIXTURE INCORRECTLY ADJUSTED
COOLING WATER LINE LEAKING OR CLOGGED
CARBON ACCUMULATION IN COMBUSTION CHAMBER
INCORRECT TYPE OF FUEL (OCTANE) USED
INCORRECT GRADE OF OIL USED
INCORRECT IGNITION TIMING DUE TO FAULTY GENERATING COIL
GROUNDED TEMPERATURE SENSOR OR WIRE
FLAME ARRESTOR CLOGGED
EXHAUST SYSTEM CLOGGED
FUEL SYSTEM
INSUFFICIENT FUEL SUPPLY TO CARBURETOR
CARBURETOR DIAPHRAGM DAMAGED
PULSE LINE LEAKING OR CLOGGED
THROTTLE VALVE NOT ABLE TO FULLY OPEN
FUEL FILTER CLOGGED
FAULTY FUEL PUMP
DAMAGED LOW SPEED SCREW
DEBRIS IN NEEDLE VALVE

CHART FIVE

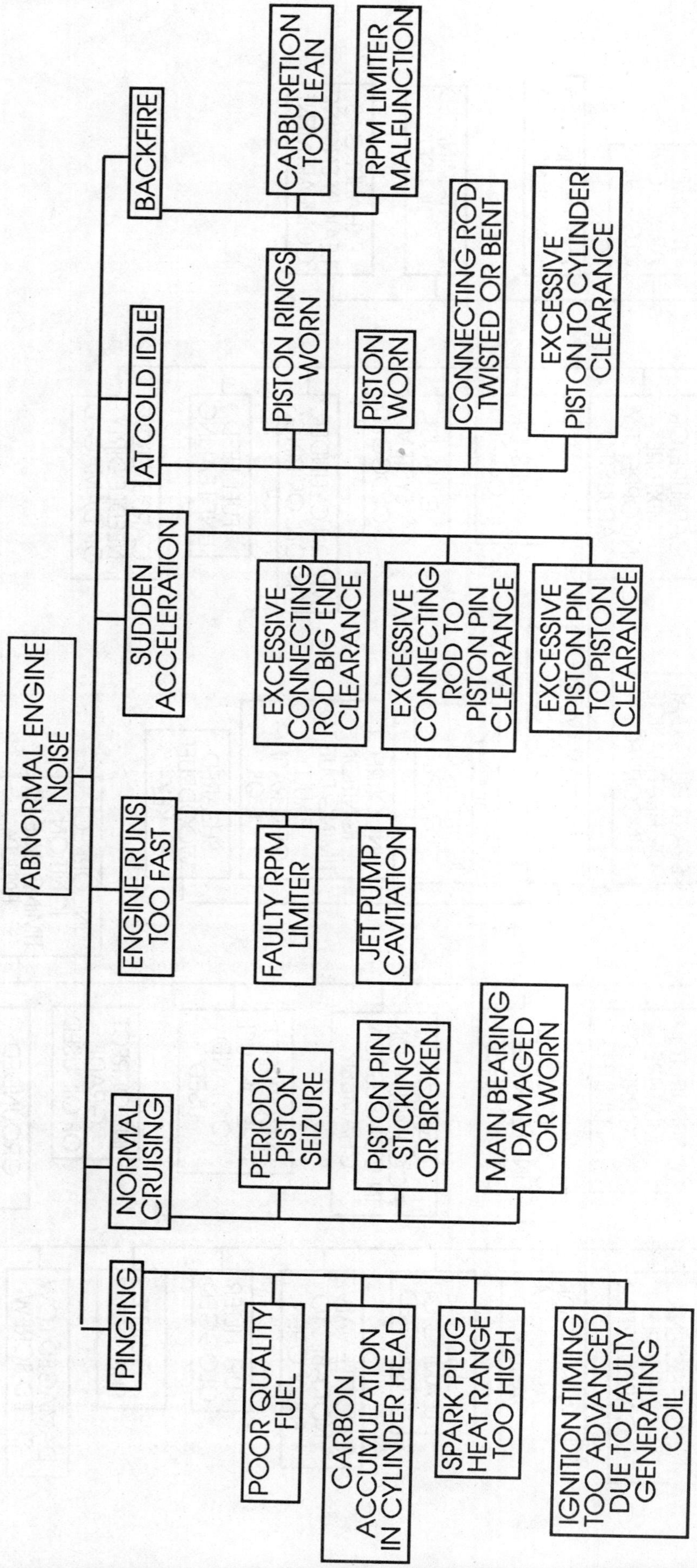
ABNORMAL ENGINE NOISE
PINGING
POOR QUALITY FUEL
CARBON ACCUMULATION IN CYLINDER HEAD
SPARK PLUG HEAT RANGE TOO HIGH
IGNITION TIMING TOO ADVANCED DUE TO FAULTY GENERATING COIL
NORMAL CRUISING
PERIODIC PISTON SEIZURE
PISTON PIN STICKING OR BROKEN
MAIN BEARING DAMAGED OR WORN
ENGINE RUNS TOO FAST
FAULTY RPM LIMITER
JET PUMP CAVITATION
SUDDEN ACCELERATION
EXCESSIVE CONNECTING ROD BIG END CLEARANCE
EXCESSIVE CONNECTING ROD TO PISTON PIN CLEARANCE
EXCESSIVE PISTON PIN TO PISTON CLEARANCE
AT COLD IDLE
PISTON RINGS WORN
PISTON WORN
CONNECTING ROD TWISTED OR BENT
EXCESSIVE PISTON TO CYLINDER CLEARANCE
BACKFIRE
CARBURETION TOO LEAN
RPM LIMITER MALFUNCTION

CHART SIX

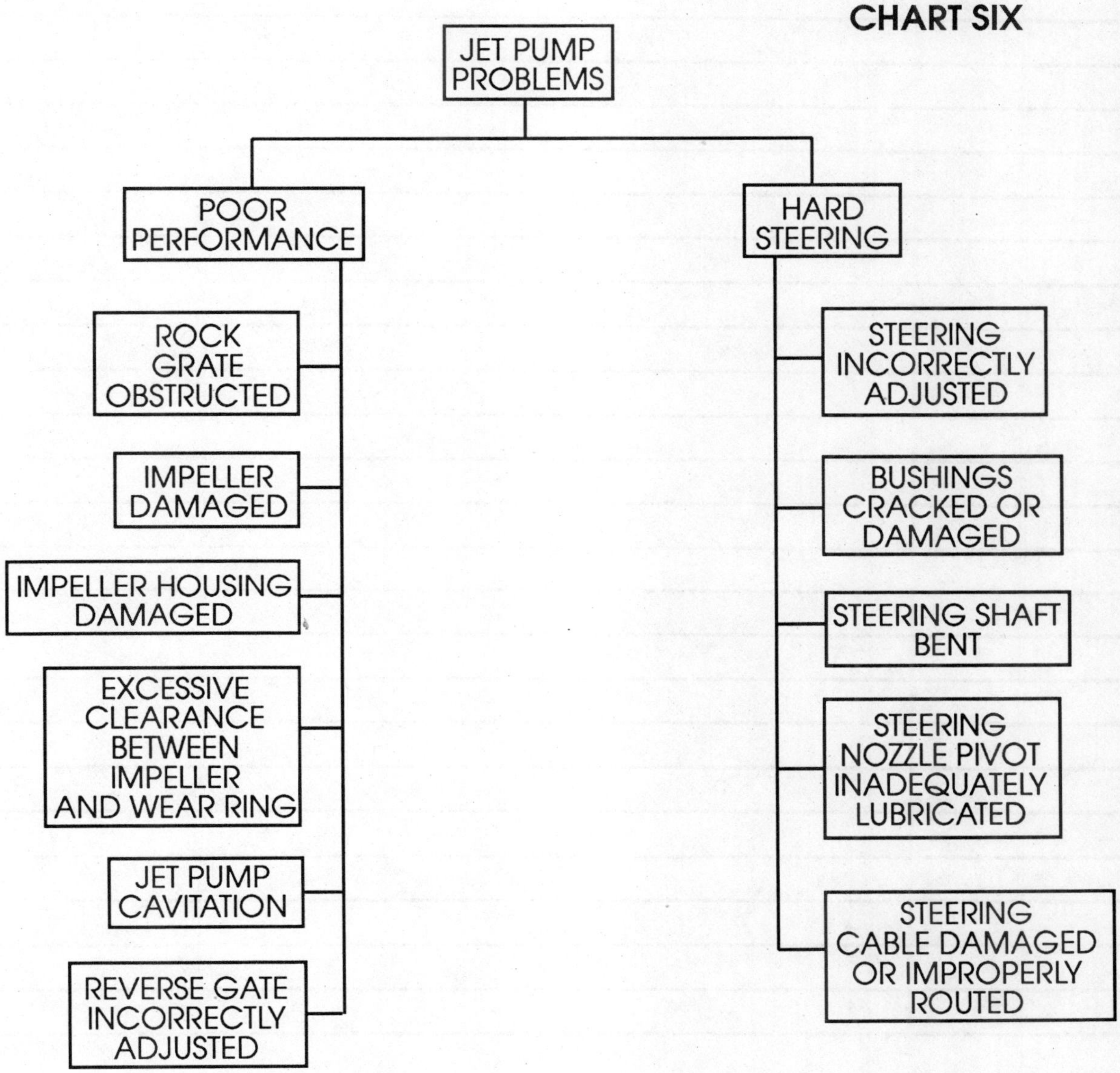
JET PUMP PROBLEMS
POOR PERFORMANCE
ROCK GRATE OBSTRUCTED
IMPELLER DAMAGED
IMPELLER HOUSING DAMAGED
EXCESSIVE CLEARANCE BETWEEN IMPELLER AND WEAR RING
JET PUMP CAVITATION
REVERSE GATE INCORRECTLY ADJUSTED
HARD STEERING
STEERING INCORRECTLY ADJUSTED
BUSHINGS CRACKED OR DAMAGED
STEERING SHAFT BENT
STEERING NOZZLE PIVOT INADEQUATELY LUBRICATED
STEERING CABLE DAMAGED OR IMPROPERLY ROUTED

NOTES & NUMBERS

6
FUEL

6-1 INTRODUCTION

The carburetion and ignition principles of two-cycle engine operation **MUST** be understood in order to perform a proper tune-up on a marine engine.

If any doubts exist concerning the reader's understanding of two-cycle engine operation, it would be best to take a few minutes to study the Introduction section in the first portion of Chapter 8 -- **ENGINE**, before tackling any work on the fuel system.

The fuel system includes the fuel tank, fuel pump, fuel filters, carburetor, and the associated parts to connect it all together. Regular maintenance of the fuel system to obtain maximum performance, is limited to changing the fuel filters at regular intervals and using fresh fuel.

If a sudden increase in gas consumption is noticed, or if the engine does not perform properly, a carburetor overhaul, including boil-out, or replacement of the fuel pump may be required.

6-2 GENERAL CARBURETION INFORMATION

The carburetor is merely a metering device for mixing fuel and air in the proper proportions for efficient engine operation. At idle speed, a marine engine requires a mixture of about 8 parts air to 1 part fuel. At high speed or under heavy duty service, the mixture may change to as much as 12 parts air to 1 part fuel.

DIAPHRAGM CARBURETOR

Description

The only carburetor used on Bombardier Sea-Doo personal watercraft, at press time, is a diaphragm type carburetor. The carburetor used is manufactured by Mikuni. All Mikuni carburetors prior to 1991 have a remote fuel pump. Since 1991 some models have an integral fuel pump.

A diaphragm carburetor does not have a float bowl or a float. Instead, a diaphragm takes the place of these items.

A hinged arm, still called a float arm, rests against the diaphragm and opens and closes the needle and seat assembly.

Operation

Following the fuel through its course, from the fuel tank to the combustion chamber of the cylinder, will provide an

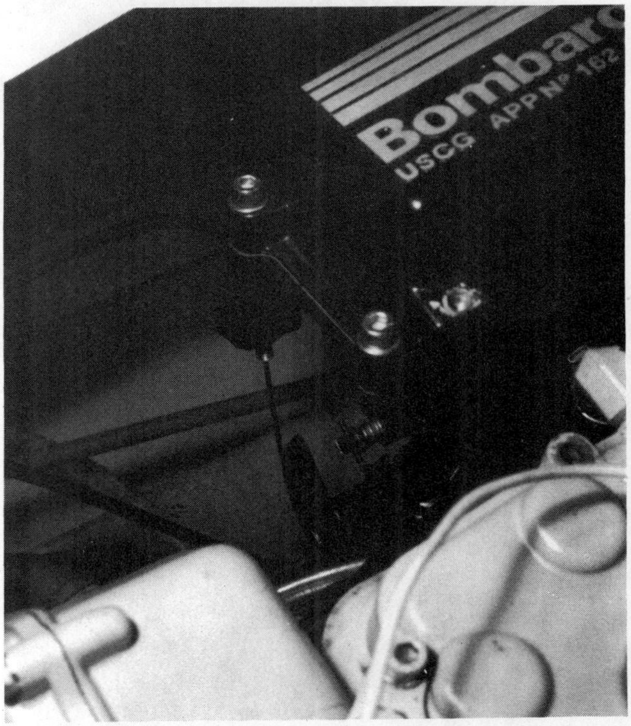

Maintenance type service of the fuel system is limited to using fresh fuel; changing the fuel filter at regular intervals; and cleaning the flame arrestor.

The Mikuni carburetor, used on all Bombardier watercraft prior to 1991. A remote fuel pump, mounted separately in the engine compartment, is used with this unit.

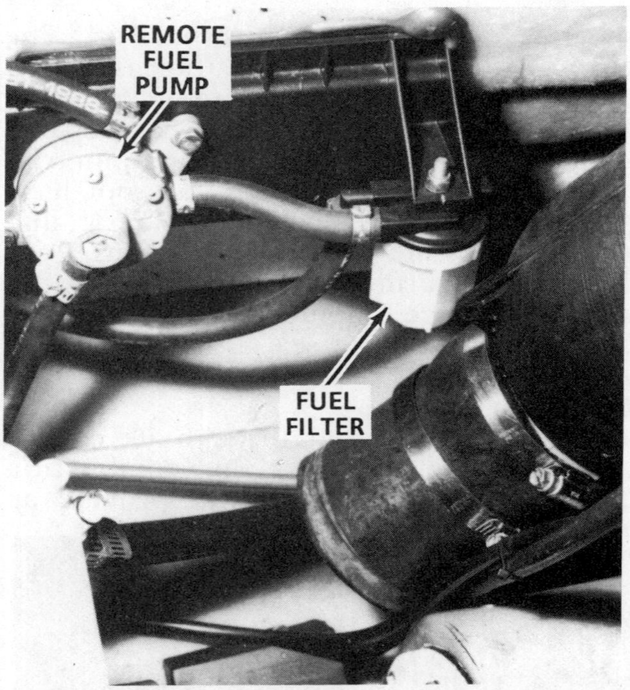

The remote fuel pump used with the Mikuni carburetor, shown above, prior to 1991. An in-line fuel filter between the fuel tank and the fuel pump is also visible.

The Mikuni carburetor, with an integral fuel pump, is used on Bombardier watercraft since 1991.

appreciation of exactly what is taking place and probably "why".

After the engine starts, the fuel passes through the pump to the carburetor. All systems have some type of filter installed somewhere in the line between the tank and the carburetor.

Fuel is delivered to an inlet valve under pressure from the integral or remote fuel pump. When the choke valve is closed and the engine is being cranked, or when the piston is on its upward stroke, the pressure in the carburetor throat orifice and the crankcase is less than atmospheric pressure. The fuel chamber above the diaphragm is connected to the crankcase through a vacuum tube. A nozzle channel connecting the carburetor throat with the fuel chamber also allows the fuel chamber to be at less than atmospheric pressure.

The chamber on the other side of the diaphragm is vented to atmosphere, and therefore, is always at atmospheric pressure.

With the two pressure differentials, the diaphragm will flex toward the lowest pressure -- toward the fuel chamber. As the diaphragm flexes, it raises the float arm, which is in constant contact with the diaphragm, because of a return spring. The float arm is hinged. As one side rises, the other side lowers and allows the inlet valve to open. At the same time, the needle valve falls away from the valve seat. As

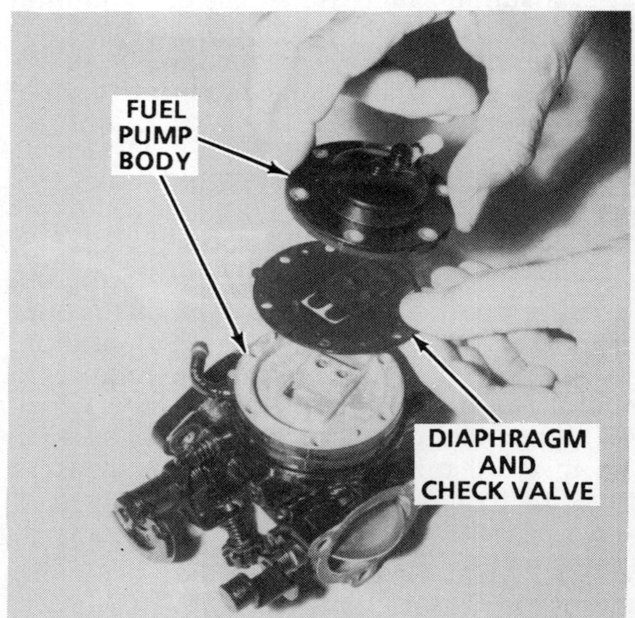

The integral fuel pump of a Mikuni carburetor, opened to show a couple of the major internal parts.

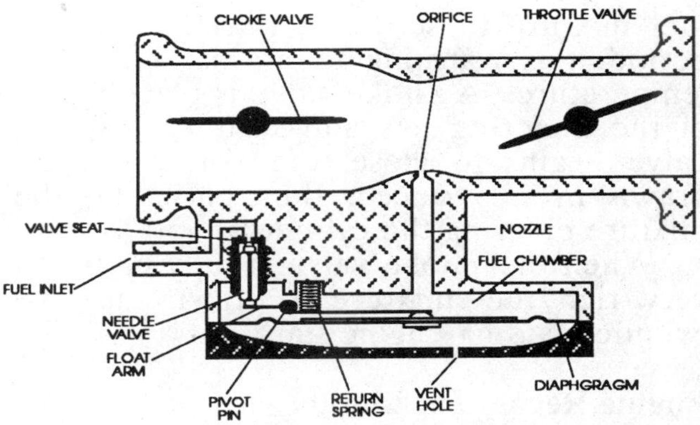

Cross-section drawing of a diaphragm type carburetor used on all series and model years covered in this manual.

soon as the inlet valve opens, fuel flows under pressure from the fuel pump into the fuel chamber.

When the piston is on its downward stroke, the crankcase, the carburetor throat, and the fuel chamber are all at a higher pressure than atmosphere. The fuel chamber is filled with fuel at a higher pressure. Therefore, the diaphragm flexes toward the side with the lower pressure -- the side vented to atmosphere.

The return spring pushes the float arm down against the diaphragm. The other side of the float arm raises up and pushes the needle valve against its seat, closing the inlet valve.

On the next piston downward stroke, the diaphragm flexes back again forcing the fuel in the fuel chamber to be ejected through a jet into the carburetor throat. On the same stroke, the fuel chamber is replenished with fresh fuel through the open needle valve. This cycle repeats itself, sending a squirt of fuel into the carburetor throat during each downward stroke of the piston.

In order to obtain the proper air/fuel mixture for all engine speeds, some models

The vent hole for the fuel chamber above the diaphragm must be kept clear to the atmosphere at all times for the diaphragm to function efficiently.

have high and low speed jets. These jets have adjustable needle valves which are used to compensate for changing atmospheric conditions. In almost all cases, the high-speed circuit has fixed high-speed jets. These fixed jets cannot be adjusted.

A throttle valve controls the flow of air/fuel mixture drawn into the combustion chambers. A cold engine requires a richer fuel mixture to start and during the brief period it is warming to normal operating temperature. A choke valve is incorporated ahead of the metering jets and venturi. As this valve begins to close, the volume of air intake is reduced, thus enriching the mixture entering the cylinders.

When this choke valve is fully closed, a very rich fuel mixture is drawn into the cylinders.

Fuel/Oil Mixture

All Bombardier Sea-Doo personal watercraft are equipped with an oil injection system.

A new or rebuilt engine must be properly "broken-in" if it is to develop maximum performance. An engine which is incorrectly "broken-in" will never be able to achieve its maximum rated horsepower. During the initial "breaking-in" period, the fuel/oil ratio should be very rich. This rich mixture may be obtained by adding an extra two ounces of oil to each gallon of fuel. The rich mixture should be used for the first ten hours of operation after the unit is

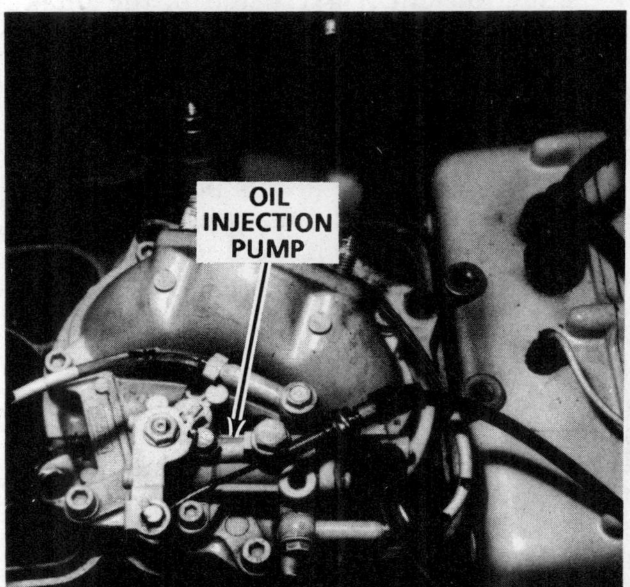

The oil injection pump is mounted on the intake manifold on the starboard side of the engine.

brought out of storage. Such a ratio will **ENSURE** adequate lubrication of moving parts, which have been drained of oil during the storage period. Operation of an engine without the proper amount of oil will quickly lead to a significant loss of horsepower.

After the first ten hours of operation, the oil injection system should take over the task of adequate automatic lubrication.

However, experienced two-stroke mechanics run their own oil injected units between 16:1 and 32:1, by adding an extra ounce of oil per each gallon of fuel. The addition of this extra oil may prevent engine seizure in the event of an oil pump failure or clog in an oil delivery line. It is far cheaper to replace a fouled spark plug than to have a complete engine overhaul with rebored cylinders, oversize pistons, and new rings.

Use only two-stroke outboard marine oil in the mixture -- **NEVER** automotive oils. Four-stroke automotive engine oil is not designed to burn completely -- only to lubricate. Therefore, automotive oil, if used in a two-stroke engine, will leave an undesirable residue in the combustion chamber.

The oil injection system is covered in Section 6-9 of this chapter, beginning on Page 6-37.

Air/Fuel Mixture

A suction effect is created in the intake manifold each time the piston moves upward in the cylinder. This suction draws air through the throat of the carburetor. A restriction in the throat, called a venturi, controls air velocity and has the effect of reducing air pressure at this point.

The difference in air pressures at the throat and in the fuel chamber, causes the fuel to be pushed out of a metering jet extending down into the fuel chamber. When the fuel leaves the jet, it mixes with the air passing through the venturi. This air/fuel mixture should then be in the proper proportion for burning in the cylinders to provide maximum engine performance.

Throttle and Choke Valves

A throttle valve controls the flow of air/fuel mixture drawn into the combustion chambers. A cold engine requires a richer fuel mixture to start and during the brief period it is warming to normal operating temperature. A choke valve is place ahead of the metering jets and venturi. As the valve begins to close, the volume of air drawn in is reduced, thus enriching the mixture entering the cylinders.

When this choke valve is fully closed, a very rich fuel mixture is drawn into the cylinders aiding engine start.

Engine Revolution Limiter

An engine revolution limiter is mounted inside the battery cover behind the cranking motor solenoid. The function of this limiter is to prevent damage to the engine by running at too high an rpm figure.

In the very early days of personal watercraft, jumping waves could cause extensive internal engine damage. When the craft left the water, unchecked engine rpm rapidly increased to as high as 12,000 rpm. As soon as the craft re-entered the water, engine rpm instantly decreased to about 6,000 rpm. This sudden increase and decrease in engine speed dramatically shortened engine life. In those days, repeated wave jumping was known to cause bent crankshafts, sheared flywheel Woodruff keys, and broken connecting rods.

Bombardier engineers developed an rpm limiter system to help prevent excessive high engine rpm and an almost "runaway" condition. The engineers answer was

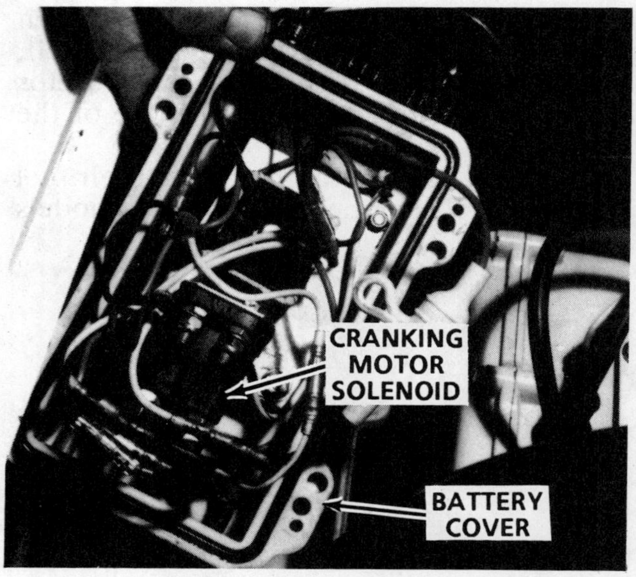

The engine revolution limiter is mounted inside the battery cover, behind the cranking motor solenoid.

an rpm limiter. This limiter restricts engine speed to about 6,500 rpm. Thus, wave jumping, suing a ramp, and a host of other almost unbelievable maneuvers could be performed without undue stress on the engine. Today, with an rpm limiter system installed, a bent crankshaft or broken connecting rod is almost non-existent.

BELIEVE THESE WORDS
Any attempt to bypass the revolution limiter could have serious and expensive consequences.

6-3 FUEL PUMP

Two designs of fuel pumps are used on the craft covered in this manual -- an integral fuel pump and a remote fuel pump.

Most models after 1991 use a carburetor with an integral fuel pump -- manufactured by Mikuni. Service procedures for this pump is included in the overhaul instructions in the Section dealing with the carburetor.

On models prior to 1991, a remote fuel pump was used. Troubleshooting, testing, and service procedures for the remote fuel pump will be found in Section 6-9, beginning on Page 6-37.

Integral Fuel Pump Description and Operation

The integral fuel pump consists of a series of diaphragms and check valves operated by crankcase pressure. The pump feeds fuel directly into the fuel chamber through the inlet valve.

As the piston moves upward, a suction, or negative pressure, is created in the crankcase. This low pressure acts upon the inner diaphragm through the carburetor. The inner diaphragm is, therefore, pulled inward. When the diaphragm moves inward the check valve opens and fuel is drawn from the bottom of the fuel pump exterior chamber into the fuel pump interior chamber.

As the piston moves downward, a positive pressure is created in the crankcase causing the inner diaphragm to be pushed outward. The delivery check valve opens under pressure and allows fuel to flow into the upper part of the exterior chamber and onto the inlet valve of the carburetor.

The function of the outer diaphragm is to absorb the pulsations of the fuel and allow a smooth uninterrupted fuel flow.

If the fuel pump diaphragm tears or is punctured, even with the tiniest hole, fuel will enter the crankcase through the pulse line causing an overrich mixture. The mixture will foul the plugs and in extreme cases flood the crankcase.

6-4 FUEL IN THE SYSTEM

The fuel system includes the fuel tank, fuel pump, fuel filters, carburetor, connecting fuel lines, and the associated parts to connect it all together. Regular maintenance of the fuel system to obtain maximum performance, is limited to changing the fuel filters at regular intervals and using **FRESH** fuel.

Even with the high price of fuel, removing gasoline that has been standing unused over a long period of time, is still the easiest and least expensive preventive maintenance possible. In most cases, old gas, even if mixed with some oil, can usually be used in an automobile, or other gas driven vehicle using regular gas, without harmful effects.

If a sudden increase in gas consumption is noticed, or if the engine does not

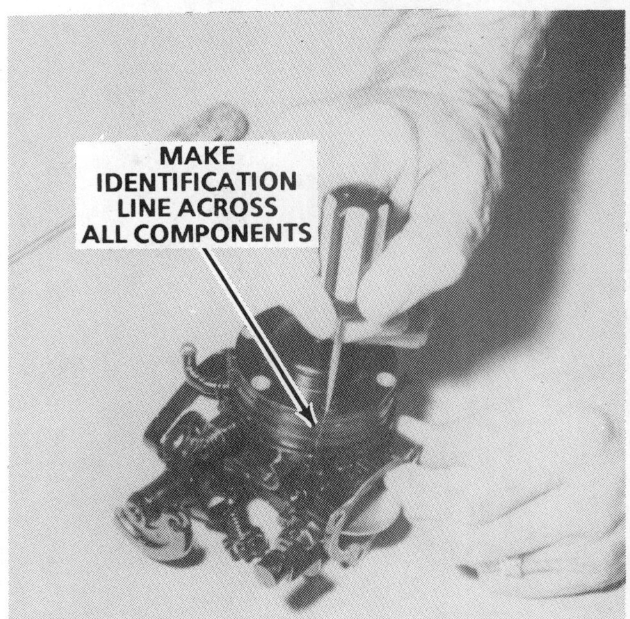

*Prior to disassembling the fuel pump, a mark should be scribed across the edge of all pump components -- eight layers in all -- gaskets, diaphragms, and check valves -- to **ENSURE** all will be assembled back in their original location and position.*

perform properly, a carburetor overhaul, including boil-out, or replacement of the fuel pump may be required.

Leaded Gasoline

The manufacturer of the units covered in this manual recommend their engines be operated using either unleaded or regular leaded gasoline with a minimum octane rating of 84 or higher.

In the United States, the Environmental Protection Agency (EPA), has slated a proposed national phase-out of leaded fuel, "Regular" gasoline, by 1993. Lead in gasoline boosts the octane rating (energy). Therefore, if the lead is removed, it must be replaced with another agent. Unknown to the general public, many refineries are adding alcohol in an effort to hold the octane rating.

Alcohol in gasoline can have a deteriorating effect on certain fuel system parts. Seals can swell, pump check valves can swell, diaphragms distort, and other rubber or neoprene composition parts in the fuel system can be affected.

A lead substitute additive -- octane booster -- can help prevent detonation when unleaded gasoline is used in a personal watercraft.

Since about 1981, every effort has been made to use materials having a high resistance to the alcohol being added to fuels for parts coming in contact with the fuel.

Fuels containing alcohol will slowly absorb moisture from the air. Once the moisture content in the fuel exceeds about 1%, the moisture will separate from the fuel taking the alcohol with it. This water/alcohol mixture will settle to the bottom of the fuel tank. The engine will fail to operate. Therefore, storage of this type gasoline for use in marine engines is not recommended for more than just a few days.

Removing Fuel from the System

For many years there has been a popular wide-spread belief that simply shutting off the fuel at the tank and then running the engine until it stops is the proper procedure before storing the engine for any length of time. Right? **WRONG!**

It is **NOT** possible to remove all the fuel in the carburetor by simply operating the engine until it shuts down from lack of fuel. Some fuel is always trapped in passages. The **ONLY** guaranteed method of removing **ALL** fuel, is to take the time to remove the carburetor, and drain the fuel.

BAD NEWS

If the engine is operated with the fuel supply shut off until it stops, the fuel and oil mixture inside the engine is removed, leaving bearings, pistons, rings, and other parts with little protective lubricant, during long periods of storage.

A water/alcohol mixture can be found in the bottom of fuel tanks, when the craft has been operated on a high alcohol content fuel.

PROPER PROCEDURE LONG TIME STORAGE

a- Shut off the fuel supply at the tank.
b- Disconnect the fuel at the tank.
c- Operate the engine **ONLY** until it begins to run **ROUGH.**
d- Stop the engine, which will leave some fuel/oil mixture inside.
e- Remove and drain the carburetor.

By disconnecting the fuel supply, all **SMALL** passages are cleared of fuel even though some fuel is left in the carburetor. A light oil should be put in the combustion chamber, as instructed in the Owner's Manual.

SHORT TIME STORAGE

For short periods of storage, the old belief of shutting off the fuel supply and running the engine until it shuts down from lack of fuel may help prevent severe gum and varnish from forming in the carburetor. This is especially true during very hot weather. Again, this procedure is only valid for short periods of time, no more than a couple weeks.

6-5 TROUBLESHOOTING

The following paragraphs provide an orderly sequence of tests to pinpoint problems in the fuel system. It is very rare for the carburetor by itself to cause failure of the engine to start.

Typical in-line fuel filter and sediment bowl installation.

FUEL PROBLEMS

Many times fuel system troubles are caused by a plugged fuel filter, a defective fuel pump, or by a leak in the line from the fuel tank to the fuel pump. A defective choke may also cause problems. **WOULD YOU BELIEVE**, a majority of starting troubles which are traced to the fuel system are the result of empty fuel tank, clogged fuel filter, or aged "sour" fuel.

Fuel Filter and Sediment Bowl Service

Some models are equipped with a fuel filter and sediment bowl mounted inside the engine compartment.

WORDS OF CAUTION

Be sure to loosen the fuel filler cap on the fuel tank to relieve any pressure built up inside the fuel system **BEFORE** removing the sediment bowl.

Have a shop cloth handy to catch any spilled fuel. Unthread the ring nut and pull down the sediment bowl with the fuel filter inside. A large **O**-ring seals the bowl to the filter cap and ring nut. The function of this **O**-ring is to prevent fuel and air leaks. If the **O**-ring is defective, the fuel pump will draw air into the system and the engine will either not start, will falter, or will die. Sometimes, this problem is not apparent with a full fuel tank, because the fuel is gravity fed to the carburetor. When the fuel level drops in the tank, the pump takes over, but its effectiveness is impaired by the air leak at the sediment bowl **O**-ring.

Clean out the sediment bowl and rinse the filter screen in solvent every 25 hours of engine operation.

"Sour" Fuel

Under average conditions (temperate climates), fuel will begin to breakdown in about four months. A gummy substance forms in the bottom of the fuel tank and in other areas. The filter screen filter screen between the tank and the carburetor and small passages in the carburetor will become clogged. The gasoline will begin to give off an odor similar to rotten eggs. Such a condition can cause the owner much frustration, time in cleaning components, and the expense of replacement or overhaul parts for the carburetor.

Even with the high price of fuel, removing gasoline that has been standing unused over a long period of time is still the easiest and least expensive preventative maintenance possible. In most cases, this old gas, even with the oil mixture, can be used without harmful effects in an automobile, or other vehicle, using regular gasoline.

A gasoline preservative additive, such as Sta-bil Fuel Conditioner and Stabilizer for 2 and 4-Cycle Engines, will keep the fuel "fresh" for up to twelve months. If this particular product is not available in your area, other similar additives are produced under various trade names.

Choke Problems

When the engine is hot, the fuel system can cause starting problems. After a hot engine is shut down, the temperature inside the fuel bowl may rise to 200oF and cause the fuel to actually boil. All carburetors are vented to allow this pressure to escape to the atmosphere. However, some of the fuel may percolate over the high-speed nozzle.

If the choke should stick in the open position, the engine will be hard to start. If the choke should stick in the closed position, the engine will flood making it very difficult to start.

In order for this raw fuel to vaporize enough to burn, considerable air must be added to lean out the mixture. Therefore, the only remedy is to remove the spark plugs; ground the leads; crank the engine through about ten revolutions; clean the plugs; install the plugs; and again attempt to start the engine. The attempt should be successful.

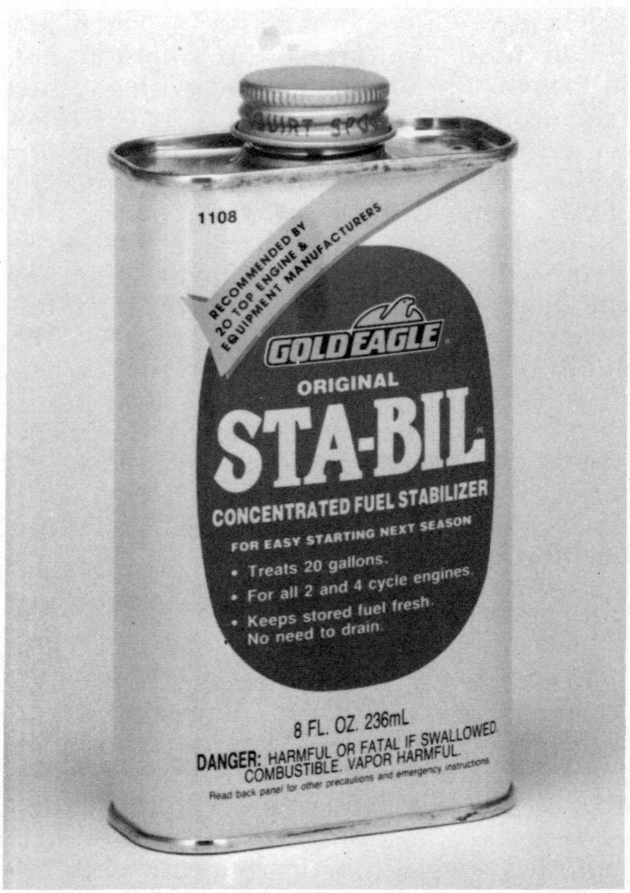

STA-BIL Gasoline Stabilizer and Conditioner -- or equivalent, may be used to prevent fuel from "souring" for up to twelve full months.

Rough Engine Idle

If an engine fails to idle smoothly, the most reasonable approach to the problem is to perform a tune-up to eliminate the possibility of fouled or faulty spark plugs.

Other problems preventing an engine from operating smoothly include: an air leak in the intake manifold or uneven compression between the cylinders.

Of course any problem in the carburetor affecting the air/fuel mixture will also prevent the engine from operating smoothly at idle speed. These problems usually include:. a weak float arm return spring; leaking needle valve and seat; defective choke; and improper adjustments for idle mixture or idle speed.

Excessive Fuel Consumption

One of two, or both conditions, may cause excessive fuel consumption. One may be inefficient engine operation and the other may be poor boating habits of the operator.

If the fuel consumption suddenly increases over what could be considered normal, then the cause can probably be attributed to the engine or watercraft and not the operator.

Check the fuel system for possible leaks. A leak between the tank and the pump many times will not appear when the engine is operating, because the suction created by the pump drawing fuel will not allow the fuel to leak. Once the engine is turned off and the suction no longer exists, fuel may begin to leak.

If a minor tune-up has been performed and the excessive fuel condition still exists, then the problem most likely is in the carburetor and an overhaul is in order.

Engine Surge

If the engine operates as if the load on the craft is being constantly increased and decreased, even though an attempt is being made to hold a constant engine speed, the problem can most likely be attributed to the fuel pump, or a restriction in the fuel line between the tank and the carburetor or between the fuel tank and the remote fuel pump, if so equipped.

6-6 MIKUNI CARBURETOR

The Mikuni carburetors installed on all Bombardier Sea-Doo watercraft are identical except for the diameter of the venturi. The venturi size does not affect overhaul procedures in any manner. The "GT" model has dual identical Mikuni carburetors except for the throttle linkage.

Complete detailed procedures for removal, disassembly, cleaning and inspecting, assembling including bench adjustments, installation, and operating adjustments for this Mikuni carburetor are presented in this section.

The Mikuni carburetor, with an integral fuel pump has been used on Bombardier craft since 1991. The carburetor used prior to 1991 is identical except the fuel pump is not an integral part of the carburetor. The procedures presented in this section are valid for both carburetors.

SERVICE MIKUNI CARBURETOR

This section provides complete detailed procedures for removal, disassembly, cleaning and inspecting, assembling including bench adjustments, installation, and operating adjustment for the Mikuni carburetor installed on the Bombardier Sea-Doo personal watercraft covered in this manual.

REMOVAL AND DISASSEMBLING

Good shop practice dictates a carburetor repair kit be purchased and new parts be installed any time the carburetor is disassembled.

Make an attempt to keep the work area organized and to cover parts after they have been cleaned before they are assembled or installed. This practice will prevent foreign matter from entering passageways or adhering to critical parts.

WORDS OF WISDOM

DO NOT attempt to remove the entire flame arrestor while the carburetor is still mounted on the intake manifold. The top cover and flame arrestor may be easily removed, without a problem. **HOWEVER**, one critical area with the bottom cover of the flame arrestor could cause a major problem. The bottom cover is secured to the carburetor with a single bolt. If this bolt should accidentally be dropped into the carburetor throat, the bolt would almost certainly find its way down through the carburetor, the intake manifold and into the crankcase. **BAD NEWS. VERY BAD NEWS.** If allowed to remain there, the bolt will reek havoc on pistons, rings, and cylinder walls. If a screw or washer is lost in this manner, the engine **MUST be** completely disassembled to locate the missing piece of hardware and avoid extensive and expensive damage. Moral to the story: **LEAVE** the bottom cover in place until the carburetor is safely on the workbench ready for disassembly.

SINGLE AND DUAL CARBURETOR REMOVAL

As the heading states, the following removal procedures cover both a single and dual carburetor installation. If the engine being serviced has only a single carburetor,

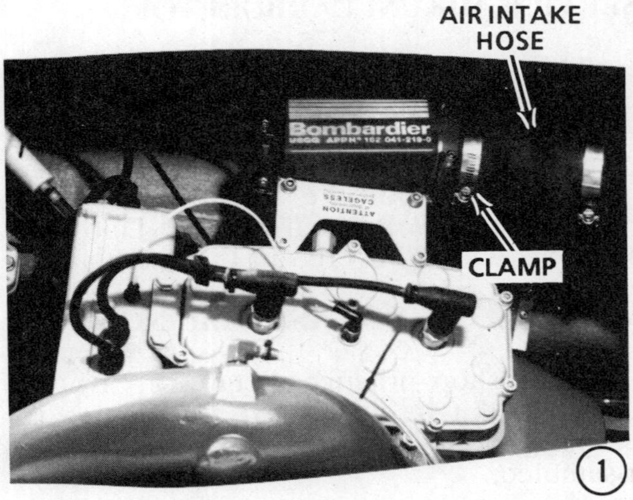

simply ignore any references to the other carburetor.

1- Rotate the fuel valve to the **OFF** position. Loosen the clamp securing the air intake hose to the flame arrestor cover and slide the clamp back along the hose. Work the hose back and free of the cover flange.

2- Obtain an M6 Allen wrench and remove the six Allen head bolts and the washers securing the support bracket to the flame arrestor and the engine. Lift off the bracket.

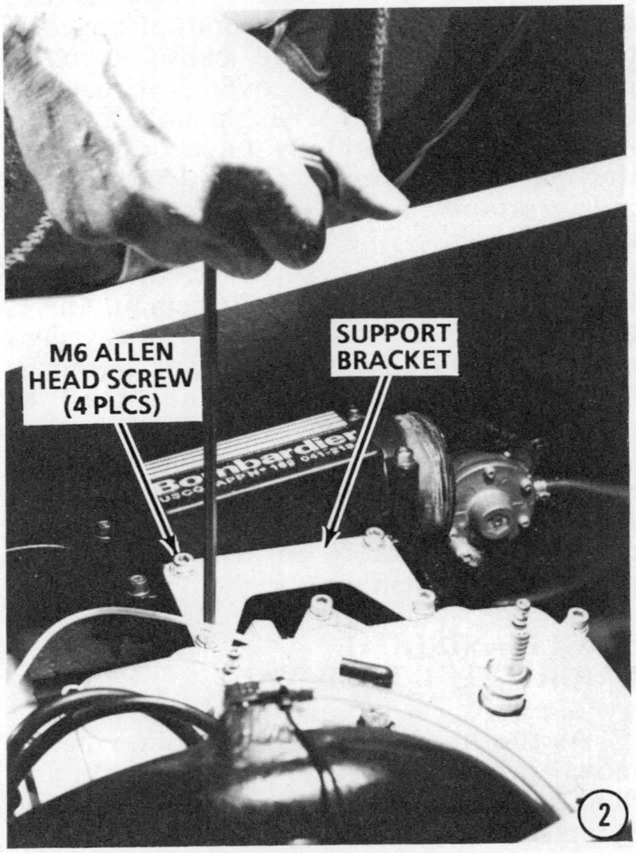

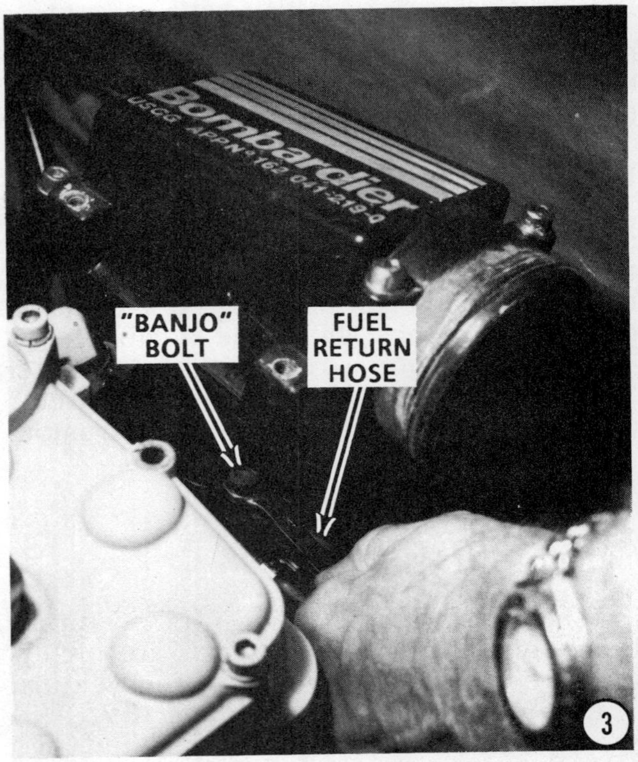

3- Remove the "banjo" bolt and fitting from the carburetor, with the fuel return hose attached. A banjo bolt is a special hollow bolt used to secure a fuel hose to the carburetor. If working on a dual carburetor installation, disconnect the fuel supply hose from the automatic shut-off valve.

4- Disconnect the throttle cable end from the curved guide on the throttle shaft. Remove the two carburetor mounting bolts.

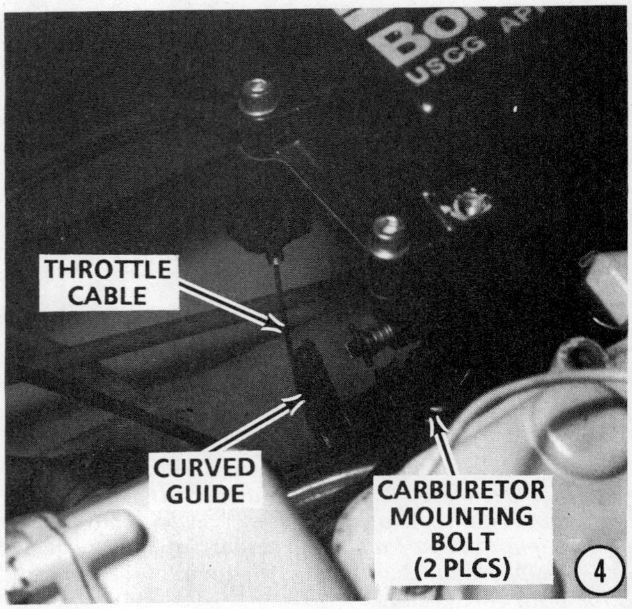

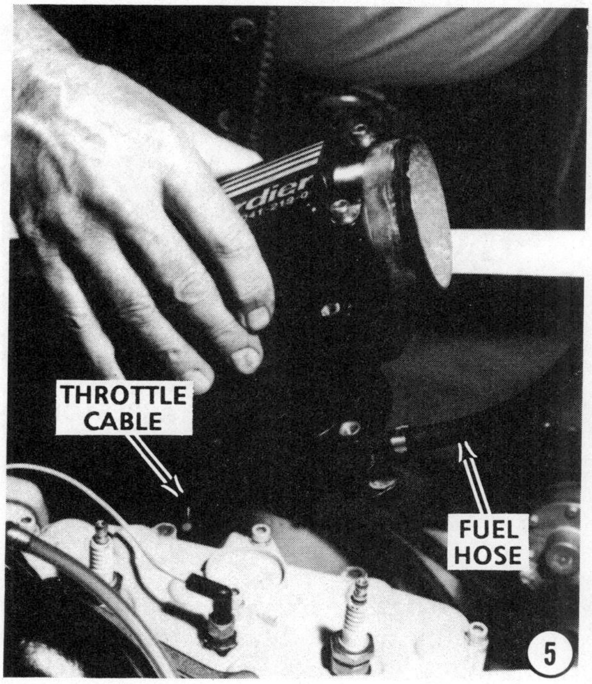

5- Lift the carburetor off the intake manifold, with the choke and throttle cables along with the fuel hose still attached. It is far easier to remove or disconnect these items after the carburetor is free of the manifold.

6- Remove all traces of carburetor gasket material from the carburetor mounting flange and the intake manifold.

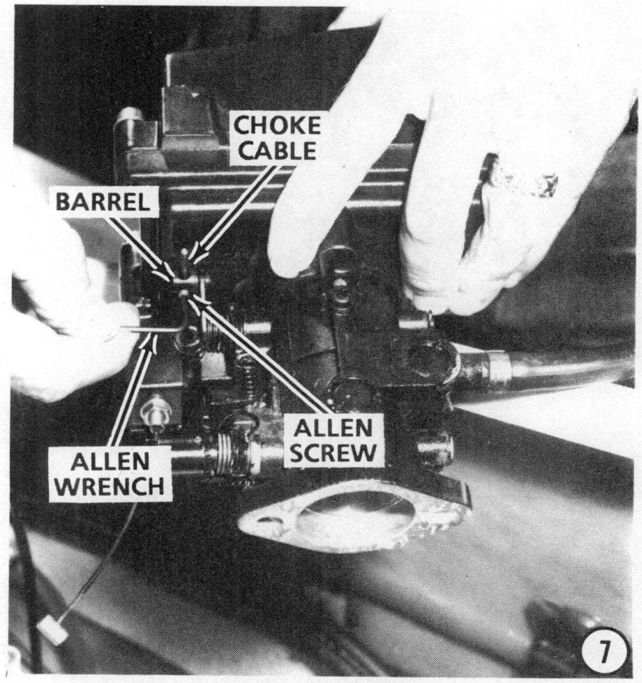

7- BEFORE disturbing the choke cable, obtain some "Whiteout", or some masking tape, or equivalent. Mark the point on the cable where it enters the barrel. Be sure the mark is fairly permanent and will not be disturbed when the cable is removed from the barrel. Marking the cable at this point, at this time, will ensure the cable can be installed back in its original position, with the same amount of cable tension and with the same choke valve setting. After the identifying mark is made, loosen the small Allen screw securing the choke cable to the barrel on the choke shaft.

8- Loosen the two nuts, one on each side, securing the choke cable to the cable

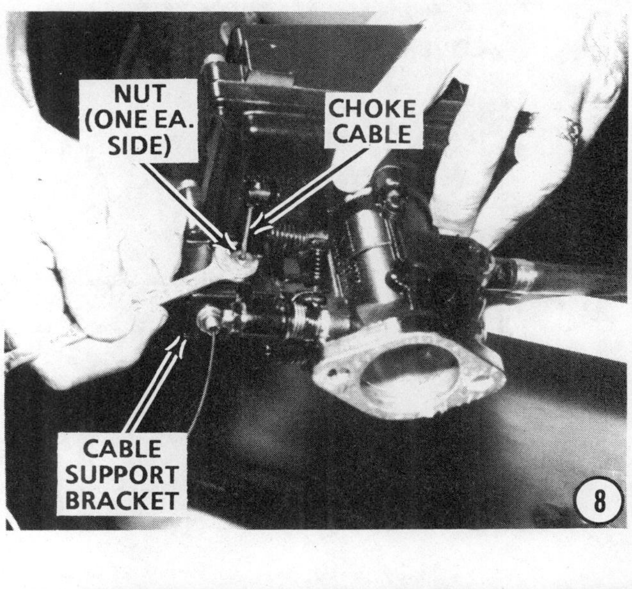

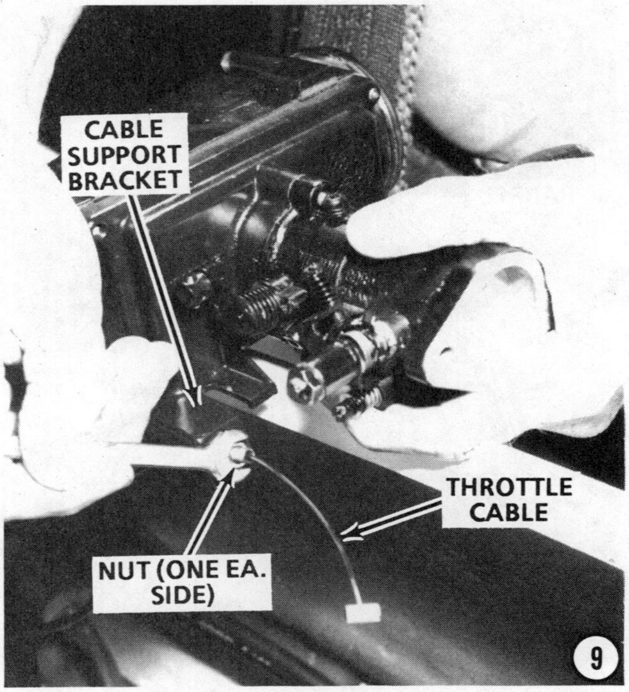

support bracket on the flame arrestor bottom cover. Pull the cable free of the slot and free of the barrel.

9- Release the throttle cable from the cable support bracket in the same manner.

10- Remove the four Allen head bolts securing the top flame arrestor cover.

11- Lift off the top cover and the flame arrestor screen.

12- Loosen the hose clamp and work the fuel supply hose free of the carburetor fitting. Move the carburetor/s to a suitable work surface. **NOW**, remove the single Allen head bolt securing the bottom flame arrestor cover to the carburetor. Lift off the cover.

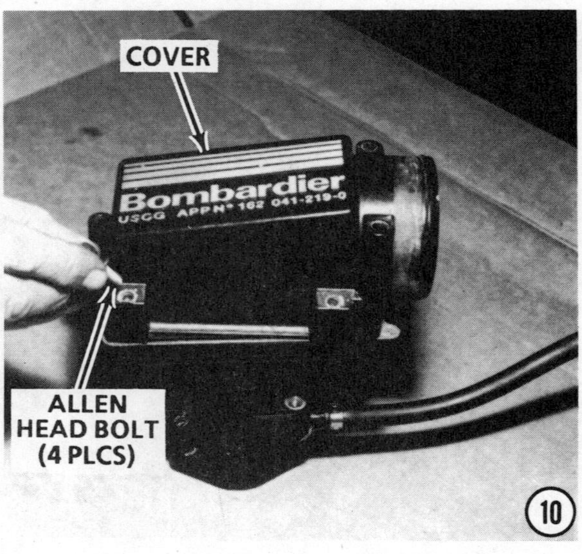

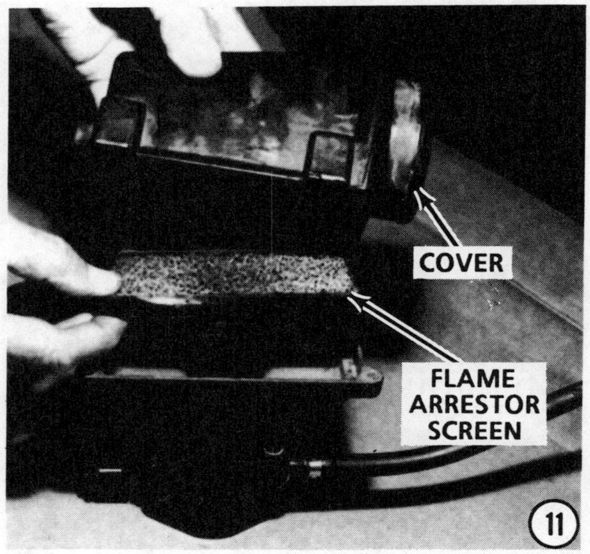

GOOD WORDS

If working with a dual carburetor installation, make an identification mark on each carburetor. The carburetor with the idle speed screw is the **AFT** carburetor. The forward carburetor does not have an idle speed screw.

Pry off the link connecting the two throttle shafts together from both ball joints. **TAKE CARE** not to alter the length of the link. Disconnect the fuel line between the two carburetors.

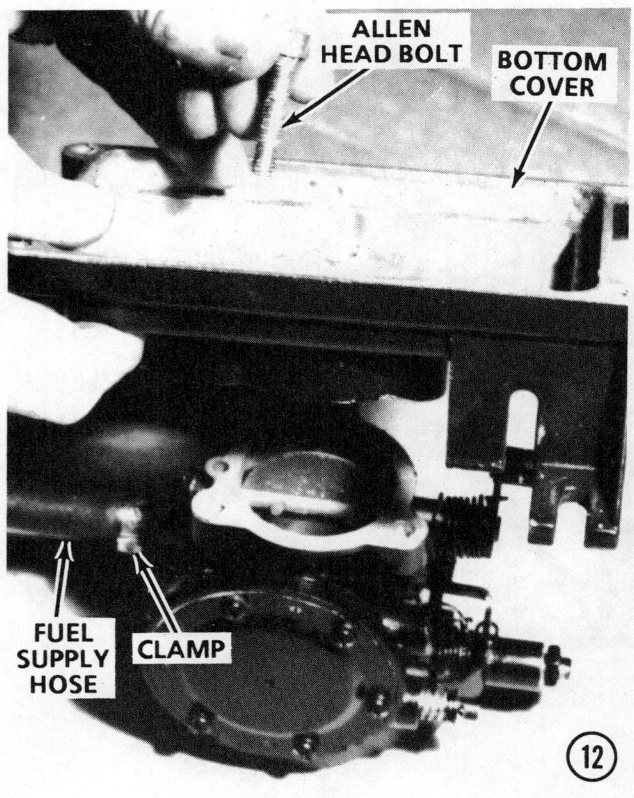

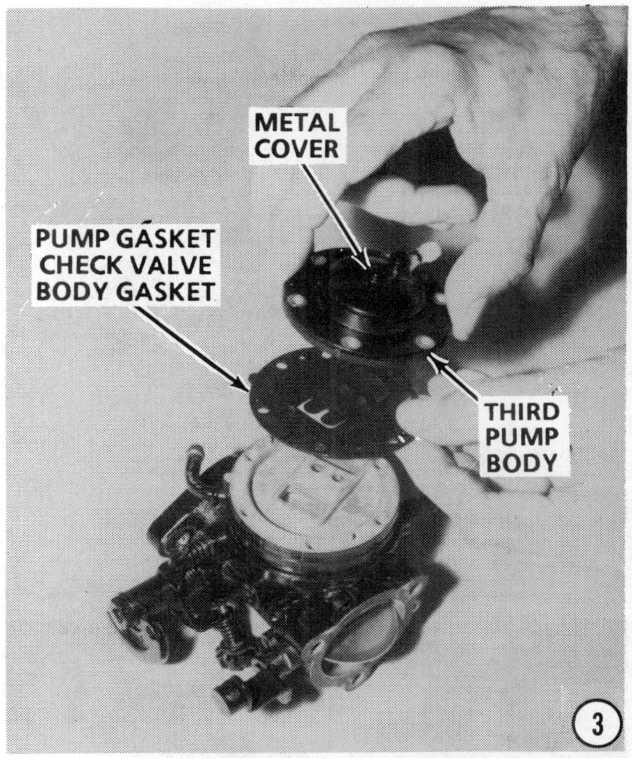

CARBURETOR DISASSEMBLING

This section covers disassembling of Mikuni carburetors with and without the integral fuel pump. Procedures to be performed on carburetors with the integral fuel pump will be clearly identified.

Carburetors WITH Integral Pump

1- Scribe a mark across the entire thickness of the fuel pump to **ENSURE** all components will be assembled in the correct position during installation.

2- Remove the six Phillips head screws and the lockwashers securing the third (outer) pump body to the carburetor.

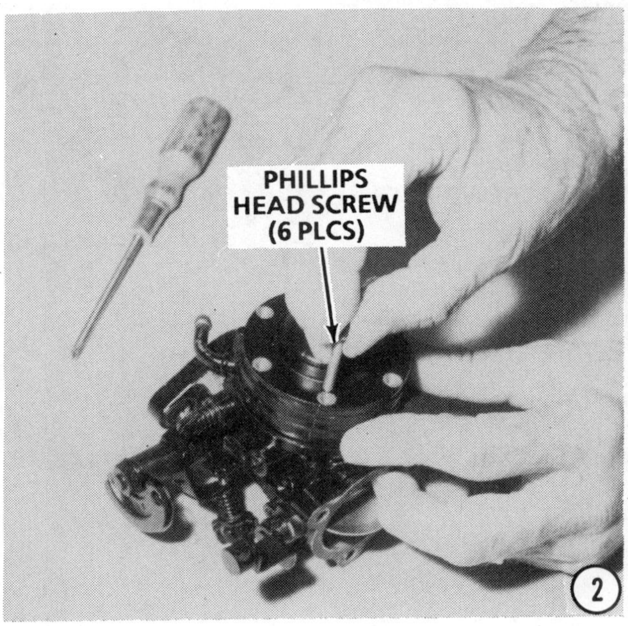

3- Lift off the following items in order: third pump body, pump gasket, check valve, and the body gasket.

Remove the Phillips head screw securing the metal cover to the third pump body. Remove and discard the gasket.

4- Lift off the second pump body, body gasket, diaphragm (plastic film), and second body gasket from the carburetor.

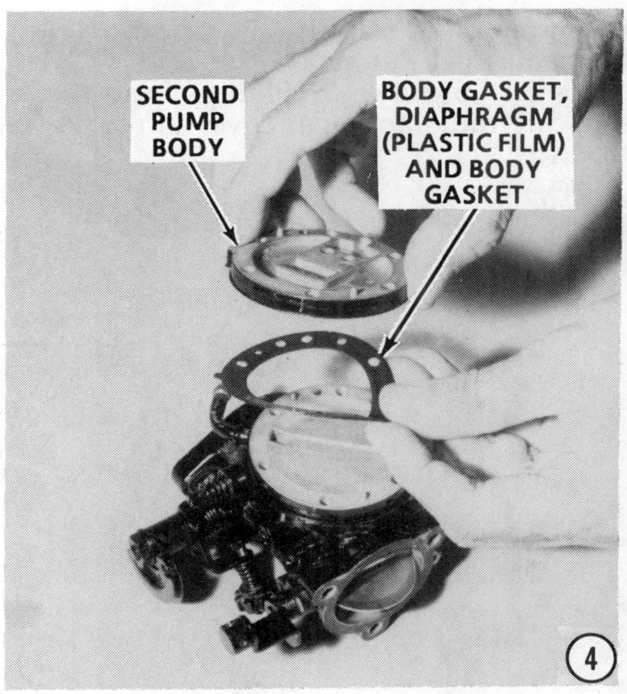

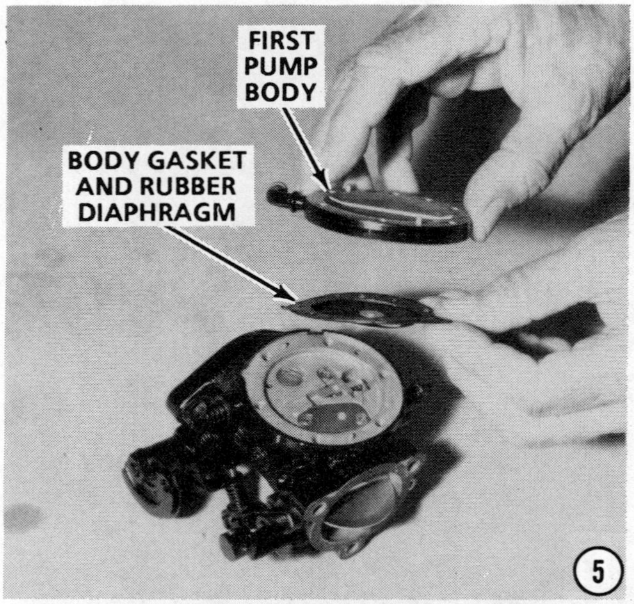

5- Remove the first pump body, body gasket, and rubber diaphragm.

Carburetors WITHOUT
Integral Fuel Pump

6- Remove the six Phillips head screws securing the diaphragm cover to the carburetor, and then lift off the cover gasket. Remove the diaphragm from the fuel chamber.

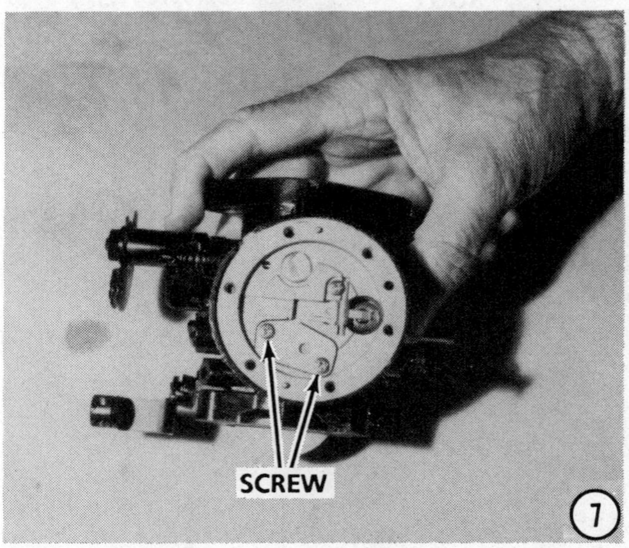

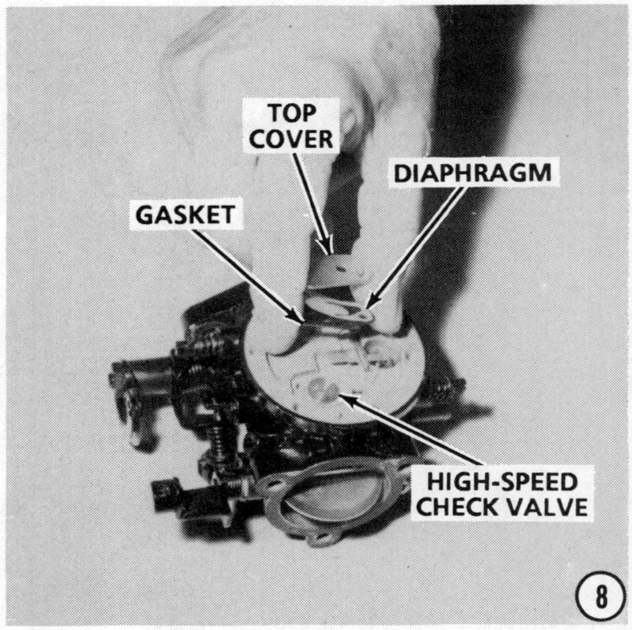

ALL CARBURETORS
Step 7 Thru Step 15

7- Remove the two small Phillips head screws from the top cover, as shown.

8- Lift off the top cover, diaphragm, and gasket from the high-speed check valve. The high-speed check valve is pressed into the fuel chamber and **CANNOT** be removed.

9- Obtain a stubby Phillips head screwdriver and a hammer. Now, tap lightly on

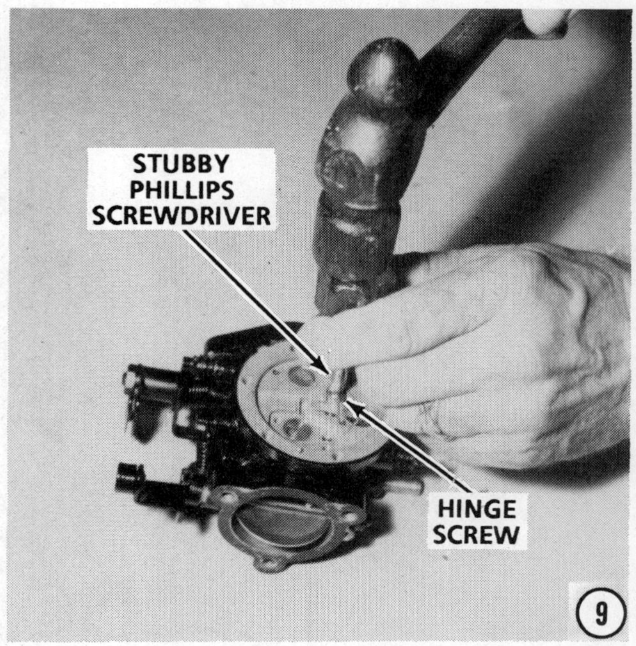

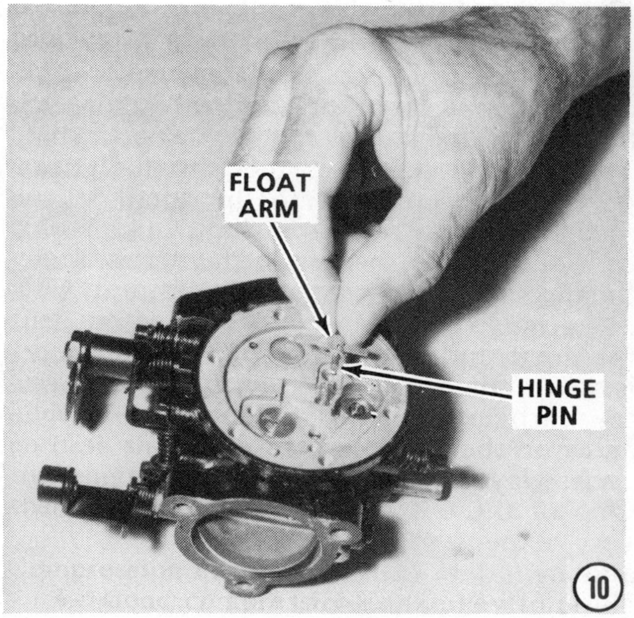

the end of the screwdriver while attempting to remove the hinge screw. **DO NOT** attempt to remove this screw "cold" -- without tapping the screwdriver -- because the head will almost certainly be stripped.

10- Lift out the hinge pin, with the float arm attached, from the groove in the carburetor surface.

11- Lift out the small spring.

12- Using a pair of needlenose pliers, remove the needle valve from the seat.

13- Obtain a narrow walled socket and remove the needle valve seat. Remove and discard the gasket under the seat.

14- Back out the main jet from the carburetor body. Remove and discard the O-ring.

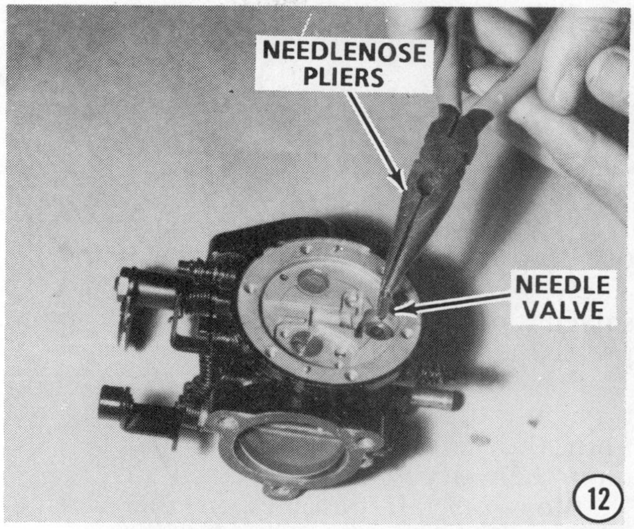

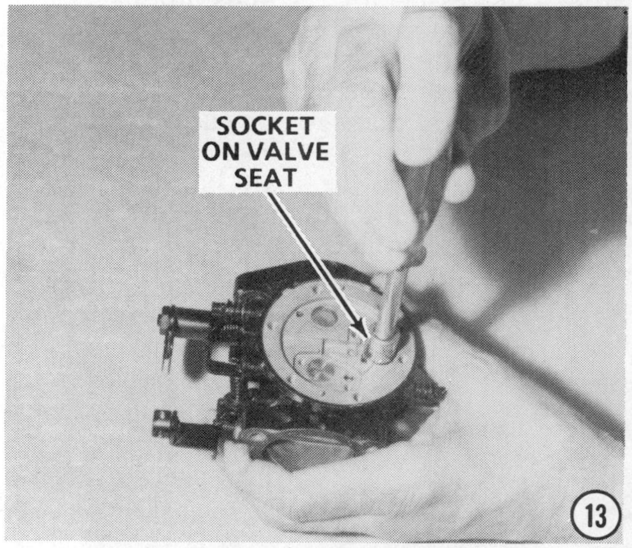

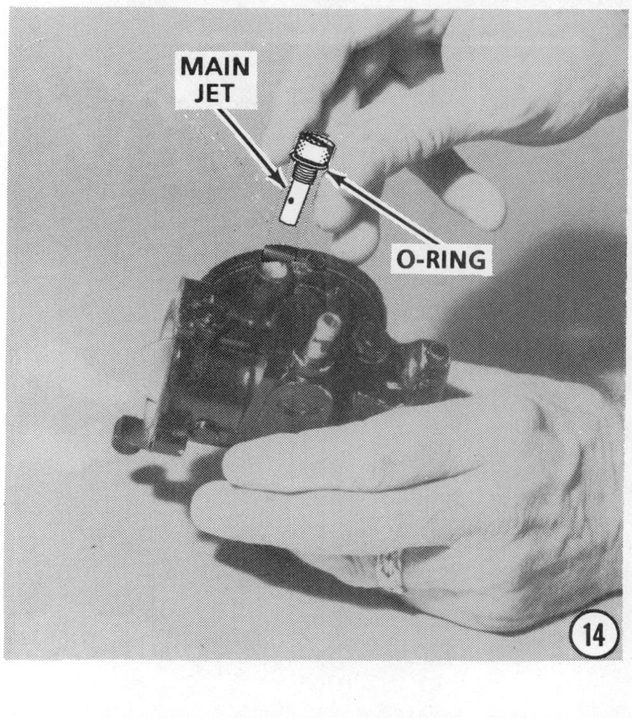

15- Back out the low-speed mixture screw.

CLEANING AND INSPECTING

NEVER dip rubber parts, plastic parts, or diaphragms in carburetor cleaner. These parts should be cleaned **ONLY** in solvent, and then blown dry with compressed air.

Place all metal parts in a screen-type tray and dip them in carburetor cleaner until they appear completely clean, then blow them dry with compressed air.

Blow out all passages in the castings with compressed air. Check all parts and passages to be sure they are not clogged or contain any deposits. **NEVER** use a piece of wire or any type of pointed instrument to clean drilled passages or calibrated holes in a carburetor.

Move the throttle shaft back and forth to check for wear. If the shaft appears to be too loose, replace the complete throttle body because individual replacement parts are **NOT** available.

CRITICAL AREA

Inspect the condition of both the choke and throttle shafts. Most shaft designs incorporate a cutout portion to enable the butterfly valve to lay flat against the shaft. Two small holes are drilled in this cutout portion to accommodate the small screws securing the valve to the shaft. The weakest point of the shaft is at the screw holes. Small cracks form at the holes due to metal fatigue. These cracks elongate the threaded holes and small screws shake loose. Once loose, they are sucked into the carburetor, then the intake manifold, and into the crankcase -- eventually finding their way into the combustion chamber. One of these small screws can then cause very expensive damage to pistons, rings, and cylinder walls. Therefore, check the cutout in the shaft **CLOSELY** for any evidence of a small crack.

Inspect the main body, airhorn, and venturi cluster gasket surfaces for cracks and burrs which might cause a leak. Check to be sure the float arm return spring has not been stretched. Check the float arm needle contacting surface and replace the diaphragm, if this surface has developed a groove.

As previously mentioned, most of the parts which should be replaced during a carburetor overhaul are included in overhaul kits available from your local dealer. One of these kits will contain a matched fuel inlet needle and seat. This combination should be replaced each time the carburetor is disassembled as a precaution against leakage.

Make a thorough inspection of the fuel chamber and diaphragm for the tiniest pin hole. If one is discovered, the hole will only get bigger. Therefore, the diaphragm **MUST** be replaced in order to obtain full performance from the engine.

Carburetor with Integral Pump

Inspect the two vent holes in the first pump body. These holes **MUST** be clean. If these two holes are not perfectly clear, the diaphragm in the fuel pump will not function properly -- full fuel volume will not be pumped.

CARBURETOR ASSEMBLING

This section covers assembly of both Mikuni carburetors -- with and without the

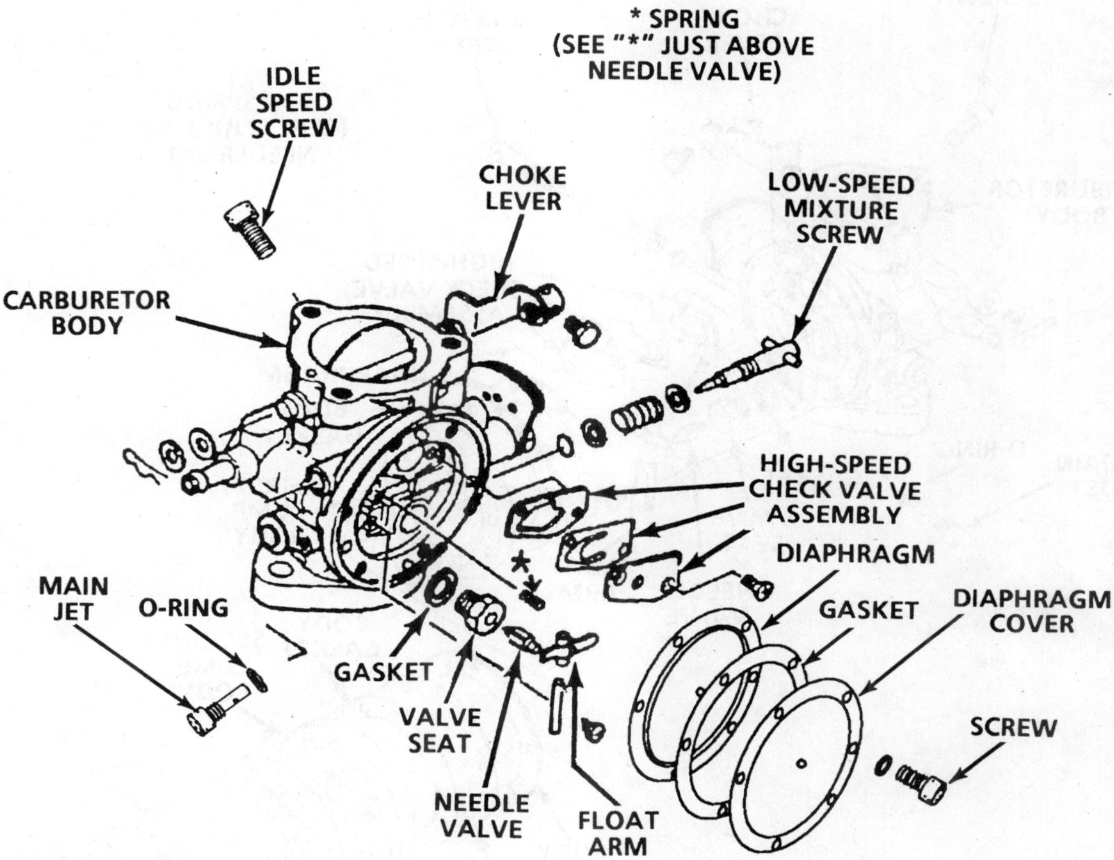

Exploded drawing of a Mikuni carburetor **WITHOUT** *an integral fuel pump. This model carburetor was installed on all Bombardier Sea-Doo watercraft prior to 1991. Major parts are identified.*

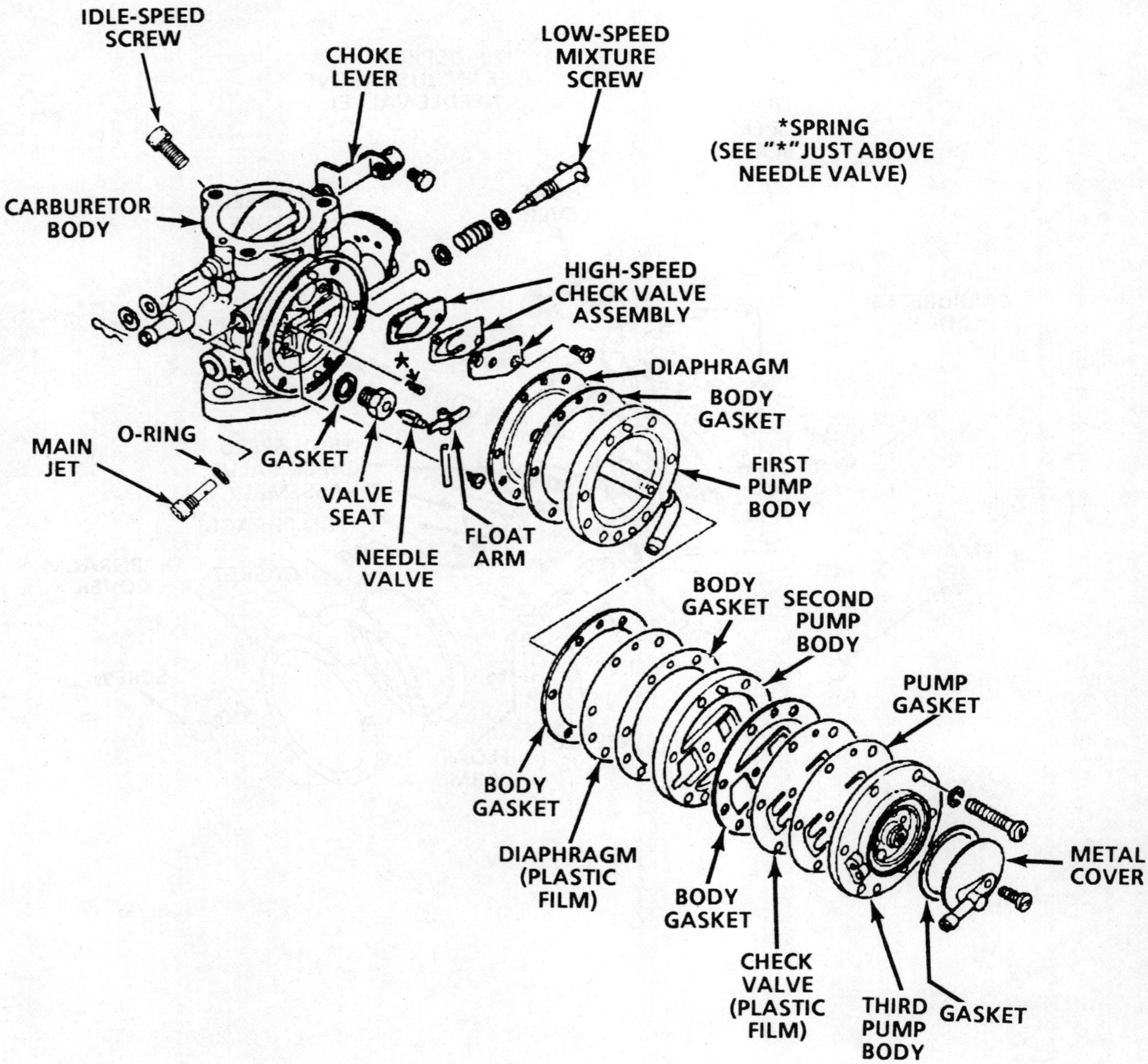

*Exploded drawing of the Mikuni carburetor **WITH** integral fuel pump. This model has been used on all Bombardier Sea-Doo watercraft since 1991. Major parts are identified.*

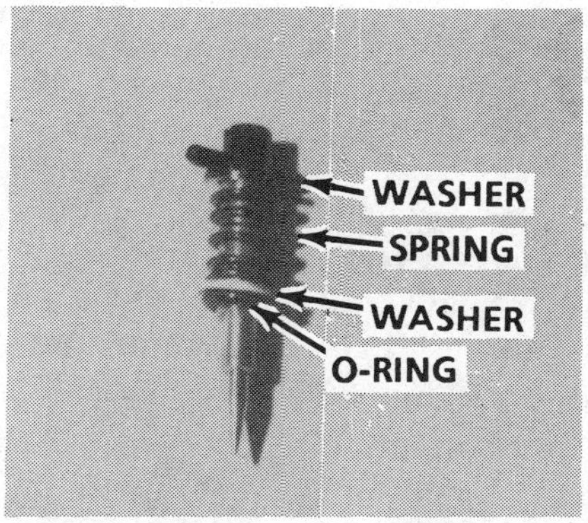

Check to be sure the low-speed adjustment screw is properly assembled, prior to installation.

integral fuel pump. Procedures for the carburetor with the integral pump are clearly identified. If servicing a carburetor without the integral fuel pump, simply ignore the instructions covering the carburetor with the pump.

All Carburetors

1- Install the low-speed mixture screw 1-1/2 turns on the linkage side of the carburetor, with the washer, spring, second washer and **O**-ring on the shaft as indicated in the accompanying illustration.

2- Install the main jet, with a new **O**-ring. Tighten the jet, just **SNUG.**

3- Place a new gasket over the needle valve seat and install the seat and gasket into the fuel chamber. Tighten the seat to a torque value of 2.9 ft lb (4Nm).

4- Lower the needle valve into the seat.

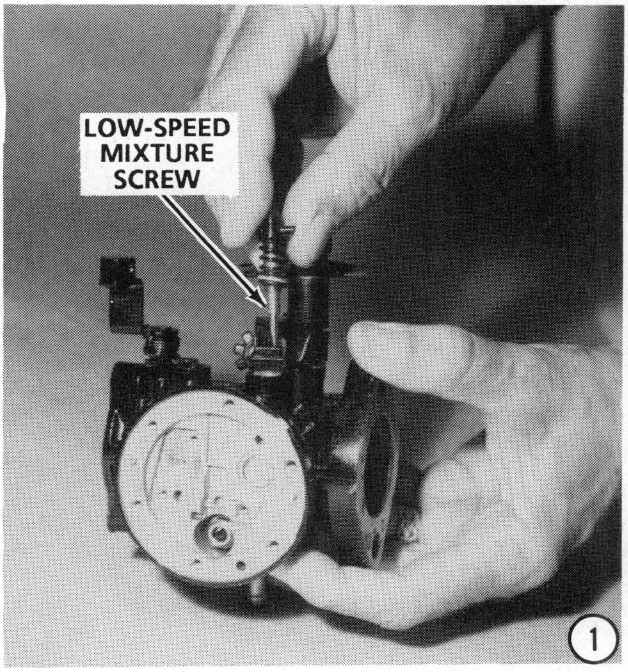

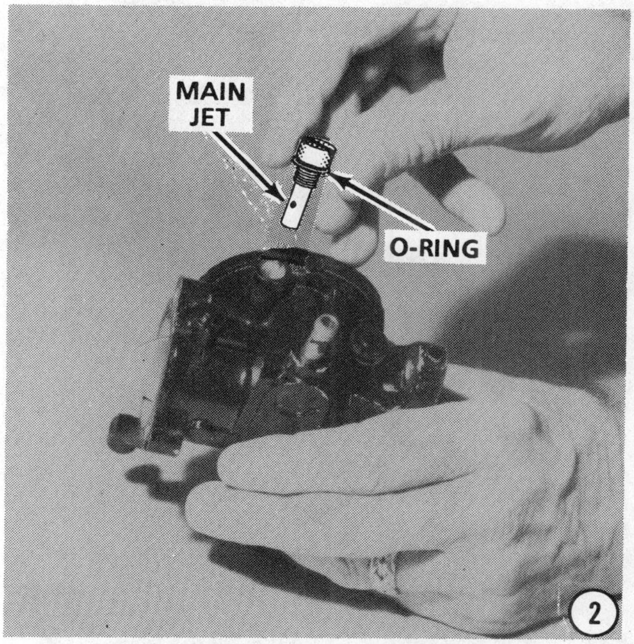

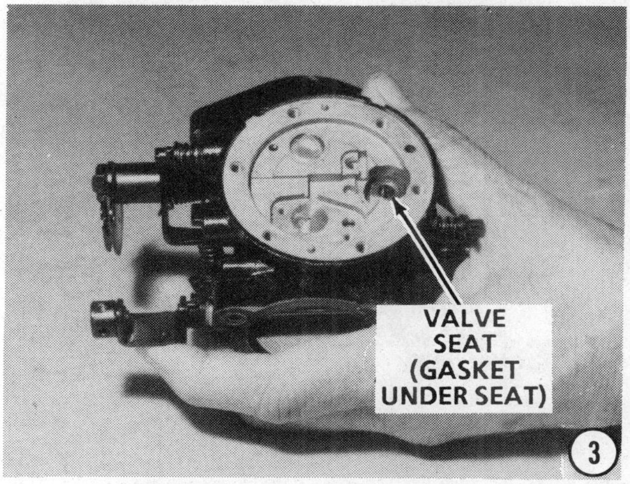

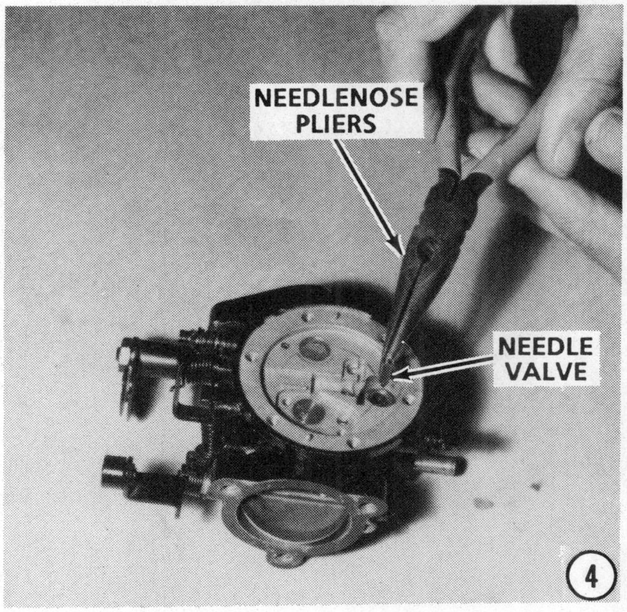

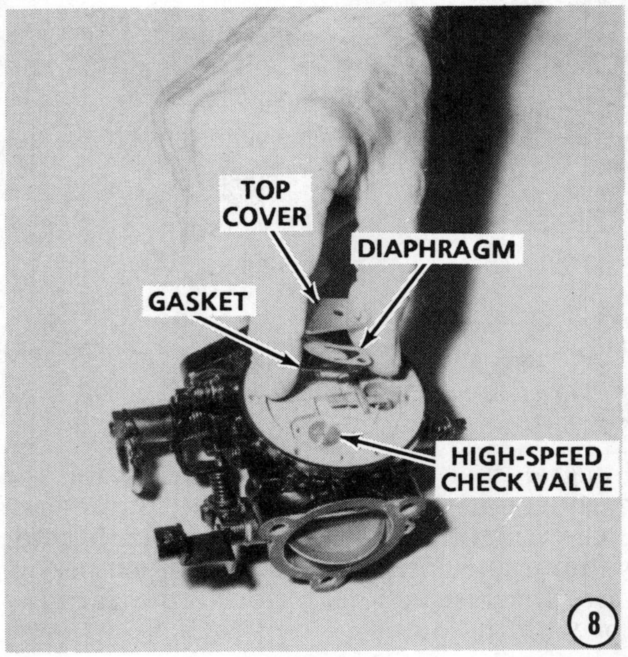

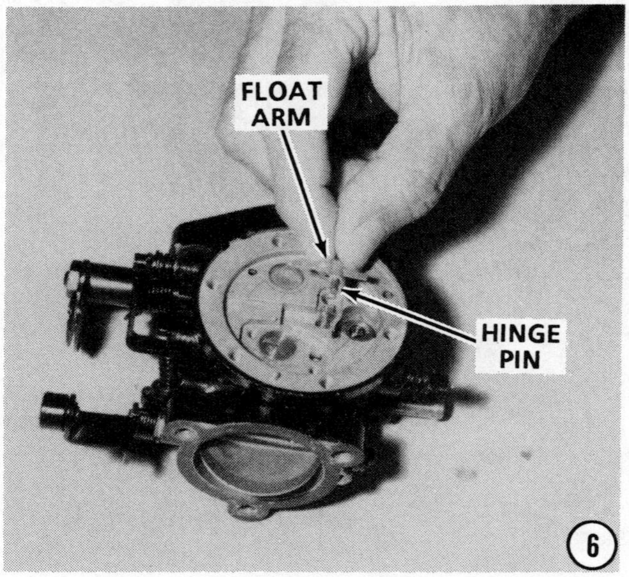

5- Slide the spring into the small hole in the fuel chamber.

6- Place the float arm at one end of the hinge pin and lower both into position over the installed spring. One end of the float arm **MUST** rest squarely on top of the needle valve, and the other end must compress the spring -- just **SLIGHTLY.**

7- Install the hinge pin retaining screw. Tighten the screw securely.

8- Position the top cover gasket, diaphragm, and top cover over the high-speed check valve.

9- Install and tighten the two small Phillips head screws securing the top cover to the fuel chamber.

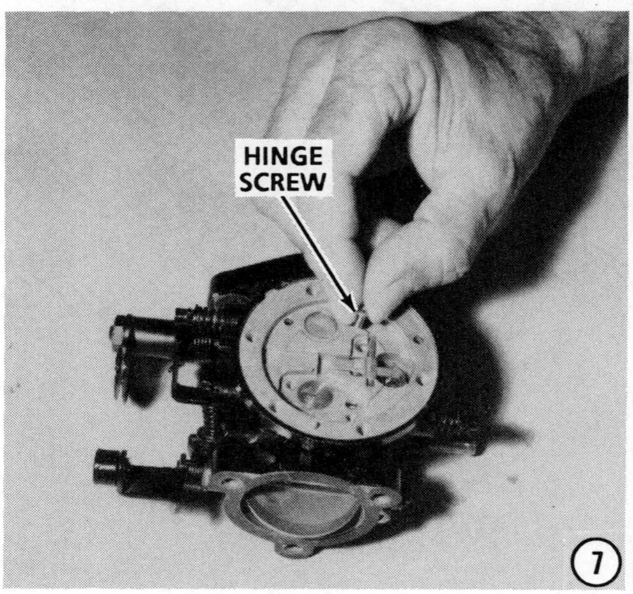

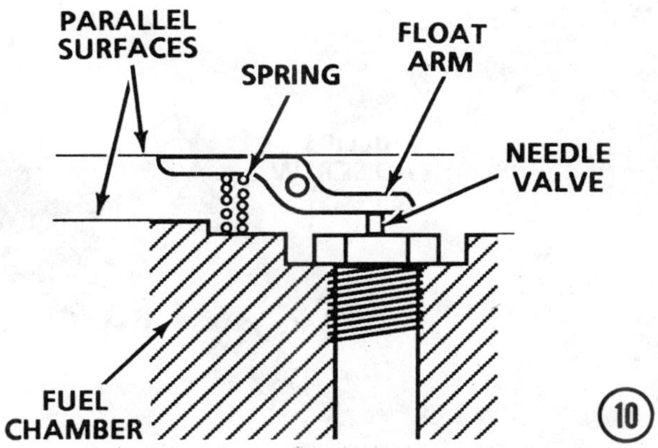

Float Arm Adjustment

10- The top of the float arm must be parallel to the fuel chamber surface and 0.024" (0.6mm) above the fuel chamber surface. If not, replace the return spring. **DO NOT** attempt to bend the float arm or to change the characteristics of the old return spring.

Carburetor With REMOTE Fuel Pump

11- Place the diaphragm over the fuel chamber with the float post on the underside facing **DOWN** to rest against the float arm. Position the diaphragm cover and gasket over the diaphragm. Secure the cover to the carburetor with the six Phillips head screws. Tighten the screws alternately and evenly to a torque value of 3 ft lb (4Nm).

FUEL PUMP ASSEMBLING

TAKE CARE to align all **EIGHT** layers of diaphragms, gaskets and check valves (made of plastic film), with their tabs indexing in the cutout in the fuel chamber base. This cutout is offset from two aligning pins. A component is correctly installed when the tab is aligned with the cut-out and the two holes are indexed over the two pins.

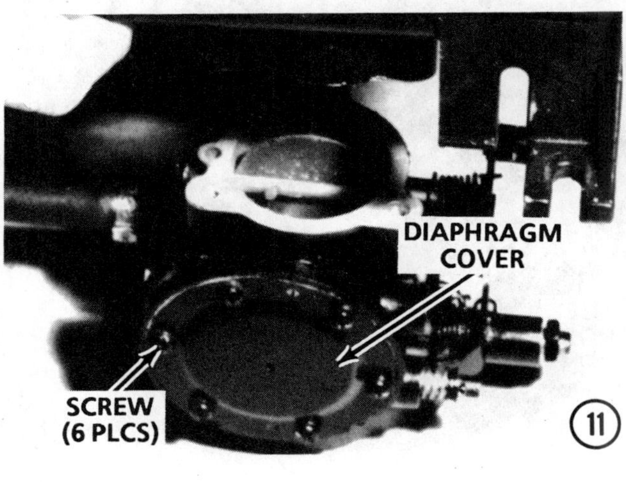

The first pump body is equipped with a vacuum fitting which takes the place of the indexing tab. The second and third pump bodies are equipped with an indexing tab. The first and second pump bodies are also equipped with a set of aligning pins. There are eleven fuel pump components to a complete assembly. All **ELEVEN** items must be carefully stacked and held together until the securing screws may be passed through the "stack" and threaded into the fuel pump body.

If a gasket, diaphragm, or check valve is askew, the slightest pressure from the end of the securing screw will tear the component and affect fuel pump performance.

12- Position the rubber diaphragm and gasket over the two indexing pins, with the tabs aligned with the cutout on the fuel pump base. Next, position the first pump body over the gasket, with the vacuum fitting aligned with the cutout on the fuel chamber base and with the two indexing pins facing **UP**.

13- Place the body gasket, diaphragm (plastic film), and second body gasket over the first pump body. Check to be sure the tabs are aligned with the vacuum fitting on the pump body and the holes in the gasket and diaphragm in place over the indexing pins. Now, place the second pump body

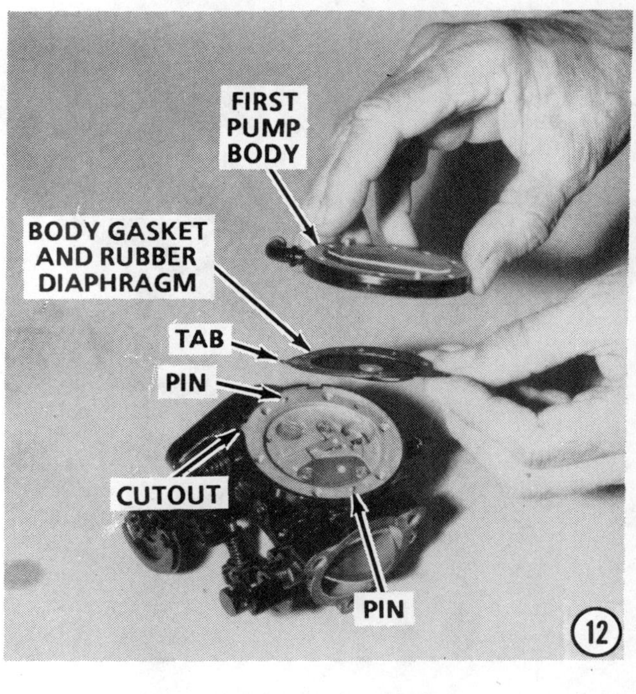

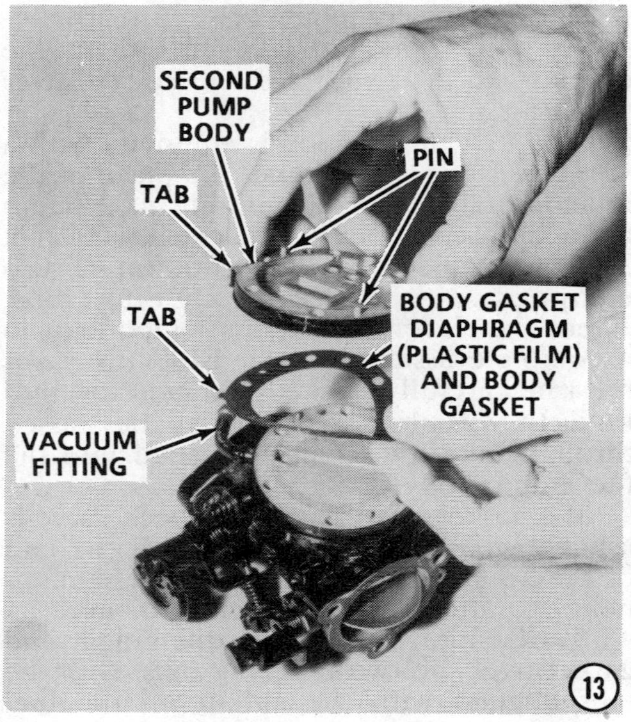

over the gasket, with the tab on the pump body aligned with the tabs on the parts already installed and the two indexing pins facing **UP**.

14- Install another body gasket, followed by the check valve (plastic film), the final body gasket and the third fuel pump body, over the two indexing pins on the

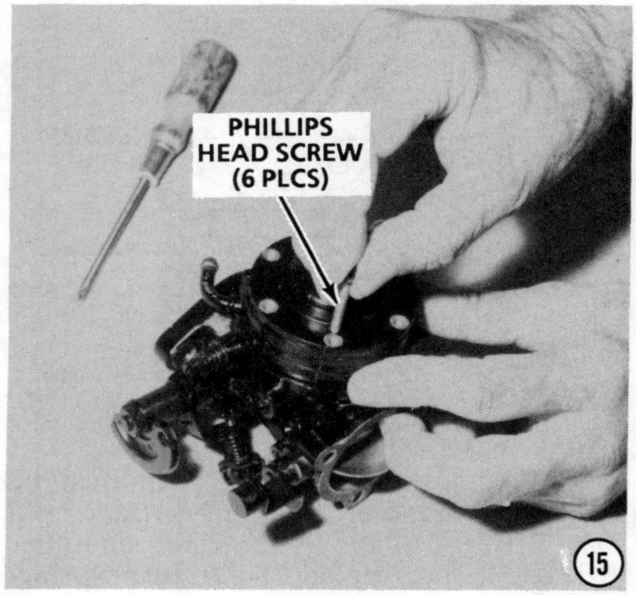

second pump body. Check to be sure all tabs are aligned.

15- Visually inspect the screws holes to verify all components are squarely in place and the path for all six screws are clear. Install the six Phillips head screws starting with two directly opposite each other. Continue to hold the stack of components tightly together until all six screws are started in the fuel chamber body. After all six screws have been started, tighten them securely -- working alternately and evenly.

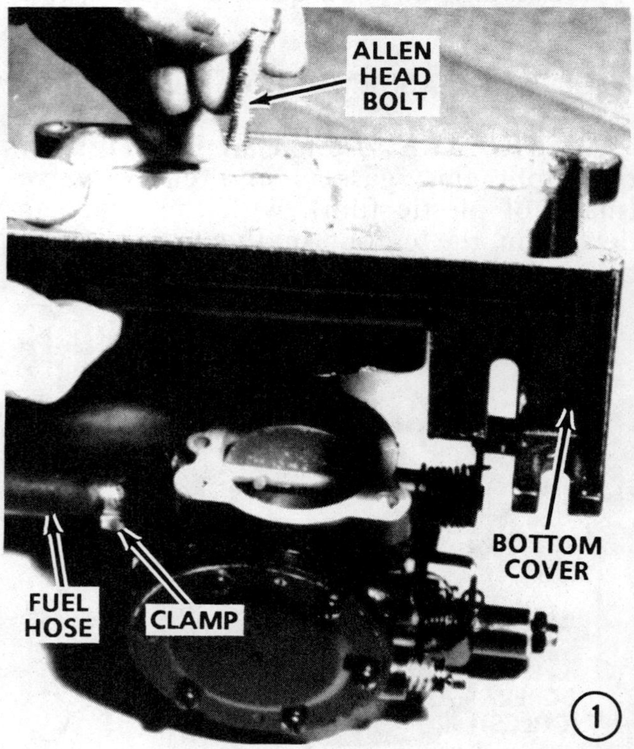

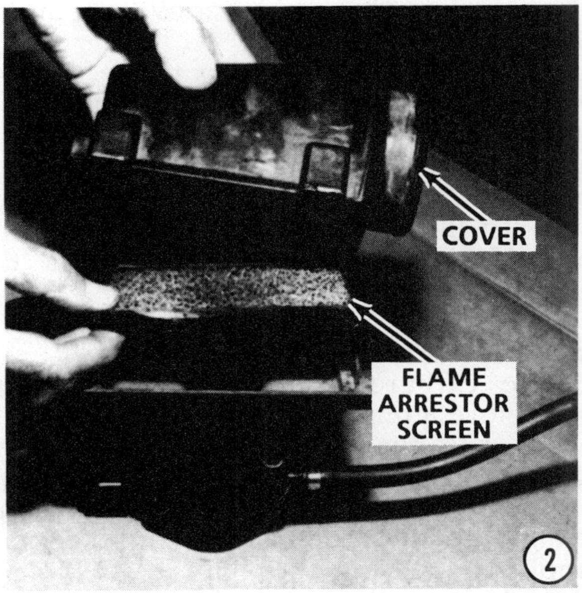

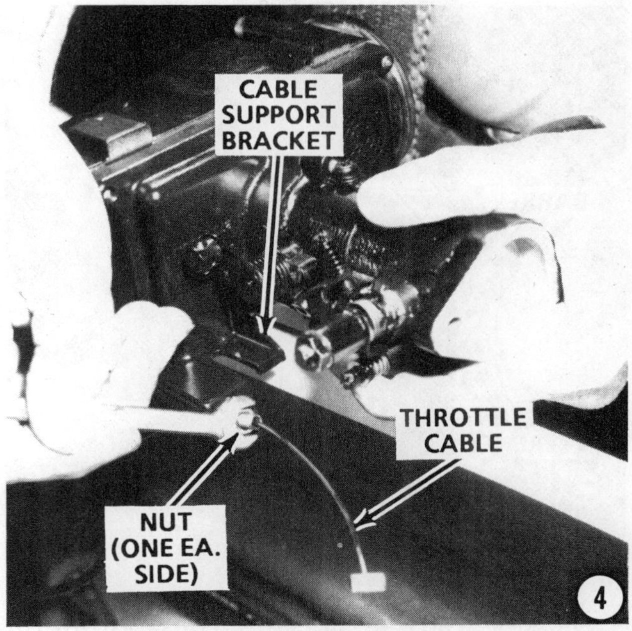

SINGLE AND DUAL CARBURETOR INSTALLATION

Preparation Tasks For Dual Installation

Place the two carburetors on the work-bench with the **AFT** carburetor -- the one with the idle speed screw -- on the left and the forward carburetor on the right. Both ball joints on the throttle levers should be facing up. Apply a dab of lubricant to both ball joints. Install the link over the ball joints. Install the short piece of fuel hose between the two carburetors.

All Carburetors

1- Work the fuel supply hose onto the proper carburetor fitting. Slide the hose clamp down over the hose and secure the hose in place, by tightening the clamp securely. Position the bottom flame arrestor cover in place over the carburetor, with the cable support bracket on the same side as the carburetor diaphragm cover. Apply a light coating of Loctite 242 onto the threads of the Allen head bolt. Install and tighten the bolt securely.

2- Place a **CLEAN** flame arrestor screen into the bottom cover and install the top cover over the flame arrestor.

3- Install and tighten the four Allen head bolts securing the two covers together. Tighten the bolts to a torque value of 8.8 ft lb (10Nm).

4- Slide the throttle cable into the slot in the cable support bracket, with a nut on each side of the slot, as shown. Tighten the two nuts against each other to secure the cable in place.

5- Slide the choke cable into the other slot on the bracket with a nut on each side

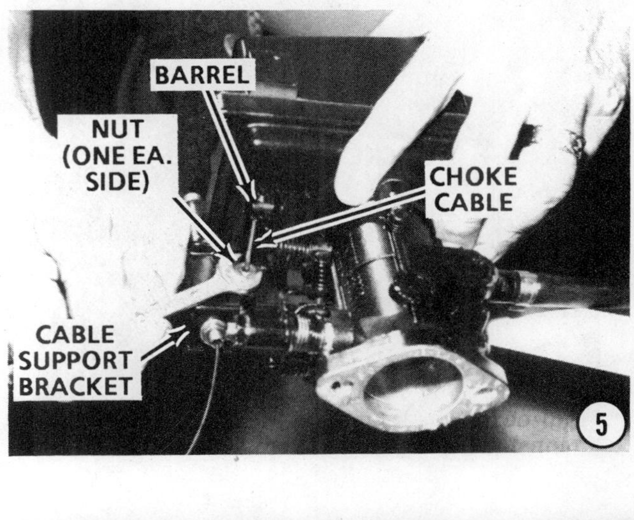

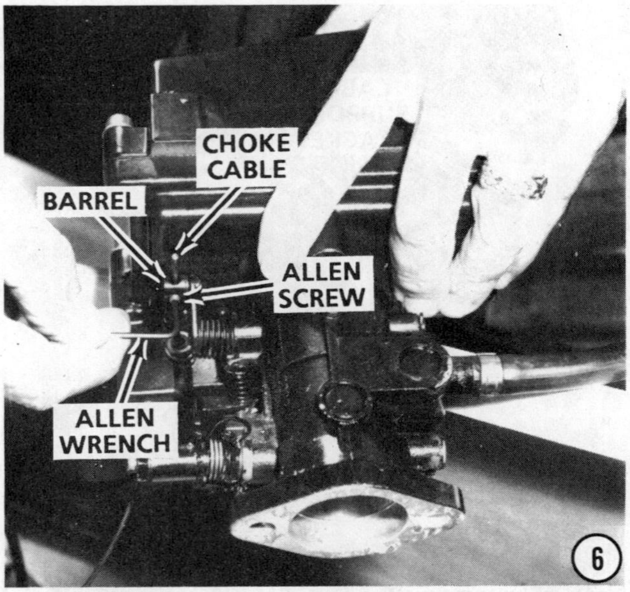

of the slot **AND** at the same time, guide the cable end into the hole in the barrel on the choke shaft. Tighten the two nuts against each other just **SNUG,** to secure the cable in place.

6- Adjust the choke cable until the mark on the cable is at the position where the cable just enters the barrel. Maintain this position and tighten the small Allen head screw on the barrel to secure the cable. If necessary, adjust the slack in the cable with the two nuts at the support bracket, and then tighten them securely.

7- Position a new carburetor mounting gasket over the intake manifold, and then

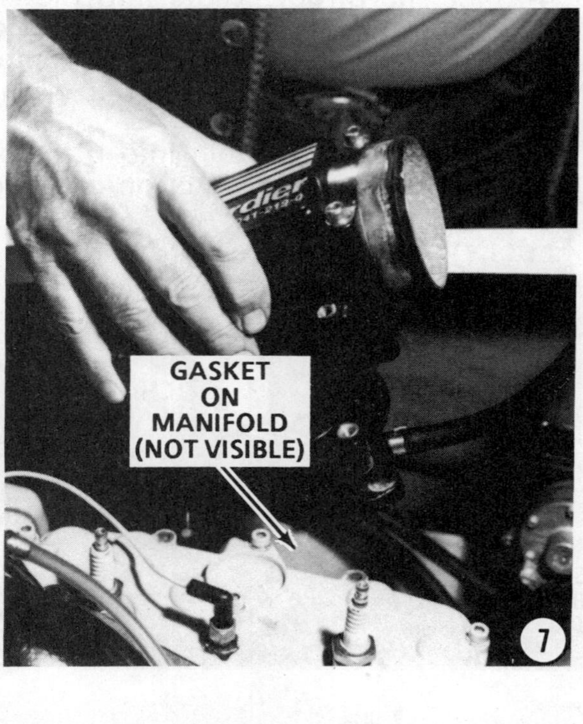

lower the carburetor/s down over the manifold studs.

8- Apply a light coating of Loctite 242 to the threads of the carburetor mounting studs. Install lockwashers, and then the nuts. Tighten the nuts to a torque value of 18 ft lb (25Nm). Insert the throttle cable end into the hole in the curved guide and route the cable into the groove in the guide.

Carburetor Priming

The carburetor must be primed with fuel before attempting to start the engine to save excessive drain on the battery and overheating of the cranking motor.

Place a shop cloth under the fuel return fitting (the fitting above the installed fuel supply line). Blow through the disconnected fuel return hose (the hose attached to the banjo fitting), until fuel begins to drip from the fuel return fitting on the carburetor.

When the fuel drips, the carburetor is full of fuel and the engine is ready for cranking after all other tasks are completed.

9- Position the "banjo" bolt, with the fuel return line attached, over the hole above the installed fuel supply hose. Install and tighten the "banjo" bolt and washer securely.

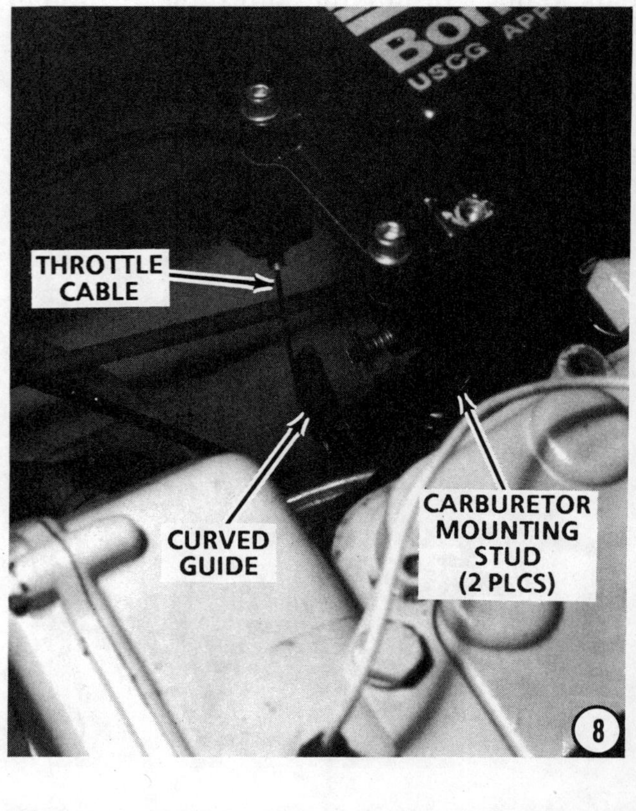

10- Install the support bracket over the flame arrestor cover and engine. Apply a light coating of Loctite 242 to the threads of the four M6 Allen head bolts. Install and tighten the four bolts securely.

11- Work the air intake hose onto the flame arrestor cover flange. Install and tighten the hose clamp to secure the hose to the flange.

Dual Carburetors

If working on a dual carburetor installation, connect the fuel supply hose to the automatic shut-off valve.

6-7 FUEL ADJUSTMENTS

Five fuel adjustments may be made on the Mikuni carburetor, for both single and twin carburetor installations, as follows:

a- Main Jet selection.
b- Throttle Cable Adjustment -- **ALWAYS** followed by oil injection synchronization
c- Low Speed Mixture Adjustment.
d- Idle Speed Adjustment -- **ALWAYS** followed by oil injection pump synchronization.
e- Choke Cable Adjustment.

Each of the listed adjustments are explained in detail in the following paragraphs.

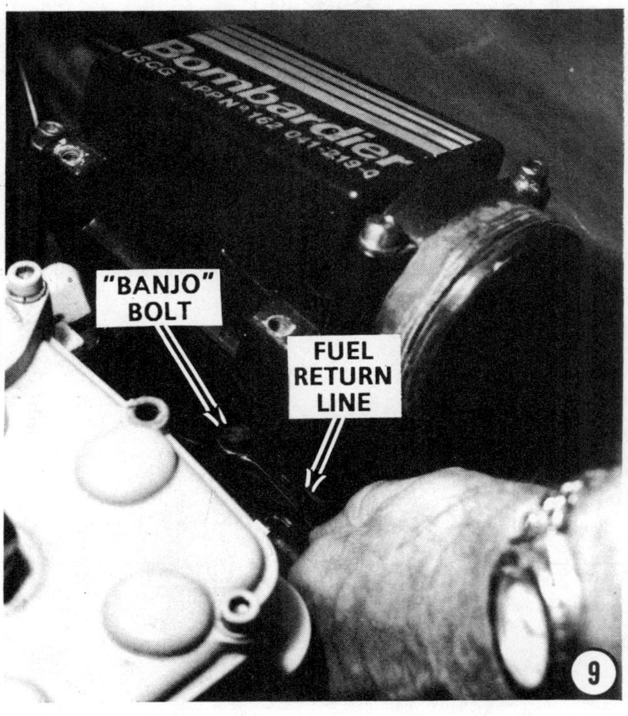

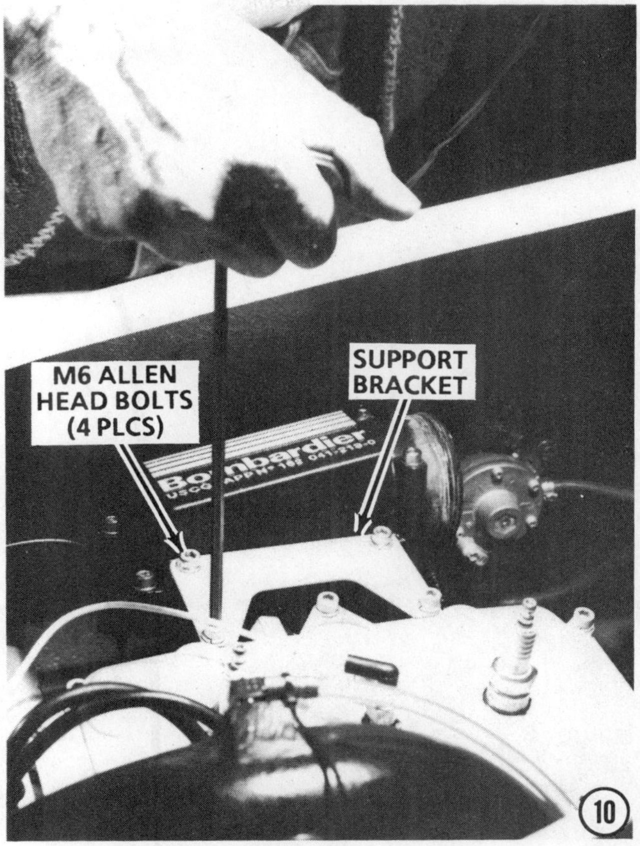

Main Jet Selection

Several main jet sizes are available to compensate for variations in altitude and temperature. The out-the-door stock main jet is intended for operation at sea level. The jet size is stamped on one face. The following main jet sizes are available: 70, 75, 80, 85, 90, 95, and 100. When operating the craft at higher altitudes, try changing the main jet in increments of 0.005" or 0.010", until an acceptable level of performance is achieved.

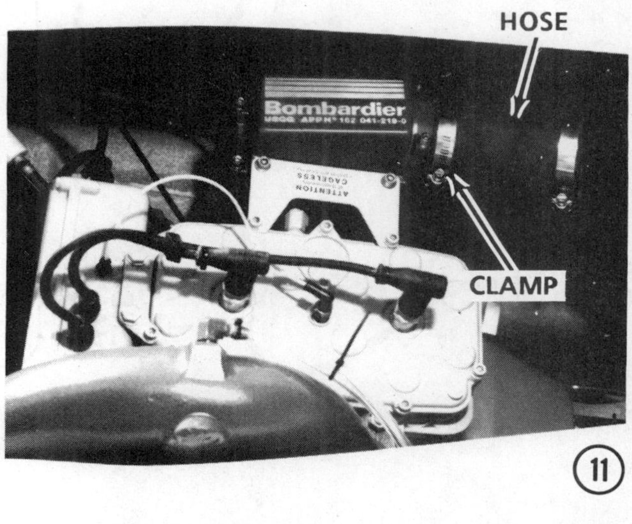

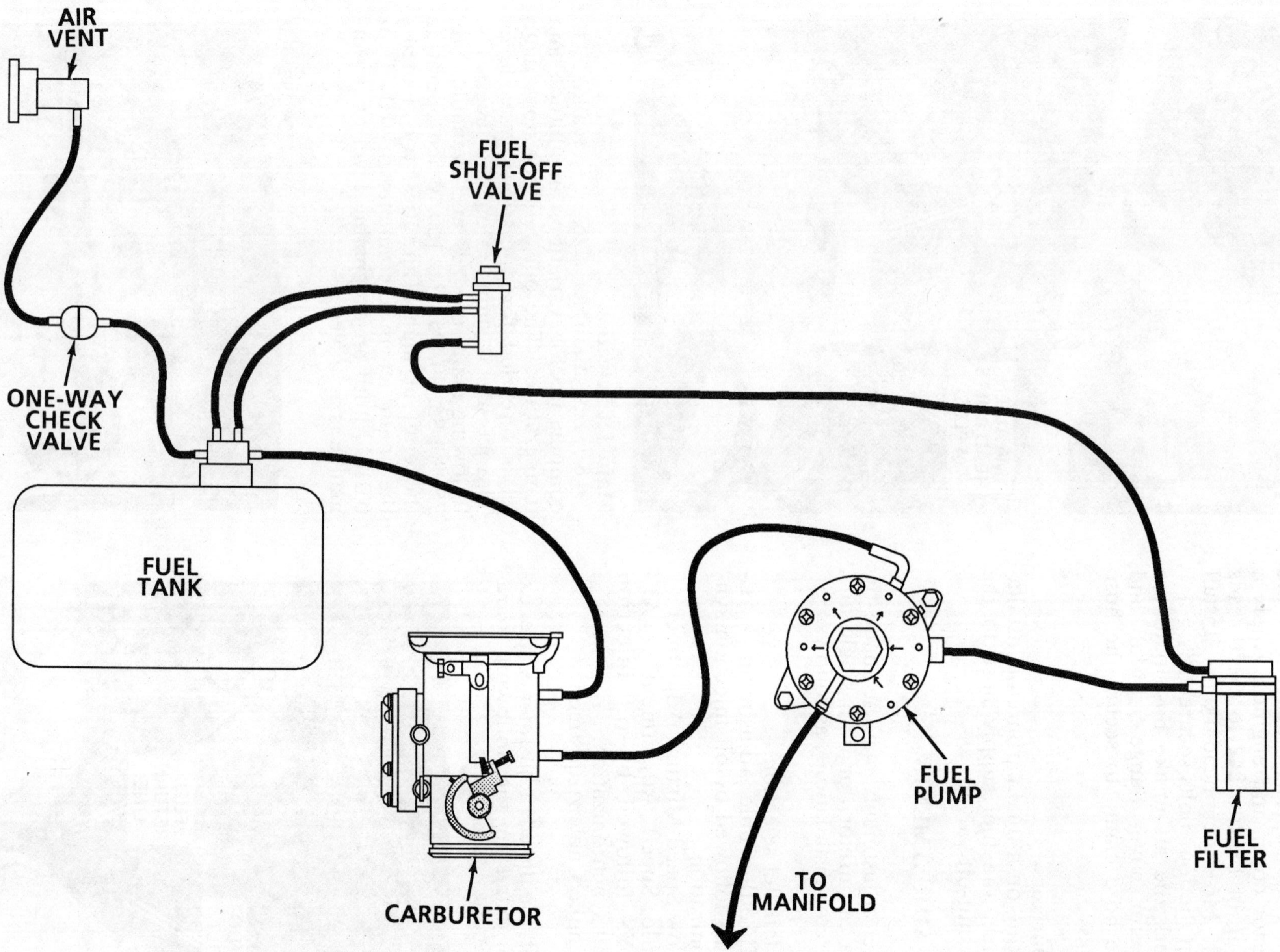

*Fuel line routing for a Bombardier Sea-Doo watercraft, **PRIOR** to 1991, equipped with a **SINGLE** carburetor, and a remote fuel pump.*

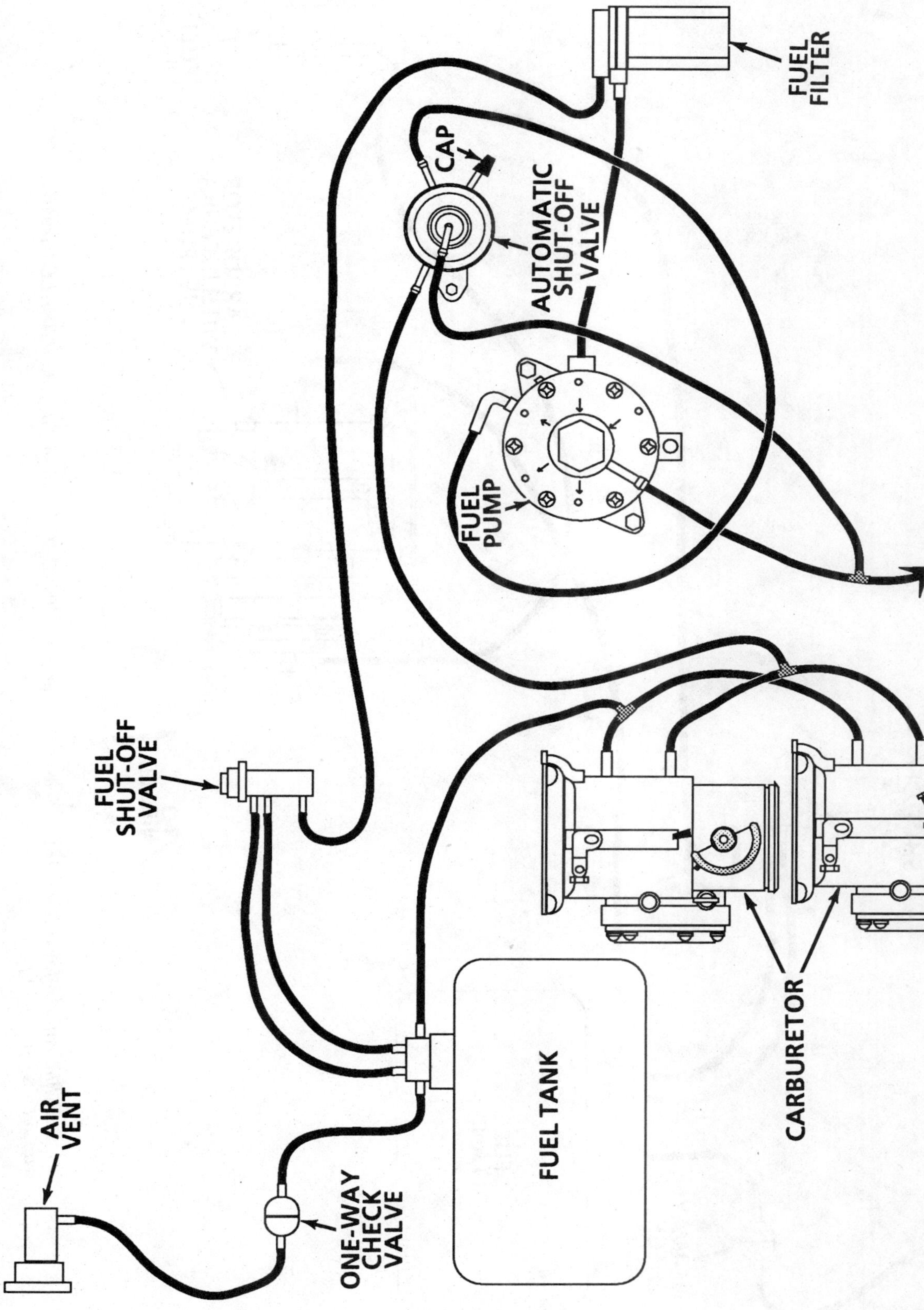

*Fuel line routing for a Bombardier Sea-Doo watercraft, **PRIOR** to 1991, equipped with **DUAL** carburetors, and a remote fuel pump.*

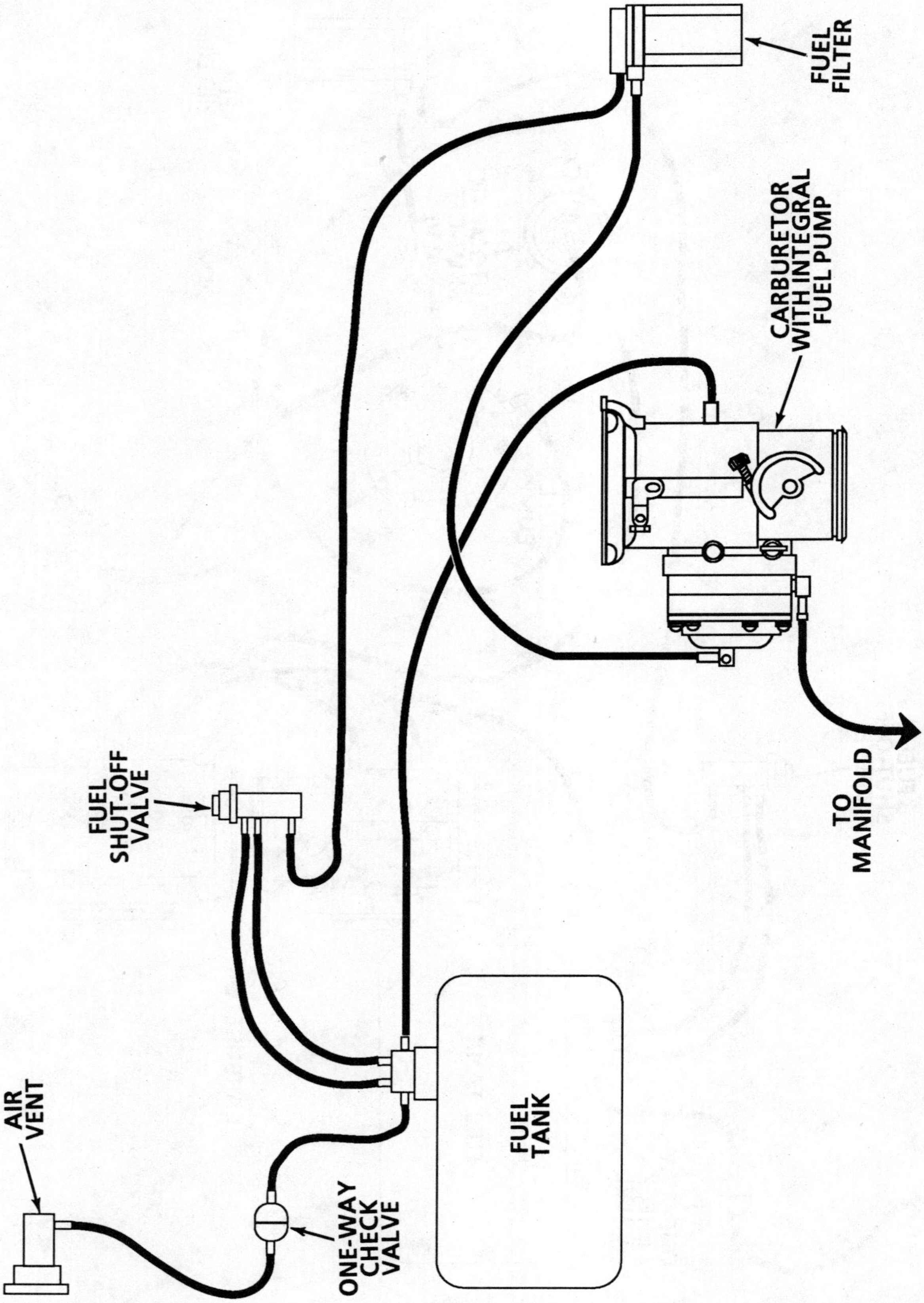

Fuel line routing for a Bombardier Sea-Doo watercraft, ***SINCE*** *1991, equipped with a* ***SINGLE*** *carburetor with integral fuel pump.*

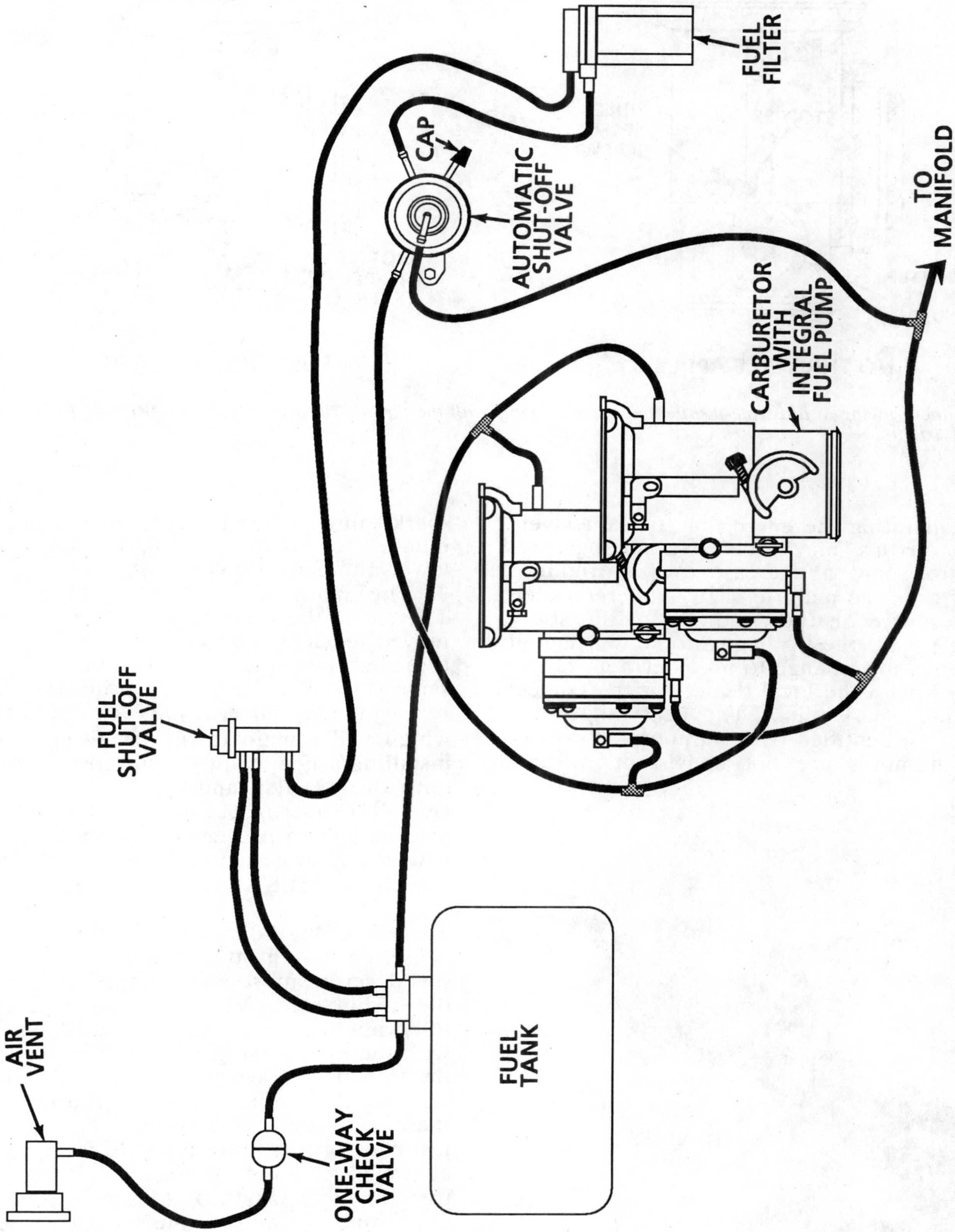

*Fuel line routing for a Bombardier Sea-Doo watercraft, **SINCE 1991**, equipped with **DUAL** carburetors with integral fuel pump.*

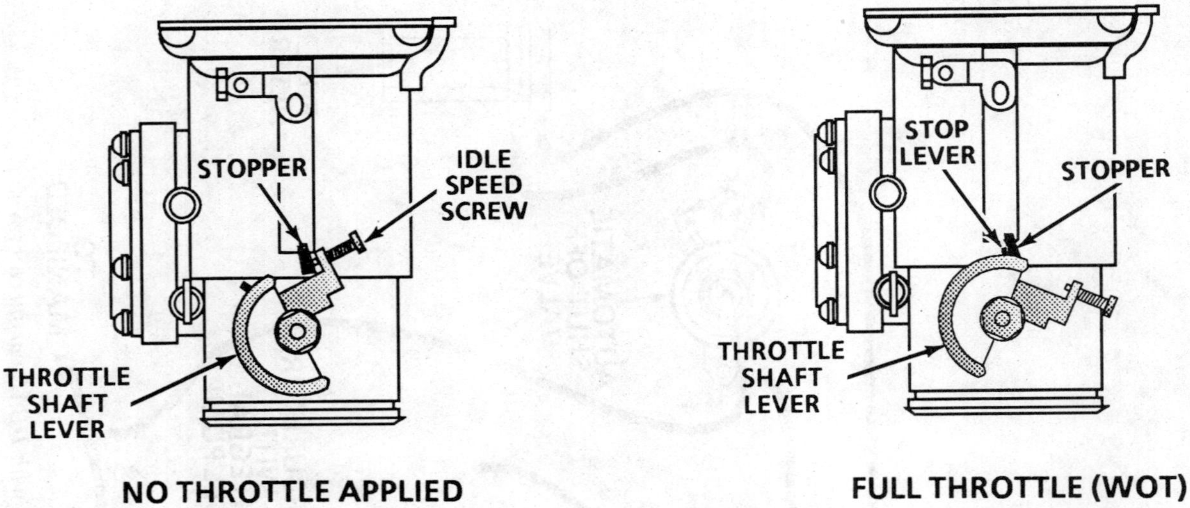

Line drawing to illustrate throttle cable adjustments with the throttle released -- closed (left), and fully open -- WOT (right).

Operating the engine on an excessively lean mixture may cause overheating, pre-ignition, and a decrease in performance. Any time the main jet size is altered, keep a close eye on the condition of the spark plug tip. If the tip begins to show signs of overheating, change to a colder heat range. Check with the local dealer for the correct number spark plug. The heat range of a spark is identified by a number. However, the numbers are not consistant. Some spark plug manufacturers use a higher number for colder plug, and others use a lower number for colder plug.

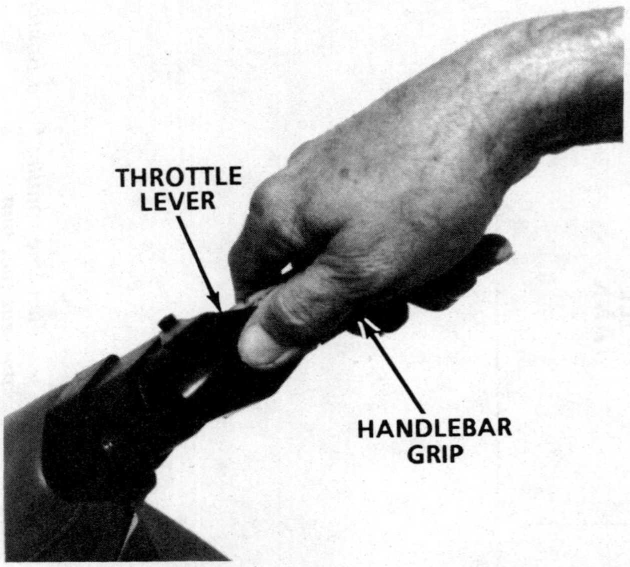

The throttle valve is wide open when the throttle lever is fully applied -- making contact with the handlebar grip.

The main jet is located externally. Therefore, the carburetor need not be removed to change a main jet. The jet is threaded into the fuel chamber, on the same side as the fuel supply hose. When removing the main jet, look for the **O**-ring, which will remain in the jet bore. When installing a new main jet, install the **O**-ring onto the jet first, and then install the jet. **DO NOT** overtighten the jet. Tighten it just enough to prevent fuel from escaping. If working on a twin carburetor installation -- both jets **MUST** be the same size.

Throttle Cable Adjustment

When no throttle is applied, the idle speed screw must contact the stopper on the carburetor. When the throttle lever is fully applied -- actually making contact with the handlebar grip -- the stop lever on the throttle linkage must contact the other side of the stopper on the carburetor. If these two conditions are not met, an adjustment on the throttle cable is necessary at the cable support bracket. Adjust the throttle cable length by loosening or tightening the two nuts -- one on each side of the support bracket.

After a throttle cable adjustment is made, **ALWAYS** perform a check to ensure the oil injection pump is synchronized with the carburetor. If the oil injection pump is not properly synchronized with the carburetor, serious damage may be caused to the engine due to lack of adequate lubrication.

Low Speed Mixture Screw

The low speed mixture screw must **NEVER** be used to alter engine rpm. Such action could result in severe engine damage. This screw is used to smooth engine operation at a set idle speed. The idle speed is changed by adjustment of the idle speed screw, covered in the following paragraph. If working on a twin carburetor installation, perform the required work on each carburetor in turn.

Now, with the engine shut down, tighten the low speed mixture screw, identified in the accompanying illustration, until the screw is **LIGHTLY** seated, and then back it out 1-1/2 to 2 full turns as a preliminary setting.

Start the engine and allow it to warm to operating temperature.

CAUTION

The following procedure MUST be followed whenever the engine is started with a hose attached for cooling the engine. Start the engine, and then IMMEDIATELY turn the water on. The engine MUST be started first to prevent water from backing up into the tuned exhaust pipe and into the engine. Water must circulate anytime the engine is operating to prevent overheating -- which could cause damage to moving parts and possibly cause engine seizure. NEVER allow the engine to operate without cooling water for more than 15 seconds. NEVER run the engine with a flushing attachment for more than 5 minutes. The driveline seal is not provided with cooling water if it is not immersed in water. To shut the engine down, first shut off the water supply, and then shut down the engine. If this sequence of tasks is not followed as stated, the engine could be severely damaged.

REMEMBER

Start the engine.
Turn the water on.
Turn off the water.
Shut down the engine.

To lean the mixture -- rotate the low speed mixture screw **CLOCKWISE.**

To enrichen the mixture -- rotate the low speed mixture screw **COUNTERCLOCKWISE.**

Rotate the low speed mixture screw in the direction which will result in the smoothest engine operation.

Idle Speed Adjustment

The idle speed adjustment screw is located on the throttle shaft linkage. If working on a twin carburetor installation, only the aft carburetor is equipped with an idle speed screw. A link rod connects the two throttle shafts together. Therefore, when an adjustment is made at the idle speed screw on the aft carburetor, the adjustment holds good for both carburetors.

When no throttle is applied, the tip of the screw rests against a stopper cast into the carburetor. To set the idle speed, obtain an induction type tachometer.

To increase engine rpm -- rotate the idle speed screw **CLOCKWISE.**

To decrease engine rpm -- rotate the idle speed screw **COUNTERCLOCKWISE.**

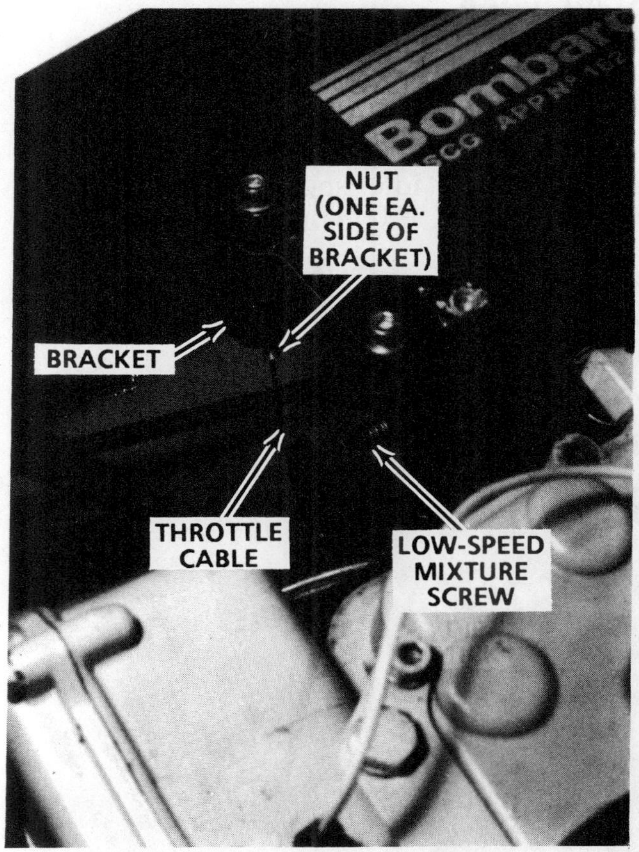

Location of the low-speed mixture screw and the throttle cable adjustment nuts.

Start the engine and allow it to warm to operating temperature.

CAUTION

The following procedure MUST be followed whenever the engine is started with a hose attached for cooling the engine. Start the engine, and then IMMEDIATELY turn the water on. The engine MUST be started first to prevent water from backing up into the tuned exhaust pipe and into the engine. Water must circulate anytime the engine is operating to prevent overheating -- which could cause damage to moving parts and possibly cause engine seizure. NEVER allow the engine to operate without cooling water for more than 15 seconds. NEVER run the engine with a flushing attachment for more than 5 minutes. The driveline seal is not provided with cooling water if it is not immersed in water. To shut the engine down, first shut off the water supply, and then shut down the engine. If this sequence of tasks is not followed as stated, the engine could be severely damaged.

REMEMBER

Start the engine.
Turn the water on.
Turn off the water.
Shut down the engine.

Rotate the idle speed screw until the engine idles at 1500 rpm -- if using a flush attachment, or 1200 rpm, if the craft is in the water.

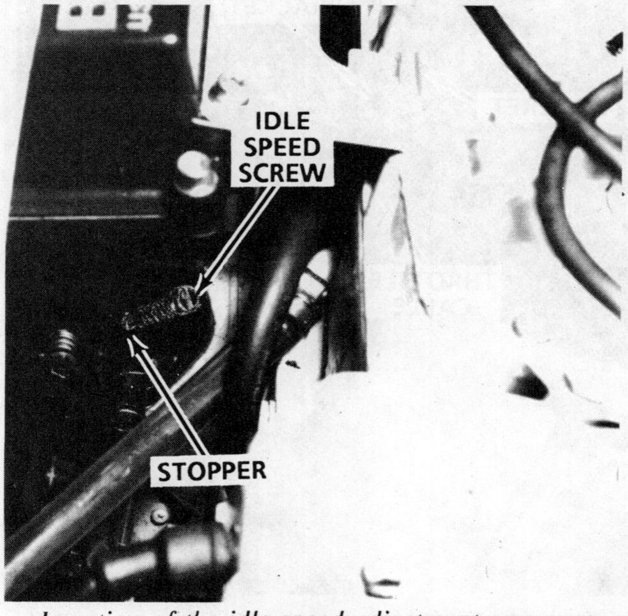

Location of the idle-speed adjustment screw, on a Mikuni carburetor.

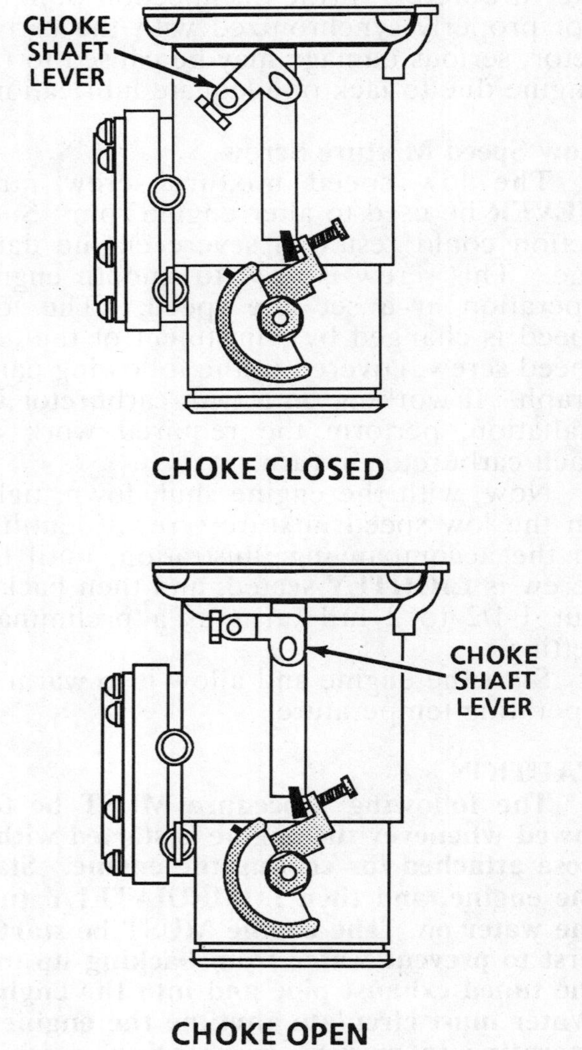

Line drawing to depict the position of the choke shaft lever when the choke is in the fully closed position (top), and in the fully open position (bottom).

After a throttle cable adjustment is made, **ALWAYS** perform a check to ensure the oil injection pump is synchronized with the carburetor. If the oil injection pump is not properly synchronized with the carburetor, serious damage may be caused to the engine due to lack of adequate lubrication.

Choke Cable Adjustment

Push the choke knob inward. Observe the choke pivot arm on the side of the carburetor. The arm should be in the position shown in the accompanying illustration, with minimum slack in the choke cable. If the choke needs adjustment: Loosen the small screw securing the choke cable end to the barrel on the choke lever. Push the choke knob inward until it rests against the control panel. Set the choke pivot arm in the fully open position, and then tighten the screw securing the cable to the barrel.

Choke Cable -- Excessive Slack

If there is too much slack in the cable, loosen the top nut around the cable at the cable support bracket and tighten the bottom nut until only a small amount of slack remains.

If the cable is tight -- with no slack -- loosen the bottom nut around the cable at the cable support bracket and tighten the top nut until there is just a small amount of slack in the cable.

If the cable has stretched, or has become kinked or frayed, or tends to stick, the entire cable should be replaced.

6-8 REMOTE FUEL PUMP

This section covers the remote type fuel pump installed on Bombardier personal water prior to 1991. Units produced since 1991 are equipped with the Mikuni carburetor having an integral fuel pump.

THEORY OF OPERATION

The next few paragraphs briefly describe, in simple terms, operation of the remote fuel pump. This description is followed by detailed procedures for testing the pressure, testing the volume, removing, servicing, and installing this remote fuel pump.

The fuel pump used is a diaphragm displacement type. The pump is secured to a bulkhead in the engine compartment. A hose from the crankcase supplies crankcase impulses to the pump. After initial engine cranking to fill the pump and engine start, the pump supplies adequate fuel to the carburetor to meet engine demands under all speeds and conditions.

The pump consists of two diaphragms, two similar check valves -- one for inlet (suction) and the other for outlet (discharge), -- a front and back cover, and a fuel pump body with a fitting for the impulse hose from the crankcase. The suction and compression created -- as the piston travels up and down in the cylinder -- causes the diaphragms to flex.

As the piston moves upward, the diaphragms will flex inward displacing volume on its opposite side to create suction. This suction will draw liquid fuel in through the inlet check valve.

When the piston moves downward, compression is created in the crankcase. This compression causes the pump diaphragms to flex in the opposite direction. This action causes the discharge check valve to lift off its seat. Fuel is then forced through the discharge valve into the fuel supply hose and delivered on to the carburetor.

Problems with the fuel pump are limited to possible leaks in the flexible neoprene suction lines; a punctured diaphragm; air leaks between sections of the pump assembly, or possibly from the check valves becoming distorted and failing to seat properly. If a cylinder indicates a wet fouled condition, as evidenced by a wet fouled spark plug, be sure to check the fuel pump diaphragms for possible puncture or leakage.

Even the tiniest pin hole in a diaphragm will cause loss of engine performance.

The remote fuel pump is mounted on a bulkhead on the starboard side of the engine compartment.

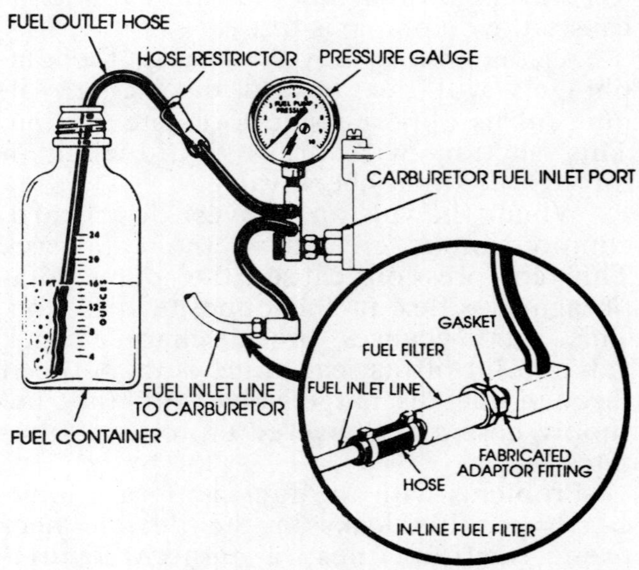

Test setup to check fuel pump pressure, as described in the text.

Pump Pressure Check

Lack of an adequate fuel supply will cause the engine to operate lean, lose rpm, or cause piston scoring.

Check the fuel strainer on the end of the pickup in the fuel tank to be sure it is of adequate size and is not clogged. Be sure to check the in-line filter at the carburetor. If this screen becomes clogged, adequate fuel cannot pass through into the carburetor to meet engine demands.

To Test Pressure

Move the craft to a body of water or a test tank. **NEVER** perform this test with a garden hose attached to the engine flush fitting.

Install a fuel pressure gauge in the fuel line between the fuel pump and the carburetor. Start the engine and allow it to warm to operating temperature, and then check the fuel pressure, as indicated by the gauge.

Operate the engine at full throttle and check the pressure reading. The gauge should indicate at least 2 psi.

Pump Volume Check

First These Words

The following test applies only to personal watercraft equipped with a remote type fuel pump separate from the carburetor.

WARNING

Gasoline will be flowing in the engine area during this test. Therefore, guard against fire by grounding the high-tension leads to prevent possible sparking, which could easily ignite fuel or fuel vapors.

A good safety procedure is to ground both spark plug leads. Disconnect the fuel line at the carburetor. Place a suitable container over the end of the fuel line to catch discharged fuel. Crank the engine and observe the flow of fuel discharged into the container.

If no fuel is discharged from the hose, check the crankcase vacuum hose connections and the fuel hose for leaks. Another possible problem may be a break or obstruction in the fuel line.

If the fuel pump is operating properly, a healthy stream of fuel should pulse from the fuel hose.

Continue cranking the engine and catching fuel for about 15 to 20 pulses to determine if the amount of fuel decreases with each pulse or if a constant amount is maintained. A decrease in the discharge indicates a restriction in the line. If the fuel line is plugged, the fuel stream may stop or never really start flowing properly .

If there is fuel in the fuel tank, but no fuel is discharged while the engine is being cranked, the problem may be in one of five areas:

1- Line from the fuel pump to the carburetor may be plugged.
2- Fuel pump may be defective.
3- Line from the fuel tank to the fuel pump may be plugged or line may be leaking air.
4- In-line fuel filter may be plugged.
5- Vacuum line from the intake manifold to fuel pump may be leaking air.

Rough Engine Idle

If an engine does not idle smoothly, the most reasonable approach to the problem is to perform a tune-up to eliminate such areas as faulty spark plugs, excessive resistance in high-tension leads, or timing out of adjustment.

Other problems contributing to rough engine idle may include:

a- Air leak in the intake manifold.
b- Uneven compression between cylinders.

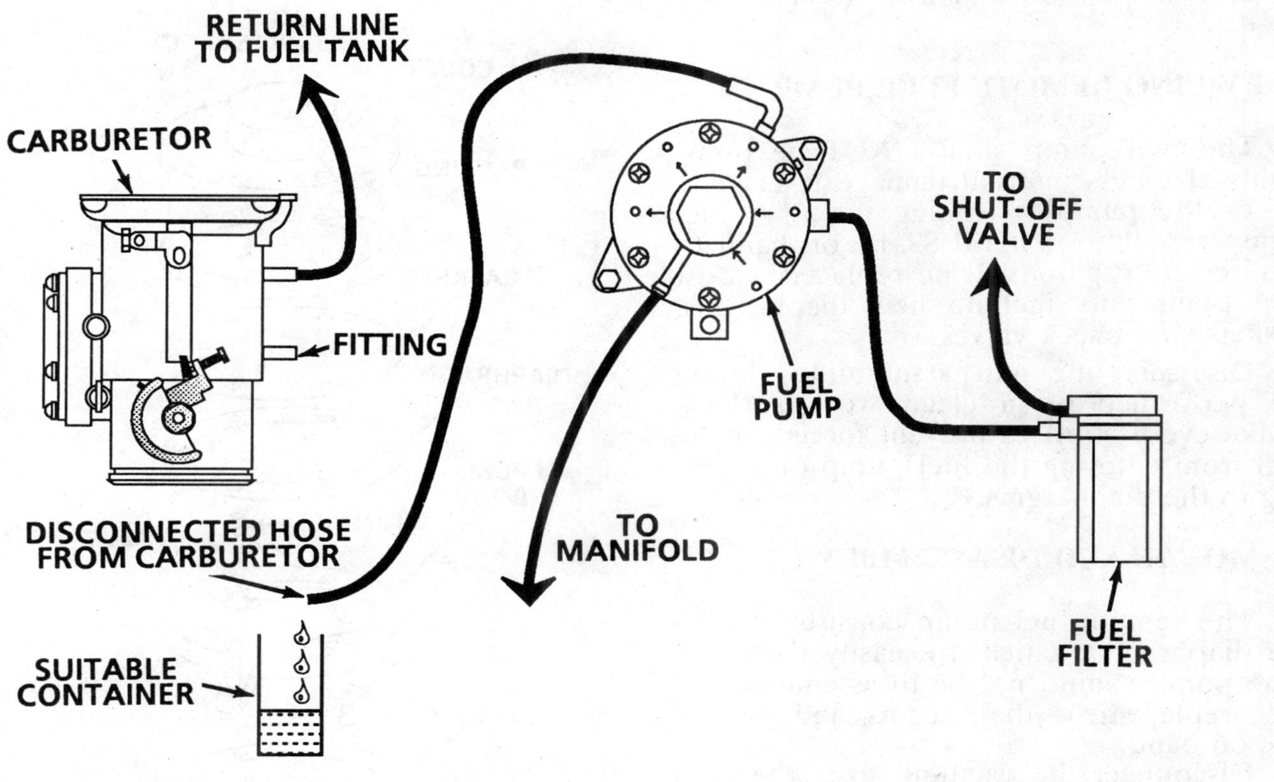

Test setup to check remote fuel pump volume for a ***SINGLE*** *carburetor installation, as explained in the text.*

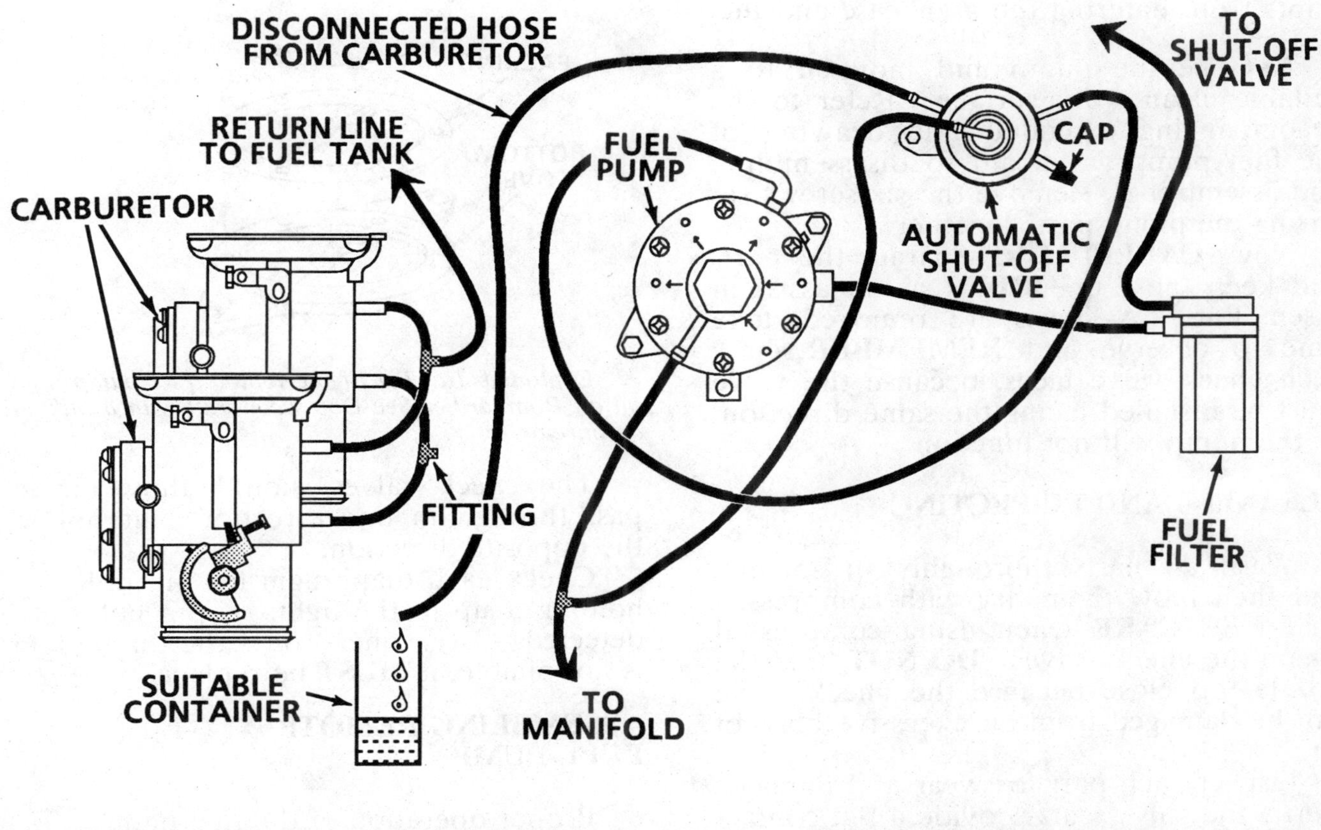

Test setup to check remote fuel pump volume for a ***DUAL*** *carburetor installation, as explained in the text.*

c- Distorted or damaged rotary valve disc.

SERVICING REMOTE FUEL PUMP

The fuel pump should **NOT** be disassembled unless internal damage is suspected or the pump is leaking fuel. A fuel pump rebuilding kit **MUST** be on hand to ensure faulty parts will be replaced. Most fuel pump kits include new diaphragms, gaskets, and check valves.

Disassembling and assembling should be performed on a clean work surface. Make every effort to prevent foreign material from entering the fuel pump or adhering to the diaphragms.

REMOVAL AND DISASSEMBLY

The remote fuel pump consists of fragile diaphragms which are easily damaged. The pump should not be disassembled unless replacement diaphragms and gaskets are on hand.

Disconnect the vacuum hose, the inlet hose and the outlet hose from the fuel pump. Take time to plug the disconnected ends of all three hoses to prevent contaminants from entering the crankcase and fuel system.

Remove the pump and move it to a suitable clean work surface. Refer to the accompanying exploded line drawing of the fuel pump as an aid to disassembling and assembling. Remove the six screws securing the pump parts together.

Now, **CAREFULLY**, separate the parts and keep them in **ORDER** as an assist in assembling. As parts are removed, take time to observe and **REMEMBER**, how each check valve faces, because the valve must be installed facing the same direction, or the pump will not function.

CLEANING AND INSPECTING

Wash all parts thoroughly in solvent, and then blow them dry with compressed air. **USE CARE** when using compressed air on the check valves. **DO NOT** hold the nozzle too close because the check valve can be damaged from an excessive blast of air.

Inspect each part for wear and damage. Verify the valve seats provide a flat contact area for the check valves.

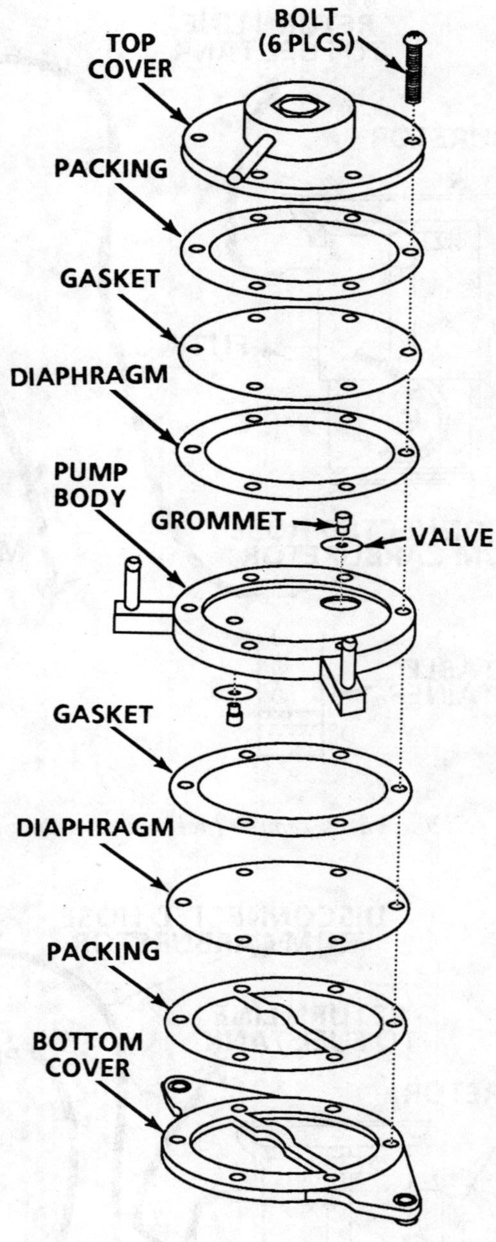

Exploded drawing of the remote fuel pump used on all Bombardier Sea-Doo watercraft with a Mikuni carburetor.

The check valves should allow air to pass through in one direction, but not in the opposite direction.

Check each diaphragm for pin holes by holding it up to the light. If pin holes are detected -- even one -- or if the diaphragm is not pliable, it **MUST** be replaced.

ASSEMBLING REMOTE FUEL PUMP

Proper operation of the fuel pump is essential for maximum engine performance.

Therefore, always use **NEW** diaphragms and gaskets.

NEVER use any type sealer on fuel pump gaskets.

TAKE CARE not to damage the very fragile flat surface of the check valve. Secure each check valve in place with a grommet.

Check to be sure the holes for the screws are all aligned. If the diaphragms are not properly aligned, a tear would surely develop when the screws are installed.

Install the six Phillips head screws through the various parts and tighten the screws securely.

Secure the pump to the bracket with the two bolts and washers.

Connect the fuel and vacuum lines, as shown in the fuel routing diagrams for single and dual carburetor installations, on Pages 6-26 and 6-27.

CAUTION

The following procedure MUST be followed whenever the engine is started with a hose attached for cooling the engine. Start the engine, and then IMMEDIATELY turn the water on. The engine MUST be started first to prevent water from backing up into the tuned exhaust pipe and into the engine. Water must circulate anytime the engine is operating to prevent overheating -- which could cause damage to moving parts and possibly cause engine seizure. NEVER allow the engine to operate without cooling water for more than 15 seconds. NEVER run the engine with a flushing attachment for more than 5 minutes. The driveline seal is not provided with cooling water if it is not immersed in water. To shut the engine down, first shut the water supply off, and then shut down the engine. If this sequence of tasks is not followed as stated, the engine could be severely damaged.

Typical installation of the remote fuel pump on the engine compartment bulkhead. Complete detailed routing of the fuel and vacuum lines for the fuel pump is shown on Pages 6-26 and 6-27.

REMEMBER

Start the engine.
Turn the water on.
Turn off the water.
Shut down the engine.

Start the engine and check the completed work. Be sure to inspect all fuel connections for leaks.

6-9 OIL INJECTION

Oil injection is standard equipment on all Bombardier Sea-Doo watercraft.

In recent years, a more sophisticated method of mixing oil with the fuel for internal lubrication has become more popular with the manufacturers of personal watercraft. This oil injection system replaces the age old method of manually

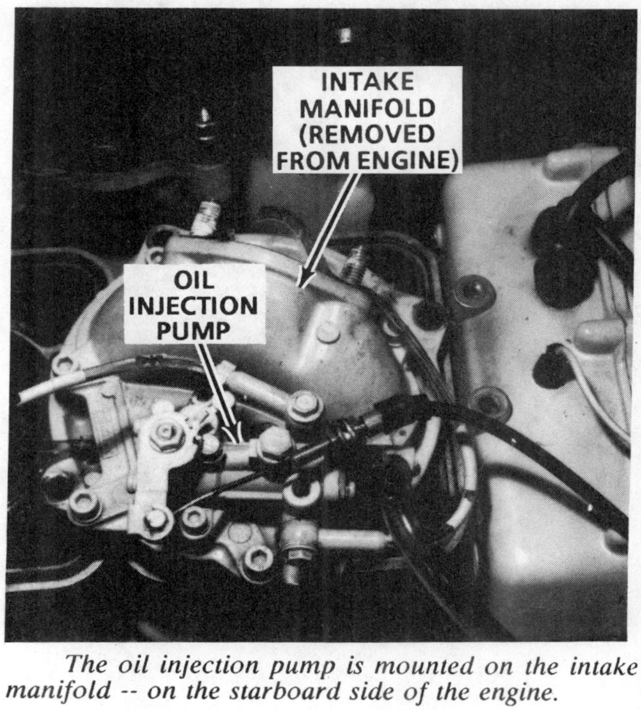

The oil injection pump is mounted on the intake manifold -- on the starboard side of the engine.

mixing oil with the fuel for lubrication of internal moving parts in the engine.

Bombardier engineers designed and developed their oil injection system as an independent unit -- separate from the fuel system until the lubricant enters the intake manifold. This system injects oil directly into the intake manifold between the carburetor and the rotary valve. Such an arrangement provides immediate oil delivery response to changes in rpm and load demands. Because the oil bypasses the carburetor, gummy deposits in the carburetor passages are avoided.

Oil Mixture

Fill the oil tank to the top line embossed on the oil tank dipstick. Use only two-cycle N.M.M.A. engine oil (or an equivalent oil with a BIA certified rating TC-W). This oil is suitable for use through a temperature range of 14°F to 140°F (-10°C to 60°C).

The rate of oil injection is controlled by engine rpm. The oil pump is directly coupled to the rotary valve shaft driven by the crankshaft. After the initial break-in period, oil is injected at a rate giving approximately a 40:1 mixture at low and idle speed. The mixture increases as the throttle is advanced to WOT (wide open throttle).

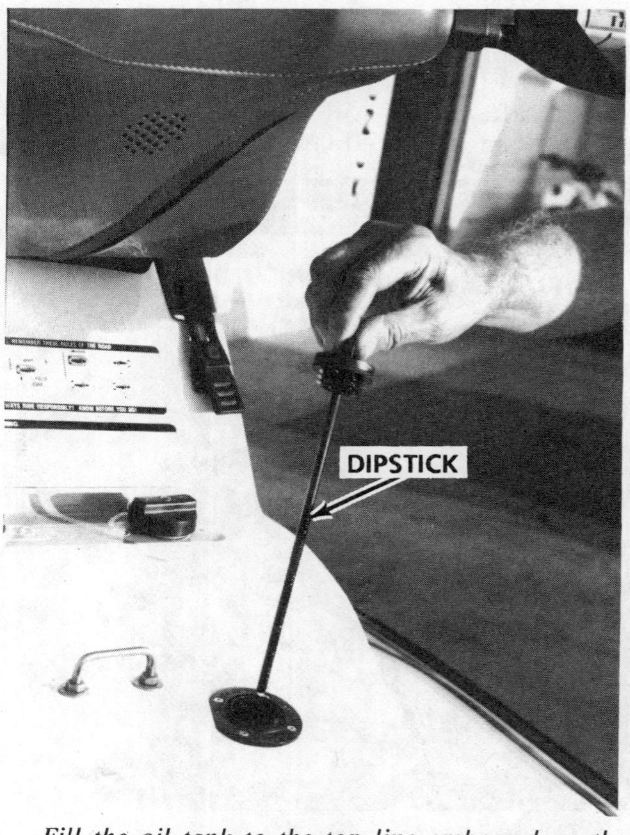

Fill the oil tank to the top line embossed on the oil tank dipstick.

In order to obtain extra lubrication for a new, rebuilt, or engine out of long storage, during the first 10 hours of break-in period, the manufacturer recommends oil to be added directly into the fuel tank to give a mixture of 25:1.

Professional mechanics, who own oil injected units, constantly add an extra ounce of oil per gallon of fuel after the break-in period. The addition of this extra oil may prevent engine seizure in the event of an oil pump failure or clog in an oil delivery line.

Mechanics are quite cost conscious. They realize all too well how much cheaper replacing fouled spark plugs can be when weighed against an engine overhaul to replace expensive parts damaged from lack of adequate lubrication.

Use only outboard marine oil in the mixture, **NEVER** automotive oils. Four-stroke automotive engine oil is not formulated to burn completely, only to lubricate. Therefore, automotive oil, if used in a two-stroke engine, will leave undesirable residue.

SYSTEM COMPONENTS

Principle components of the oil injection system include:

The oil tank is mounted inside the engine compartment next to the fuel tank. The capacity of the tank is 95 fl oz (2.8L).

Main oil tank, with dipstick.

A one-way check valve.

An oil injection pump.

An in-line oil filter.

A network of connecting hoses.

Oil Tank

The oil tank is mounted inside the engine compartment, with the filler easily accessible when the engine cover is removed. The capacity of this tank is 95 fl oz (2.8L).

Oil supply to the oil injection pump is gravity fed from the oil tank. Therefore, a one-way check valve is mounted on top of the tank. The other end of the valve is open to atmosphere. The check valve allows air to enter the tank replacing oil used, but will not allow oil to flow out of the tank -- thus the tank is "spill-proof".

Oil Injection Pump

The oil injection pump is a positive displacement type unit driven by the rotary valve shaft. With each revolution of the crankshaft, a plunger cam moves up and down, pumping oil through two transparent hoses to the intake manifold.

The in-line oil filter is located in the supply line between the tank and the oil injection pump.

System Inspection

Remove the oil tank filler cap. Inspect the condition of the seal beneath the cap. Replace the seal if it is worn or cracked.

Disconnect the ventilation hose from the fitting on the oil tank.

To test operation of the check valve:

Pull the two hoses free of the fittings. Now, blow through one of the fittings, and then the other fitting. Air should pass through the valve **ONLY** in one direction.

A faulty check valve will starve the engine of lubrication and quickly cause extensive and expensive damage.

Troubleshooting

Unfortunately, most problems with the oil injection system are only discovered after it is too late. If either of the two hoses becomes clogged, kinked, or restricted, oil delivery to the cylinders will be reduced. Insufficient oil delivery to the engine will cause the oil level to drop slower than normal; the engine will overheat due to inadequate lubrication; and moving internal parts will wear more quickly or be severely damaged.

Excessive oil delivery to the cylinders will cause the oil level to drop at a much faster rate than normal. The engine will smoke -- especially at idle speed -- and the spark plugs will become fouled, causing the engine to misfire.

First Checks
Oil Delivery

If excessive or reduced oil delivery occurs, as described above, one or more of the following areas may require immediate attention:

Defective oil pump.

Clogged oil tank check valve.

Kinked oil hose.

Blocked oil hose.

Air in the system.

Leaking oil hose.

Improper idle fuel adjustment.

Poor quality fuel.

Poor quality two-stroke oil.

BAD NEWS

An oil pump gasket set is the only kit available to overhaul an oil pump. No spare parts are sold for the oil pump used on these units. Therefore, a defective oil pump **MUST** be replaced.

Systematically check each oil line and connection. A free flow of oil through an unrestricted line is **CRITICAL** for adequate engine lubrication.

CAUTION

Any time an oil tank hose is disconnected, the oil injection pump **MUST** be purged ("bled") of any trapped air. Failure to "bleed" the system could lead to engine seizure due to lack of adequate lubrication.

Instruction for "bleeding" -- purging -- air from the system follows.

Purging Air from Oil Injection System

The following procedures are to be performed any time the oil injection system has been opened (other than to add oil to the tank), and air has entered the system.

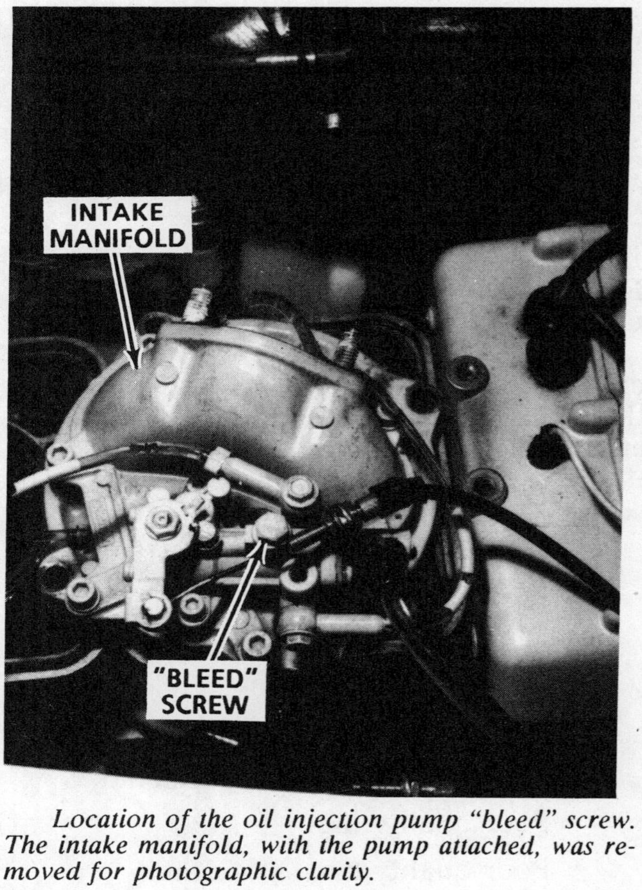

Location of the oil injection pump "bleed" screw. The intake manifold, with the pump attached, was removed for photographic clarity.

Prepare about a ten minute supply of 40:1 premix using two-stroke oil. "Jerry-rig" a setup to provide this mixture to the fuel pump during this purging procedure. The oil/fuel mixture prepared will provide adequate lubrication while the system is being purged. Check the level of oil in the oil tank and replenish as necessary.

CAUTION

The following procedure MUST be followed whenever the engine is started with a hose attached for cooling the engine. Start the engine, and then IMMEDIATELY turn the water on. The engine MUST be started first to prevent water from backing up into the tuned exhaust pipe and into the engine. Water must circulate anytime the engine is operating to prevent overheating -- which could cause damage to moving parts and possibly cause engine seizure. NEVER allow the engine to operate without cooling water for more than 15 seconds. NEVER run the engine with a flushing attachment for more than 5 minutes. The driveline seal is not provided with cooling water if it is not immersed in water. To shut the engine down, first shut the water supply off, and then shut down the engine. If this sequence of tasks is not followed as stated, the engine could be severely damaged.

REMEMBER

Start the engine.
Turn the water on.
Turn off the water.
Shut down the engine.

SPECIAL WORDS

Because the fuel and oil premix is being fed through the carburetor, in addition to oil from the oil injection system also being fed directly into the manifold, the engine will be operating on a heavy mixture of oil. This condition will cause the engine to smoke heavier than normal.

Obtain a small container and have a shop cloth handy to catch any oil drips. Loosen the bleed screw on the oil pump a few turns and wait until a slow pulsing flow of oil emerges without any air bubbles. Observe the flow of oil for four full minutes to verify the pump is functioning properly and all air has been expelled from the system.

The two small transparent hoses between the pump and the intake manifold **MUST** be full of oil. If not, operate the engine at idle speed while holding the pump lever in the fully open (**UP**) position. **DO NOT** activate the throttle lever at the handlebar until the two hoses are full of oil. Tighten the bleed screw securely.

Replenish the oil tank with two-stroke oil.

Oil Injection Pump Synchronization

Oil injection pump adjustment is dependent on throttle cable position. Therefore, any time the throttle cable is adjusted, the oil injection pump **MUST** be synchronized. The throttle cable from the handlebar splits into two cables. One cable controls engine rpm at the carburetor. The other cable is attached to the oil injection pump. With this arrangement, the rate of oil injected into the intake manifold is controlled by engine rpm -- as it should be. The higher rpm -- the more oil is needed.

Observe the mark on the pump lever and the mark on the pump boss. These two marks **MUST** be aligned for the pump to be properly synchronized.

Adjust the length of the throttle cable to bring these two marks into alignment. This is accomplished using the two adjustment nuts -- one on each side of the slot in the cable support bracket on the pump. A small mirror would be most helpful to see these marks, because there is very little clearance between the oil injection pump and the starboard engine compartment bulkhead.

Oil Injection Pump Operational Test

Remove the four bolts and washers securing the rotary valve housing to the upper and lower crankcase. The intake manifold and the oil pump are attached to the valve housing and will come away with the housing.

Remove the six Allen head bolts and the washers securing the intake manifold to the rotary valve housing. The oil injection pump will remain on the manifold.

Lift the intake manifold free of the rotary valve housing. **TAKE CARE** not to lose the single dowel pin from one of the mating surfaces. The dowel pin may remain in the housing or in the manifold.

Obtain Gear Holder P/N 290-277-905 and use it to prevent the oil pump gear from rotating while the nut is loosened. If this tool is not available, hold the gear with a shop cloth, as shown. Remove the nut and washer, gear, and another washer.

Obtain an electric drill capable of **COUNTERCLOCKWISE** rotation (reverse). Insert the oil pump shaft into the drill chuck and tighten the chuck just **SNUG** -- not enough to damage the shaft threads.

Place a shop cloth inside the rotary valve housing under the discharge ports of the two transparent hoses. Check to be sure the drill selector is in the **REVERSE** position. Now, hold the pump lever in the fully open position and at the same time turn the drill motor on. Oil should drip from both discharge ports. If not, the pump is defective and **MUST** be replaced.

GOOD WORDS

Individual replacement parts are **NOT** available. However, if the pump is leaking, the condition may be corrected by installing a gasket kit, which is available. If the gasket set does not solve the problem, the pump **MUST** be replaced. Removal procedures follow.

Oil Pump Removal

First, remove the two "banjo" bolts and fittings from the pump, with the oil hoses

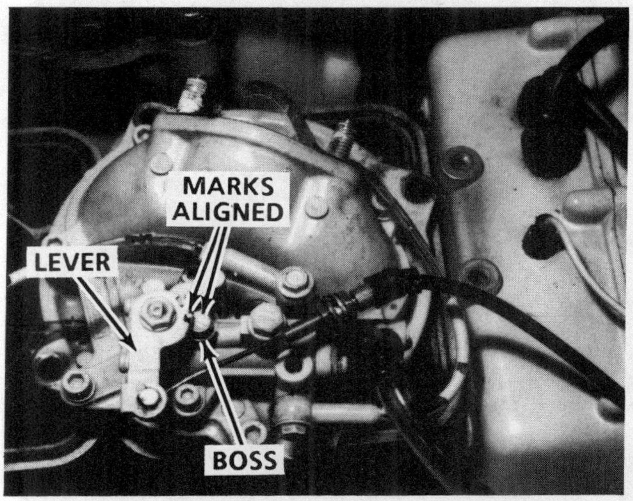

The text explains in detail, how and why the two marks -- one on the pump lever and the other on the pump boss -- MUST be aligned, for proper pump synchronization with the carburetor.

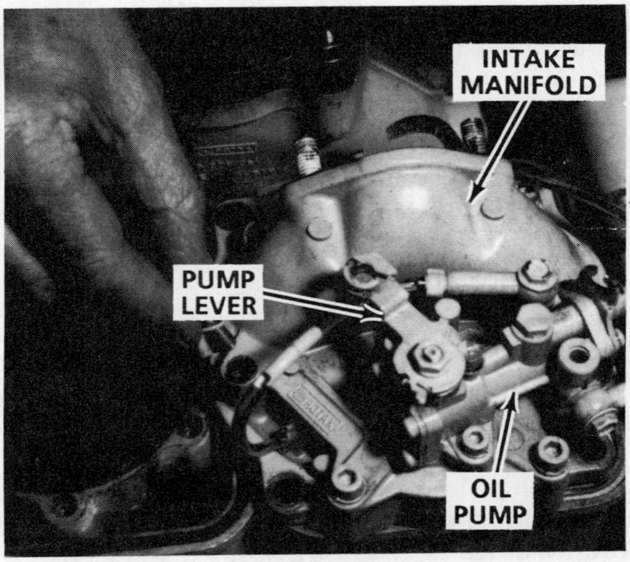

*Securing the oil injection pump to the intake manifold. The throttle cable is installed into the pump lever **AFTER** the rotary valve housing has been installed on the engine.*

attached. Remove the two mounting bolts, and then lift the pump clear of the intake manifold. Remove and discard the **O**-ring.

Oil Pump Installation

Install a new **O**-ring over the oil pump shaft. Install the pump into the intake manifold, with the two bolts and washers. Tighten the bolts to a torque value of 53 in lb (5Nm). Install the two "banjo" bolts and fittings to the pump, with the oil hoses attached. Apply Loctite 242 to the pump shaft threads.

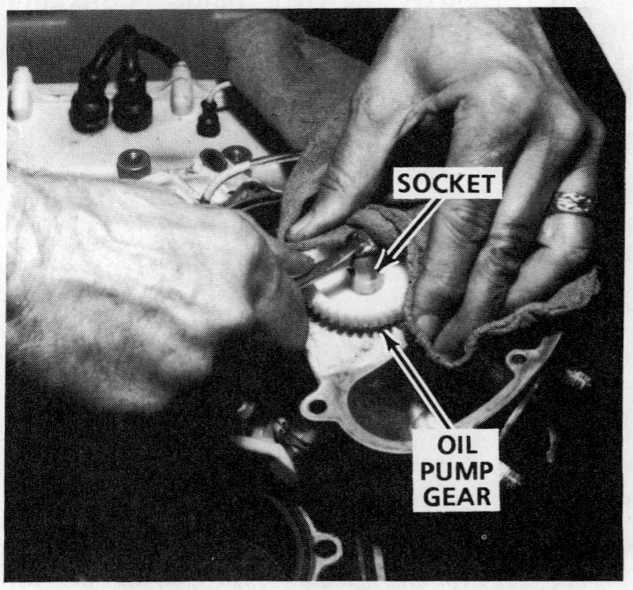

Installation of the washer, gear, another washer, and nut onto the pump shaft from inside the intake manifold.

Inside the intake manifold: Install the washer, gear, another washer, and then the nut onto the oil pump shaft. Obtain Gear Holder P/N 290-277-905 or a shop cloth. Hold the gear firmly to prevent it from rotating while the nut is tightened to a torque value of 89 in lb (10Nm).

Install the intake manifold onto the rotary valve cover, with the manifold indexed over the single dowel pin in the cover. Install the six bolts and washers securing the intake manifold to the rotary valve cover. Tighten the bolts alternately and evenly to a torque value of 80 in lb (9Nm).

Oil Pump Synchronization

Observe the mark on the pump lever and the mark on the pump boss. These two marks **MUST** be aligned for the pump to be properly synchronized.

Adjust the length of the throttle cable to bring these two marks into alignment. This is accomplished using the two adjustment nuts -- one on each side of the slot in the cable support bracket on the pump. A small mirror would be most helpful to see these marks, because there is very little clearance between the oil injection pump and the starboard engine compartment bulkhead.

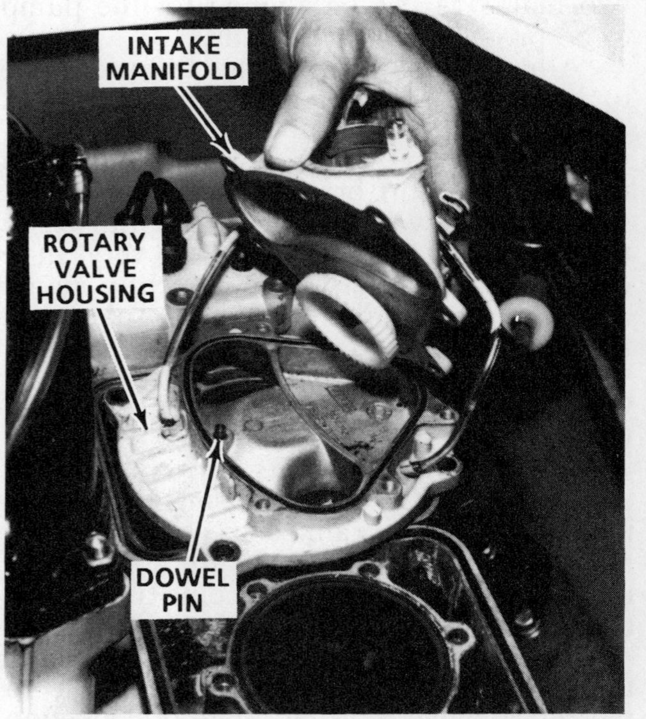

Installation of the intake manifold, with the oil injection pump attached onto the rotary valve housing. A hole in the intake manifold must index over a single dowel pin in the rotary valve housing.

7
IGNITION

7-1 INTRODUCTION AND CHAPTER COVERAGE

The less a marine engine is operated, the more care it needs. Allowing a marine engine to remain idle will do more harm than if it is used regularly. To maintain the engine in top shape and always ready for efficient operation at any time, the engine should be operated every 3 to 4 weeks throughout the year.

The carburetion and ignition principles of two-cycle engine operation **MUST** be understood in order to perform a proper tune-up on a marine engine.

If you have any doubts concerning your understanding of two-cycle engine operation, it would be best to study the operation theory section in the first portion of Chapter 8 -- **ENGINE**, before tackling any work on the ignition system.

A Capacitor Discharge Ignition (CDI) is used on all engine models covered in this manual.

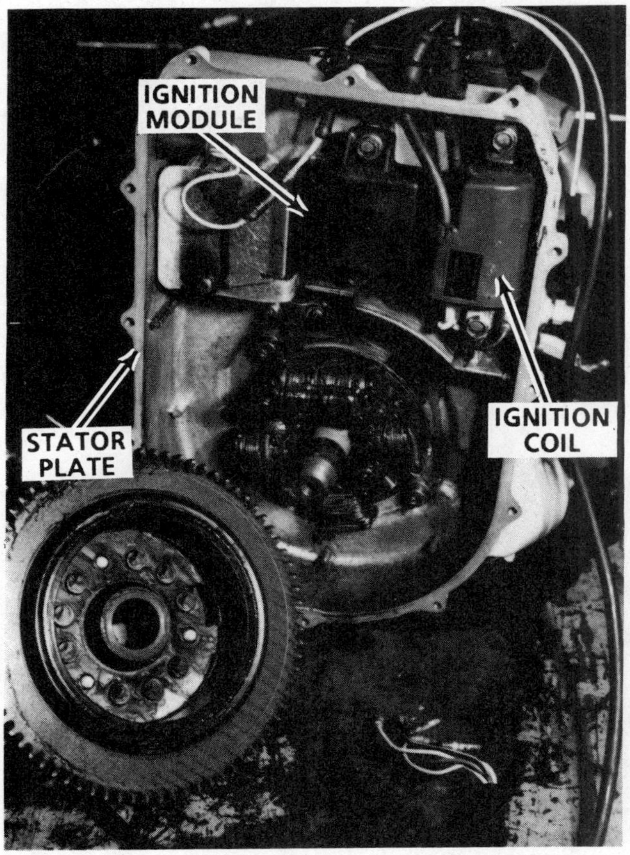

The ignition module and the ignition coil are mounted on the stator plate above the flywheel.

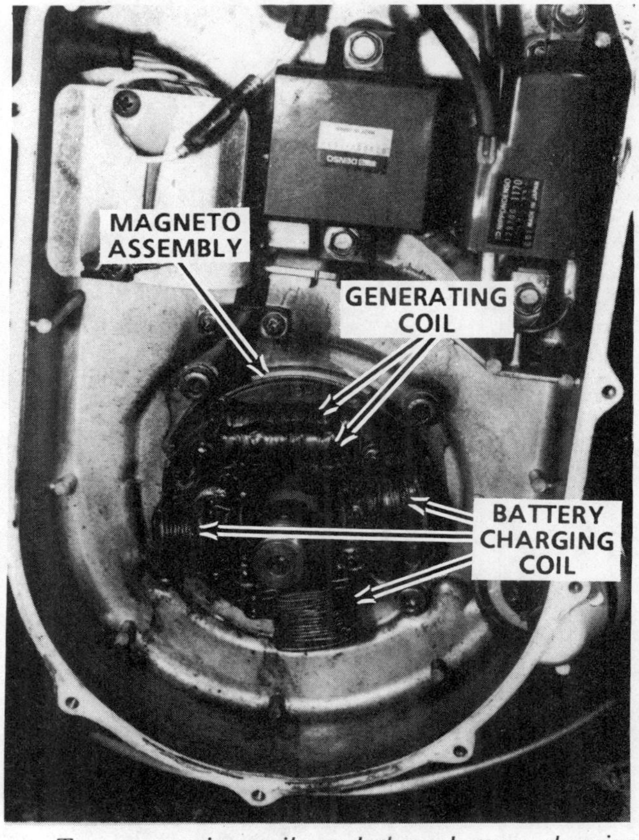

Two generating coils and three battery charging coils are mounted on the magneto assembly behind the flywheel.

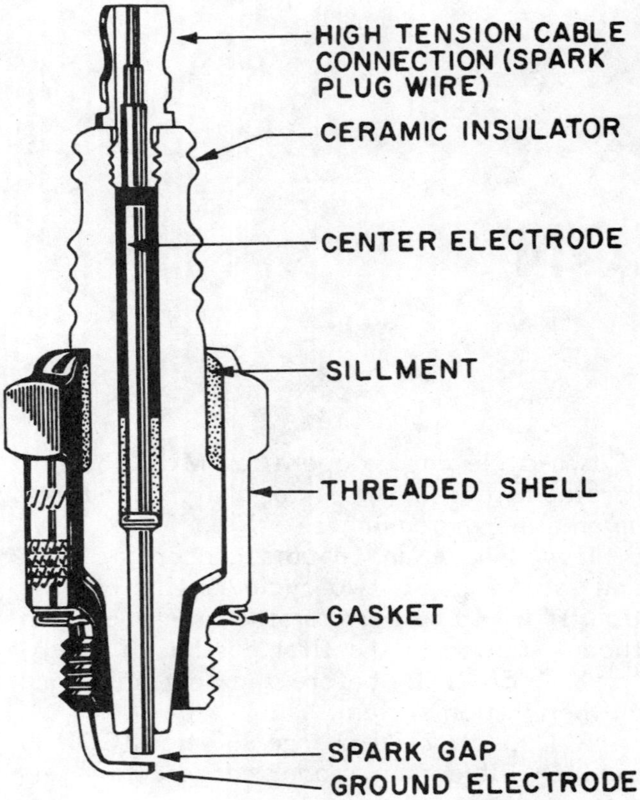

Cross-section drawing of a typical spark plug, with major parts identified.

7-2 SPARK PLUG EVALUATION

Removal

Remove the spark plug leads by pulling and twisting on only the molded cap. **NEVER** pull on the lead or the connection inside the cap may become separated or the boot damaged. Remove the spark plugs. **TAKE CARE** not to tilt the socket as you remove the plug or the insulator may be cracked.

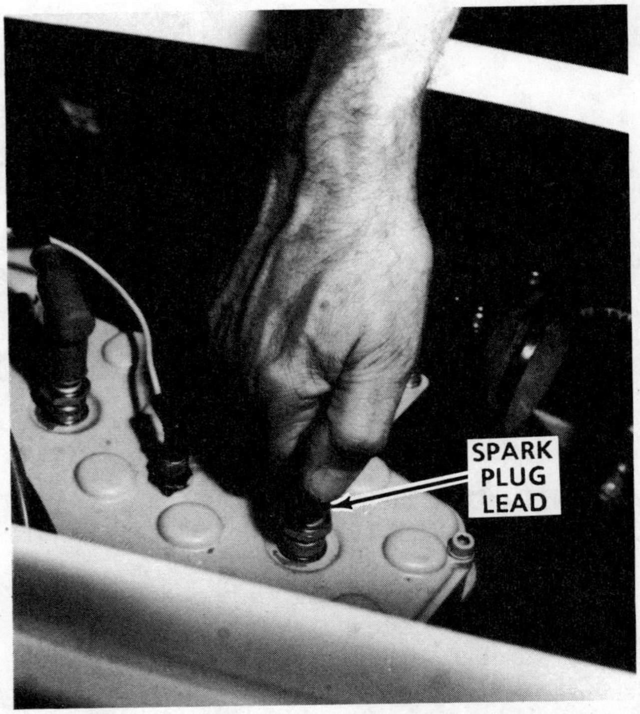

ALWAYS *remove the spark plug high tension lead by pulling and twisting with a good grip on the molded cap.*

Examine

Carefully examine both plugs to determine the firing conditions in the cylinder. If the side electrode is bent down onto the center electrode, the piston is traveling too far upward in the cylinder and striking the spark plug. Such damage indicates the piston pin or the rod bearing is worn excessively.

In most cases, an engine overhaul is required to correct the condition. To verify the cause of the problem, rotate the flywheel by hand. As the piston moves to the full up position, push on the piston crown

The spark plug is foul from operating with an overrich air/fuel mixture, possibly caused by an improper carburetor adjustment.

This spark plug has been operating too-cold, because it is rated with a too-low heat range for the engine.

with a screwdriver inserted through the spark plug hole, and at the same time rock the flywheel back-and-forth. If any play in the piston is detected, the engine must be rebuilt.

Correct Color

A proper firing plug should be dry and powdery. Hard deposits inside the shell indicate too much oil is being mixed with the fuel. The most important evidence is the light gray to tan color of the porcelain, which is an indication this plug has been running at the correct temperature. This means the plug is one with the correct heat range and also that the air/fuel mixture is correct.

Rich Mixture

A black, sooty condition on both spark plug shells and the porcelain is caused by an excessively rich air/fuel mixture, both at low and high speeds. The rich mixture lowers the combustion temperature so the spark plug does not run hot enough to burn off the deposits.

Deposits formed only on the shell is an indication the low-speed air/fuel mixture is too rich. At high speeds with the correct mixture, the temperature in the combustion chamber is high enough to burn off the deposits on the insulator.

Too Cool

A dark insulator, with very few deposits, indicates the plug is running too cool. This condition can be caused by low compression or by using a spark plug of an incorrect heat range. If this condition shows on only one plug it is most usually caused by low compression in that cylinder. If both plugs have this appearance, then it is probably due to the plugs having a too-low heat range.

Fouled

A fouled spark plug may be caused by the wet oily deposits on the insulator shorting the high-tension current to ground inside the shell. The condition may also be caused by ignition problems which prevent a high-tension pulse from being delivered to the spark plug.

Carbon Deposits

Heavy carbon-like deposits are an indication of excessive oil in the fuel. This condition may be the result of worn piston rings or excessive ring end gap.

Overheating

A dead white or gray insulator, which is generally blistered, is an indication of overheating and pre-ignition. The electrode gap wear rate will be more than normal and in the case of pre-ignition, will actually cause the electrodes to melt. Overheating and pre-ignition are usually caused by overadvanced timing, detonation from using too-low an octane rating fuel, an excessively lean air/fuel mixture, or problems in the cooling system.

*Damaged spark plugs. Notice the broken electrode on the left plug. The missing part **MUST** be found and removed before returning the engine to service. If overlooked, the part could cause extensive and expensive damage to internal parts.*

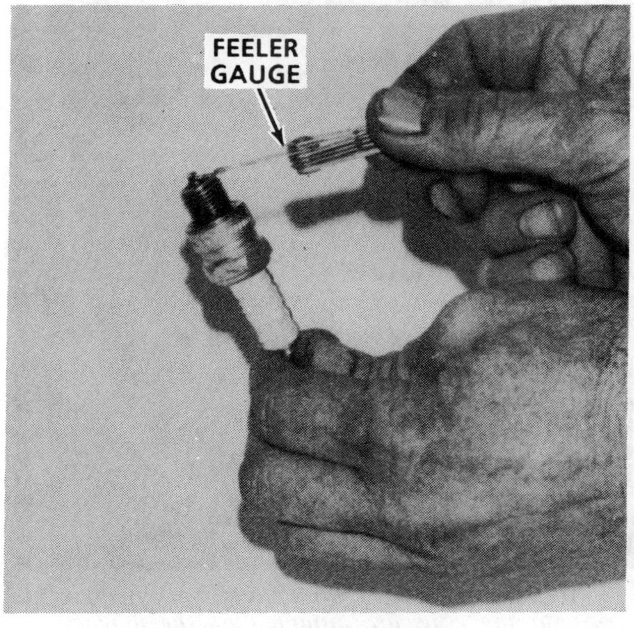

The spark plug gap should always be checked with a wire-type feeler gauge before installing new or used spark plugs.

Electrode Wear

Electrode wear results in a wide gap and if the electrode becomes carbonized it will form a high-resistance path for the spark to jump across. Such a condition will cause the engine to misfire during acceleration. If both plugs are in this condition, it can cause an increase in fuel consumption and very poor performance at high-speed operation. The solution is to replace the spark plugs with a rating in the proper heat range and gapped to specification.

Red rust-colored deposits on the entire firing end of a spark plug can be caused by water in the cylinder combustion chamber. This can be the first evidence of water entering the cylinders through the exhaust manifold because of an accumulation of scale. This condition **MUST** be corrected at the first opportunity. Refer to Chapter 8 -- **ENGINE.**

7-3 CDI (CAPACITOR DISCHARGE IGNITION) AND CHARGING SYSTEM

A Capacitor Discharge Ignition (CDI) system is used on all Bombardier Sea-Doo

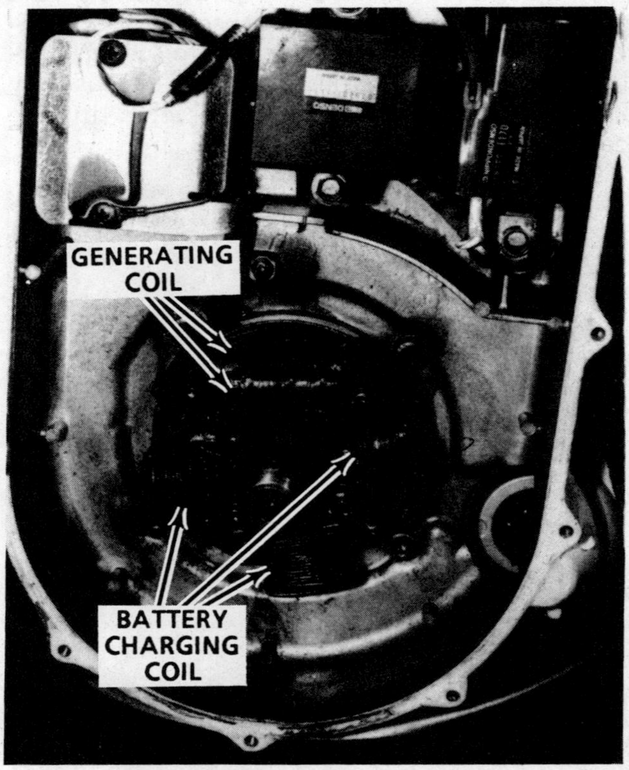

Although both generating coils and all three battery charging coils are mounted on the magneto assembly behind the flywheel, the flywheel does not have to be "pulled" in order to test these coils, as explained later in the text.

Personal Watercraft covered in this manual. This CDI system incorporates an automatic electronic spark advance, which should not require adjustment **UNLESS** the magneto assembly becomes misaligned or an ignition component is replaced.

About half of the ignition circuit is housed and well protected behind the flywheel. The remainder of the system components are mounted inside the battery cover in the engine compartment, or on the stator plate behind the flywheel.

SPECIAL WORDS

Five coils are mounted on the magneto assembly. A group of three coils together are the battery charging coils. The other two coils, stacked one on top the other, are the generating coils. The magneto assembly is mounted on the stator plate behind the flywheel. Therefore, the charging system is presented in this chapter and in this section because once the flywheel is removed the components for both systems are exposed.

DESCRIPTION AND OPERATION IGNITION CIRCUIT

The following paragraphs describe -- in layman's language -- the CDI system and how it functions. Every effort has been made to keep "technical" terms to a minimum.

Three major components are used in the CDI system.

1- Two stacked generating coils, are mounted on the magneto assembly behind the flywheel. These coils are sometimes called trigger coils, depending on the manufacturer of the ignition system. The generating coils provide a source of AC current and also time the electrical pulses.

2- The ignition module, mounted on the stator plate, is used to convert AC to DC current. The module stores the DC current, then releases it and automatically controls the timing.

3- The ignition coil, also mounted on the stator plate next to the ignition module, boosts boots the DC voltage instantly to approximately 20,000 volts to the spark plugs.

Two basic circuits are used with the CDI system -- a charge circuit and an ignition circuit.

Charge Circuit

The charge circuit consists of the flywheel magnet, battery charging coils, a diode, and a rectifier/regulator. As the flywheel magnet passes the charging coils, a voltage is induced in the coils. As the flywheel continues to rotate and the coils are no longer influenced by the magnet, the magnetic field collapses. Therefore, an alternating current is produced at the charging coils. This AC current is changed to DC by diodes inside the rectifier/regulator.

Ignition Generating Circuit

The ignition generating circuit has its own flywheel magnet, a generating coil, a diode, and a thyristor. A thyristor is a solid state electronic switching device which permits voltage to flow only after it is triggered by another voltage source.

At this point in time, when the ignition timing marks align, an alternating current is induced in the generating coil, in the same manner as previously described for the charging coil. This current is then passed to a second diode located in the CDI unit where it becomes DC current and flows on to the thyristor. This voltage triggers the thyristor to permit the voltage stored in the capacitor to be discharged. The capacitor voltage passes through the thyristor and on to the primary windings of the ignition coil.

In this manner, a spark at the plug may be accurately timed by the timing marks on the flywheel relative to the magnents in the flywheel and to provide as many as 100 sparks per second for an engine operating at 6,000 rpm.

Timing Advance

All Bombardier Personal Watercraft covered in this manual are equipped with an automatic advance type CDI system with no moving or sliding parts. Therefore, adjustment is not required. Instead, ignition advance timing is accomplished electronically and signal output increases proportionately to engine speed increase.

The timing is correctly set at the factory. As mentioned earlier, the timing should never require adjustments **UNLESS** the magneto assembly is misaligned or an ignition component has been replaced.

Detailed ignition timing procedure is presented in Section 7-7 of this chapter, beginning on Page 7-19.

Charging System

The charging system consists of the flywheel magnets, three battery charging coils, a rectifier/regulator, and the battery. As the flywheel magnets rotate, voltage is induced in the charging coil mounted on the magneto assembly under the flywheel. The alternating current passes through a series of four diodes inside the rectifier/regulator and is transformed to DC current.

The rectifier/regulator allows a maximum of 14 volts to pass on to the battery. Any excess voltage is passed to a SCR gate and back through the charging coil. In this manner, overcharging the battery is avoided. The life of components in the charging circuit is greatly extended due to the absence of power surges.

The rectifier/regulator is a sealed "black box" type unit. The regulator circuit contains four diodes. If any one of the diodes becomes defective, the complete rectifier/-regulator must be replaced.

TROUBLESHOOTING IGNITION CDI SYSTEM

Always attempt to proceed with the troubleshooting in an orderly manner. The "shot in the dark" approach will only result in wasted time, incorrect diagnosis, replacement of unnecessary parts, and frustration.

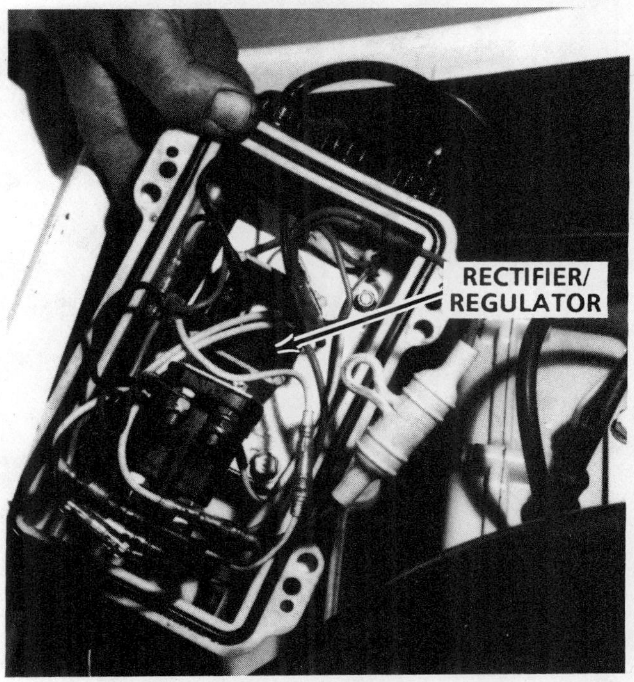

The rectifier/regulator is mounted inside the battery cover where it is well protected from damage and corrosion.

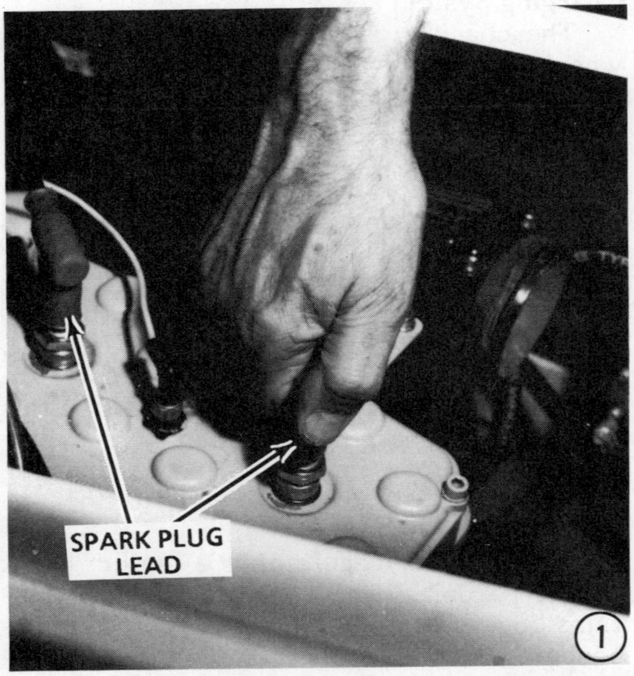

Begin the ignition system troubleshooting with the spark plug and continue through the system until the source of trouble is located.

Spark Plugs

1- Check the plug leads to be sure they are properly connected. Check the entire length of the leads from the plugs to the magneto assembly on the stator plate. If the leads are to be removed from the spark plug, **ALWAYS** use a pulling and twisting motion as a precaution against damaging the connection.

Attempt to remove the spark plug by hand. This is a rough test to determine if the plug is tightened properly. The attempt to loosen the plug by hand should fail. The plug should be tight and require the proper socket size tool. Remove the spark plug and evaluate its condition as described in Section 7-2.

2- Use a spark tester and check for spark. If a spark tester is not available, hold the plug wire about 1/4" (6.4mm) from the engine. Crank the engine through a few revolutions using the cranking motor and check for spark. A strong spark over a wide gap must be observed when testing in this manner, because under compression a strong spark is necessary in order to ignite the air/fuel mixture in the cylinder. This means it is possible to think a strong spark is present, when in reality the spark will be too weak when the plug is installed. If there is no spark, or if the spark is weak, the trouble is most likely behind the flywheel in the magneto assembly.

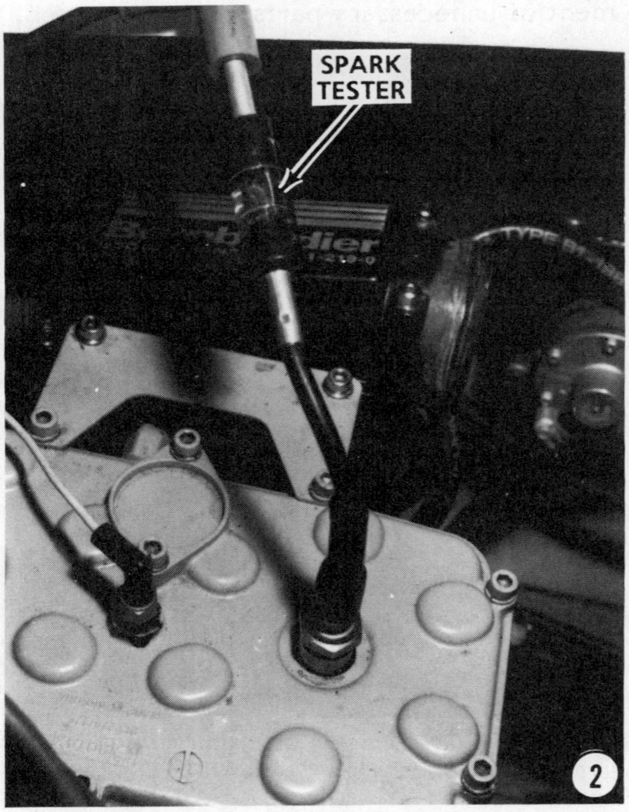

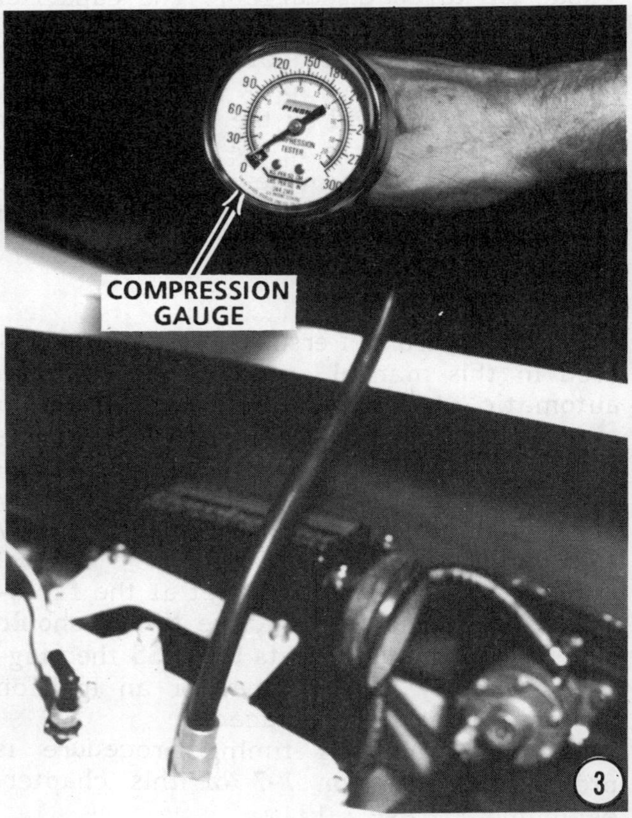

Compression

3- Before spending too much time and money attempting to trace a problem to the ignition system, a compression check of both cylinders should be made. If a cylinder does not have adequate compression, troubleshooting and attempted service of the ignition or fuel system will fail to give the desired results of satisfactory engine performance.

Remove the spark plug lead by pulling and twisting **ONLY** on the molded cap. **NEVER** pull on the lead because the connection inside the cap may be separated or the boot may be damaged. Remove the spark plug. Insert a compression gauge into the cylinder spark plug hole. Ground the spark plug leads to prevent damage to the ignition coil. If a lead is not grounded, the coil will attempt to match the demand created by the spark trying to jump from the electrode to the nearest ground.

Crank a cold engine through several revolutions with the electric cranking motor. Note the compression reading.

An acceptable pressure reading for an engine covered in this manual is 142 psi (981 kPa) or more. The readings between cylinders should not vary by more than 10%.

TROUBLESHOOTING CHARGING SYSTEM

As mentioned earlier, the charging coil is mounted on the magneto assembly located under and protected by the flywheel. The charging coil seldom causes problems in the charging circuit. Most problems can be traced to the rectifier/regulator, the battery, or to the connecting wiring.

The following three conditions will cause damage to the rectifier/regulator.

a- Battery cables connected in reverse.

b- Operating the engine with one or both battery cables disconnected.

c- A loose or broken connection in the charging circuit.

The voltage generated by the charging coil is intended to charge the battery. Anytime the circuit is incomplete, the voltage has actually no where to go. As a result, the voltage will actually reverse itself. Such reversed voltage will damage the internal structure of a diode in the circuit. The diode will then have a scorched or burned appearance.

Perform a visual inspection of the following areas before making any resistance tests on charging circuit components.

a- Check to be sure the battery cables are connected to the correct terminals.

b- Check condition of the battery and recharge, if necessary.

c- Inspect the wiring and the connections between the charging coil, the rectifier/regulator, and the battery. Replace any damaged or deteriorated cable insulation. Pay particular attention for loose, corroded, or disconnected terminals or connections.

7-4 TESTING IGNITION COMPONENTS

General Information

Due to the complex nature of the circuitry and high energy output pulses of microsecond duration, conventional testing devices such as a Volt/Ohm/Ammeter will not measure electrical output with the degree of accuracy required. A Bombardier Ignition Tester, as used by the local Bombardier dealer is an electrical device capable of measuring the peak energy output of capaci-

NEVER attempt to verify the charging circuit by operating the engine with the battery disconnected. Such action will damage the rectifier/regulator.

tor discharge ignition units, as well as the charging and generating coils. This instrument was designed to troubleshoot the CDI ignition system. Unfortunately this tester may not be easily accessible. Therefore, this chapter includes only those tests which may be performed with instruments normally and easily available.

WORDS ON TESTERS

Resistance tests performed with a Volt/-Ohmmeter are not as accurate or as conclusive as tests performed using a Bombardier tester.

To replace a component, the flywheel must be "pulled" -- see Section 7-6, this chapter.

WORDS FROM EXPERIENCE

During the tests, if a reading is slightly different from the specifications, but the engine still operates, then there is no real need to replace the affected component, until it actually fails. Bear in mind, in **MOST** cases electrical components are not returnable, once the item leaves the store.

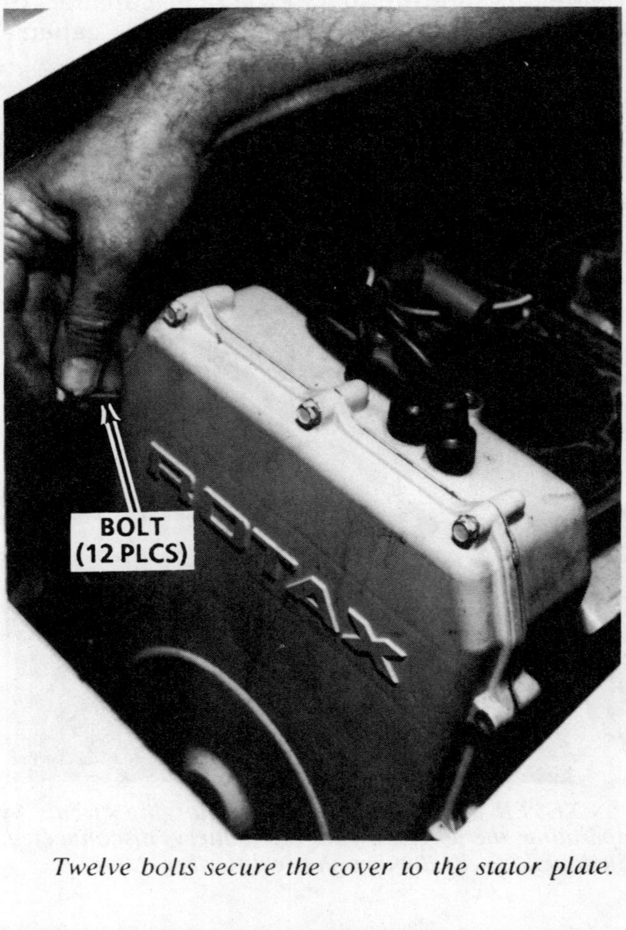

Twelve bolts secure the cover to the stator plate.

Therefore, make every attempt to avoid the **BUY** and **TRY** method of troubleshooting. Such a practice will lead only to wasted time, unneeded cash outlay, and frustration.

Intermittent or Multiple Problems

Many ignition problems occur only during engine operation -- when the component is subject to vibration, the engine is operating under a load, and/or there is a marked increase in temperature -- from the engine operating.

SPECIAL WORDS ON IGNITION AND CHARGING CIRCUIT TESTING

Most of the components in the ignition and charging circuits have leads with quick-disconnect fittings or harness plugs. Some components in these circuits are housed inside the battery cover -- others are housed on the magneto assembly behind the flywheel, or on the stator plate above the flywheel. If a component inside the battery cover or on the stator plate, is found to be defective, it may be replaced immediately -

*After the stator plate cover has been removed, work the sealing ring out of the groove, and then discard the ring. This type sealing ring will **NEVER** seal properly after it has been used once.*

- usually through attaching hardware. If a component behind the flywheel -- battery charging or generating coil -- is found to be defective, the flywheel must be "pulled" in order to remove and replace the item. See Section 7-6, beginning on Page 7-12 for instuctions to "pull" the flywheel.

CRITICAL WORDS

Never attempt to verify the charging circuit by operating the engine with the battery disconnected. Such action would force current -- normally directed to charge the battery -- back through the rectifier and damage the diodes in the rectifier.

Magneto Assembly Resistance Tests

Remove the 12 bolts and washers securing the stator plate cover to the plate. **TAKE CARE** not to damage the sealing ring.

Generating Coil Test

Two generating coils -- one a high-speed coil, the other a low-speed coil -- are stacked on top of each other and secured to the magneto plate behind the flywheel. It is not

Once the stator plate cover is removed, the leads from all the ignition and charging components are easily accessible.

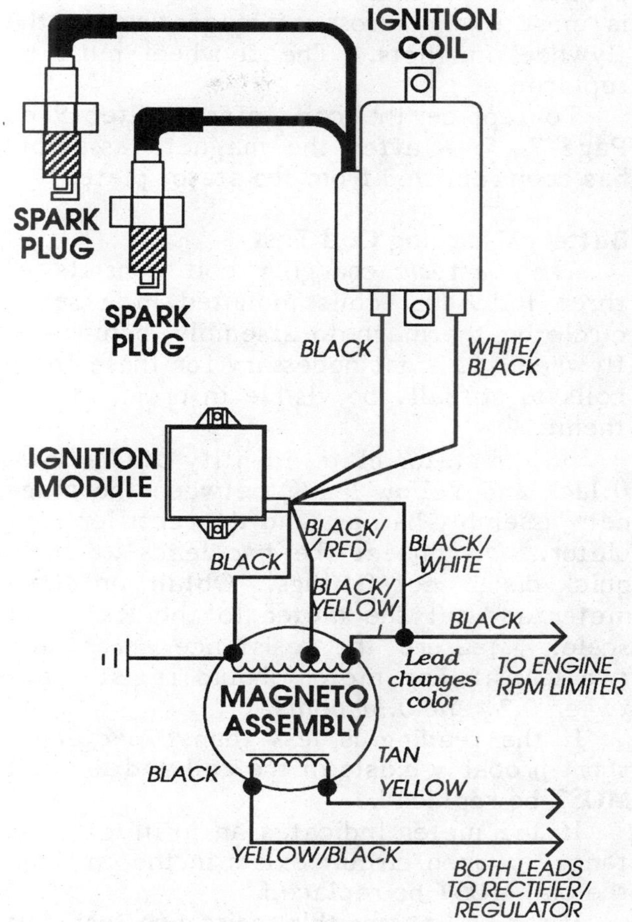

Wire identification and routing for the stator plate and the magneto assembly circuits.

necessary for these two coils to actually be visible in order to test them.

Inside the stator plate: identify the Black/White, Black/Red, and Black leads between the magneto assembly harness --coming out from behind the flywheel -- and the ignition module. Disconnect the four leads at their quick disconnect fittings. Obtain an ohmmeter and set the meter to the Rx1 ohm scale. Measure the resistance across the Black/White and Black/Red lead. If the meter registers between 2.8 and 4.2 ohms, the high-speed generating coil is satisfactory. Next, set the meter to the Rx100 ohm scale. Measure the resistance between the Black/Red and Black leads. If the meter registers between 1.2 and 1.8 ohms (120-180), the low-speed generating coil is satisfactory. If either coil fails the resistance test, both coils **MUST** be replaced, because they share a common lead. If the coil passes the resistance test, but the problem is still determined to be in the ignition portion of the magneto assembly, the cause is most likely a loss of magnetism in the flywheel magnets. The flywheel must be replaced.

To replace the coil refer to Step 9 on Page 7-15 -- after the magneto assembly has been removed from the stator plate.

Battery Charging Coil Test

The battery charging coil consists of three individual coils mounted in a semi-circle on the magneto assembly, behind the flywheel. It is not necessary for these three coils to actually be visible in order to test them.

On the stator plate, identify the Yellow-/Black and Yellow leads, between the magneto assembly harness and the rectifier/regulator. Disconnect the two leads at their quick disconnect fittings. Obtain an ohmmeter and set the meter to the Rx1 ohm scale. Measure the resistance across the two leads. The meter should register between 0.21 and 0.31 ohms.

If the reading is less than expected, a short probably exists in the coil and the coil **MUST** be replaced.

If the meter indicates an infinite resistance, an open circuit exists in the coil and the coil **MUST** be replaced.

If the coil passes this resistance test, but the problem is still determined to be in the charging portion of the magneto assembly, the cause is most likely a loss of magnetism in the flywheel magnets. If such is the case, the flywheel **MUST** be replaced.

To replace the battery charging coil, see Step 9 on Page 16 -- after the magneto assembly has been removed from the stator plate.

Ignition Coil Test

The ignition coil is mounted on the stator plate. To check **PRIMARY** resistance: first, identify the White/Blue and the Black leads between the ignition coil and the ignition module. Disconnect the leads at their quick disconnect fitting. Obtain an ohmmeter and set the meter to the Rx1 ohm scale. Measure the resistance between the two leads. The meter should register between 0.23 and 0.43 ohms.

To check the **SECONDARY** resistance: set the meter scale to the Rx1000 ohm scale. Insert the meter probes into the high tension lead boots and make contact with the spark plug caps. The meter should

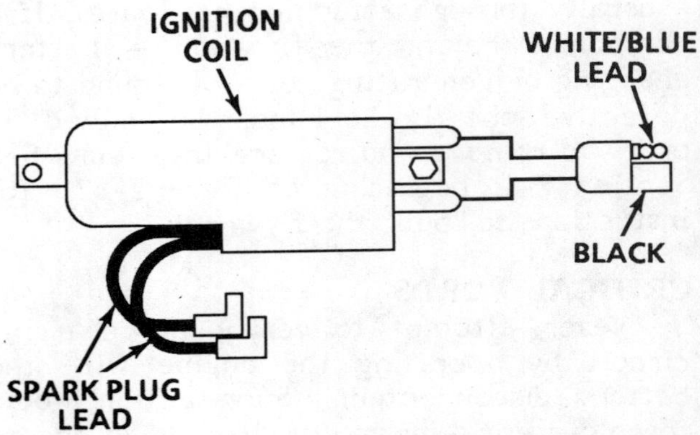

Detailed instructions are presented in the text for testing the primary and secondary resistance of the ignition coil.

register between 5.85 and 6.50 ohms (5,850 and 6,500 ohms).

If the ignition coil fails either test, it is defective and **MUST** be replaced.

If the reading is questionable, but the engine operates, install a spark plug tester and check the strength of the spark.

If the spark is acceptable, it is not necessary to replace the ignition coil. However, keep this area in mind for possible future trouble.

If the spark is weak, it will get weaker. Therefore, replace the ignition coil.

7-5 CHARGING CIRCUIT

DESCRIPTION AND OPERATION

The charging circuit consists of the following items:

- Permanent magnets -- attached to inside perimeter of the flywheel.
- Charging coil -- installed on the magneto assembly.
- A rectifier/regulator -- located inside the battery case cover.
- An external battery.
- Necessary wiring -- to connect the components together.

The **NEGATIVE** side of the rectifier/regulator is always grounded. The **POSITIVE** side is always connected to the battery. The negative terminal on the battery must be connected to a good ground on the engine.

The alternating current generated in the charging coil windings passes to the rectifier/regulator. The rectifier portion changes the alternating current (AC) to direct current (DC) to charge the 12-volt battery.

A second function of the rectifier is to smooth the pulses of current to form a steady DC current flow.

The regulator portion of the rectifier/-regulator limits the current produced and sends any excess current back through the charging coil windings.

Electronic components in the charging circuit, including the battery, will last much longer if subjected to a low steady voltage. A sudden voltage surge could overheat and damage charging circuit components.

The magneto assembly is located behind and protected by the flywheel. Therefore, the magneto assembly, including the charging coil, seldom causes problems in the charging circuit. When problems do occur, they can usually be traced to the rectifier/-regulator or to the battery.

If either the charging coil or the rectifier/regulator fails troubleshooting tests, the defective unit cannot be repaired, it **MUST** be replaced.

Testing Charging Circuit Output

Before testing the charging circuit, verify the condition of the battery -- see Chapter 9 -- **ELECTRICAL.**

Move the craft to a test tank or secure it in a body of water next to a float or in a "slip". Secure the craft fore and aft to prevent movement during the following tests.

CAUTION

Water must circulate through the jet pump -- to and from the engine, anytime the engine is operating. Circulating water will prevent overheating -- which could cause damage to moving engine parts and possible engine seizure.

NEVER allow the engine to operate without cooling water for more than 15 seconds.

Amperage Output Test

Obtain an ammeter and set it to the highest DC amp scale. The current flow is much higher when the engine is first started and could very well damage the meter if it were set directly to the operating range scale.

Disconnect the positive battery cable and connect the meter between the disconnected lead and the battery --positive meter terminal to the positive battery terminal and the negative meter terminal to the disconnected battery lead. Connect a tachometer to the engine.

CAUTION

The following tests **CANNOT** be made with a flush attachment connected to the unit. The craft **MUST** be moved to a body of water or to a test tank.

Start the engine and allow it to warm to operating temperature, with the rpm rate stabilized. Now, select the Ax1 amp scale on the meter. Advance the throttle to 5,500 rpm. The meter should register approximately 5 amps, if the rectifier/regulator is functioning correctly. If the rectifier/regulator is faulty, the engine may not be able to reach the required rpm and the ammeter will fail to indicate 5 amps. If such is the case, replace the rectifier/regulator.

Shut the engine down and disconnect the ammeter.

Voltage Output Test

With the craft in a test tank or body of water, obtain a voltmeter capable of indicating up to 25 DC volts. Connect the voltmeter across the battery terminals, negative-to-negative -- positive-to-positive. Connect a tachometer to the engine.

Start the engine and allow it to warm to operating temperature with the rpm rate

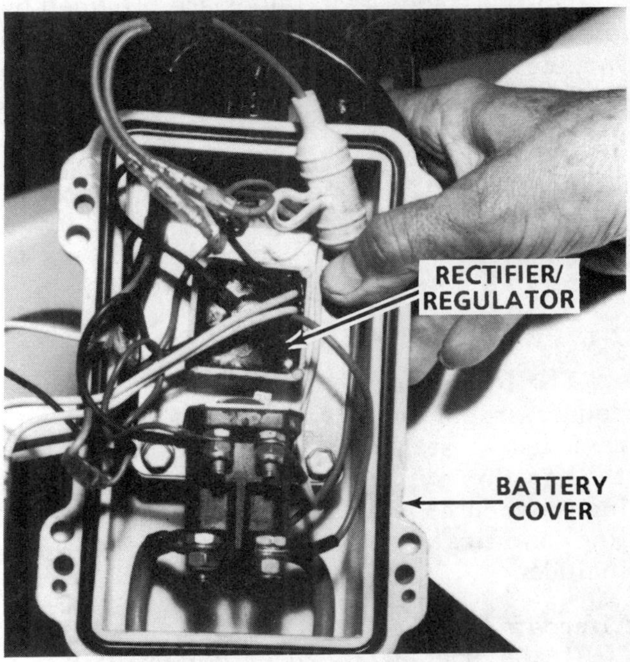

If the rectifier/regulator is identified as a problem area during troubleshooting, the text explains how the unit may be tested for amperage and voltage output.

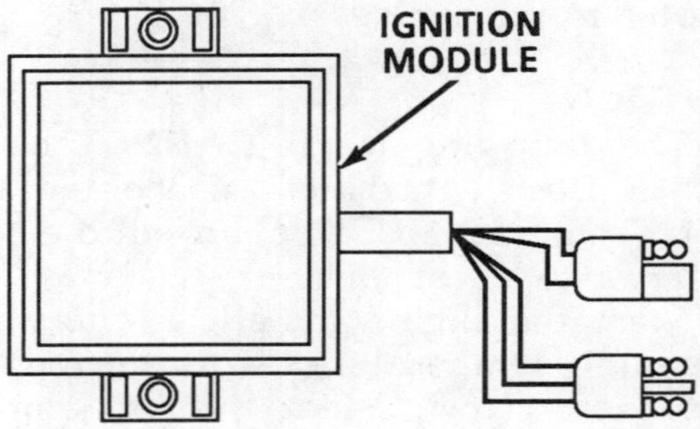

Unfortunately, the manufacturer has failed to provide resistance values as guides for testing the ignition module. However, if troubleshooting leads to this unit as a possible problem area, replacing the module with a known good one will eliminate suspicion for the present and immediate future.

stabilized. Now, advance the throttle to 5,500 rpm. If the meter registers over 15 VDC, the rectifier/regulator is defective and **MUST** be replaced. If the meter registers less than 12 VDC, inadequate voltage is being produced to charge the battery.

Check the battery charging coil behind the flywheel -- see Page 7-10 for resistance tests. If the coil is satisfactory, replace the rectifier/regulator.

Ignition Module Test

No resistance test values are provided by the manufacturer for testing the ignition module. If all ignition components are found good, and an ignition malfunction is still suspected, then the module should be replaced.

Replacing a questionable component with a known good one will eliminate a possible problem area, which is difficult to trace, in a defective ignition system.

7-6 FLYWHEEL AND MAGNETO

The following short section lists the procedures required to "pull" the flywheel and remove the stator plate in order to service the ignition system. Cleaning and Inspecting procedures in addition to proper assembling and installation instructions are also included.

Alternate Flywheel Holder

If the special flywheel holder listed is not available, a tool may be fabricated from a metal bar and two long bolts. The longer the bar, the more leverage gained, but still

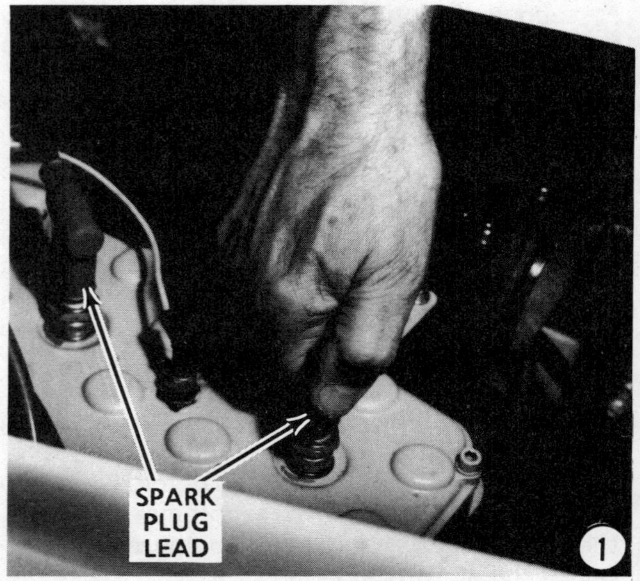

limited by the confines of the engine compartment. The bolts should be of sufficient length to allow a socket and a torque wrench to be placed over the flywheel nut and still clear the bar. Drill two holes, large enough for the bolts to pass through, at the end of the bar, with the same distance between as the two holes in the flywheel. Secure the bolts to the bar.

Alternate Flywheel Puller

If the special flywheel puller listed is not available, a similar puller, shown in the accompanying illustration, may be used **PROVIDED** it will pull from the bolt holes in

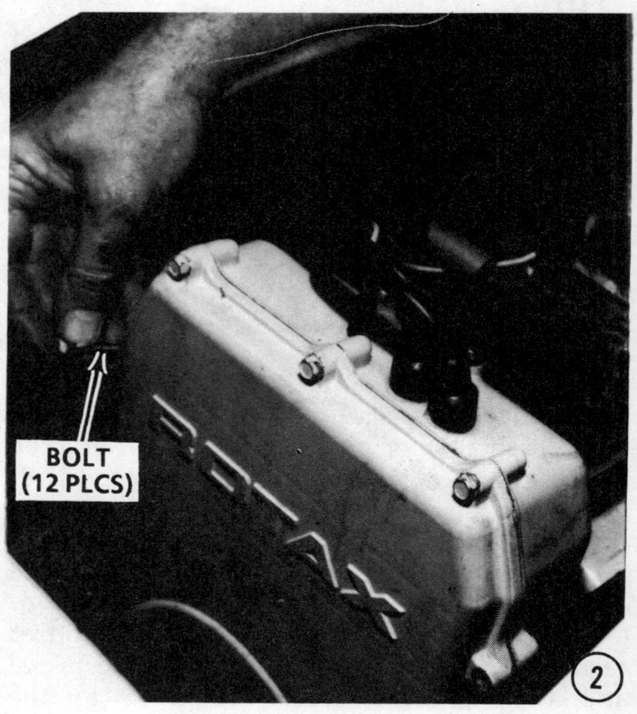

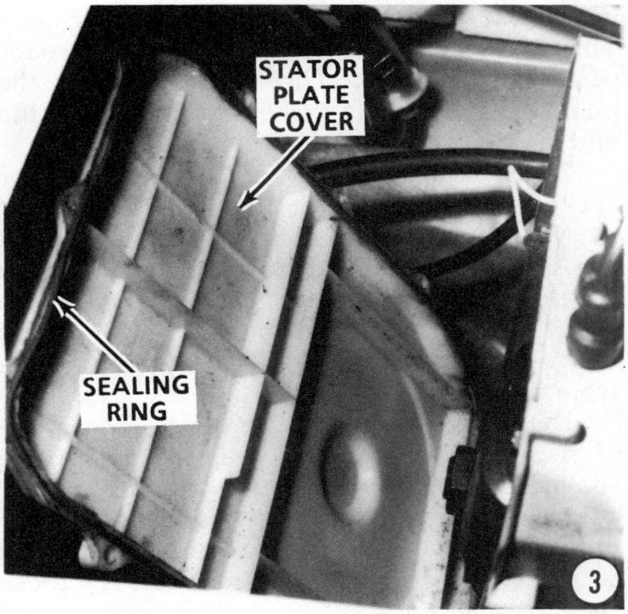

the flywheel and **NOT** from around the perimeter of the flywheel. **NEVER** attempt to use a puller which would put a strain on the outside edge of the flywheel.

"PULLING" THE FLYWHEEL

1- Remove the spark plugs.

AUTHORS' WORDS

For photographic clarity, the illustrations in this section were made while the work was being performed on a work bench, The confined area of the engine compartment would not allow the necessary details to be shown. The steps are valid and may be performed with the engine in place in the craft.

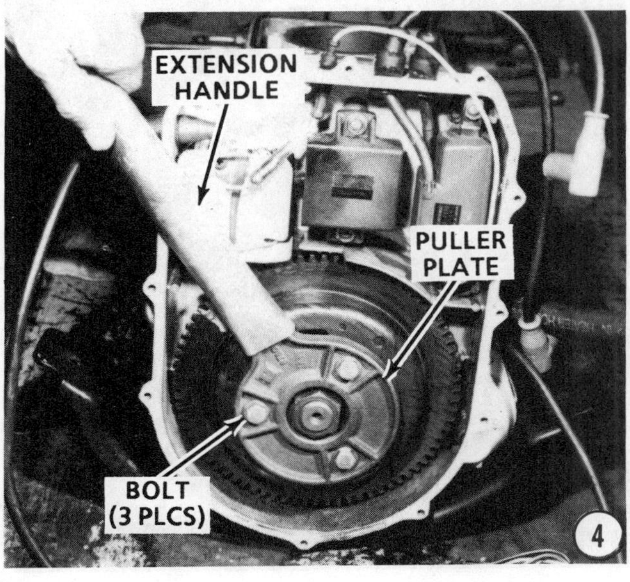

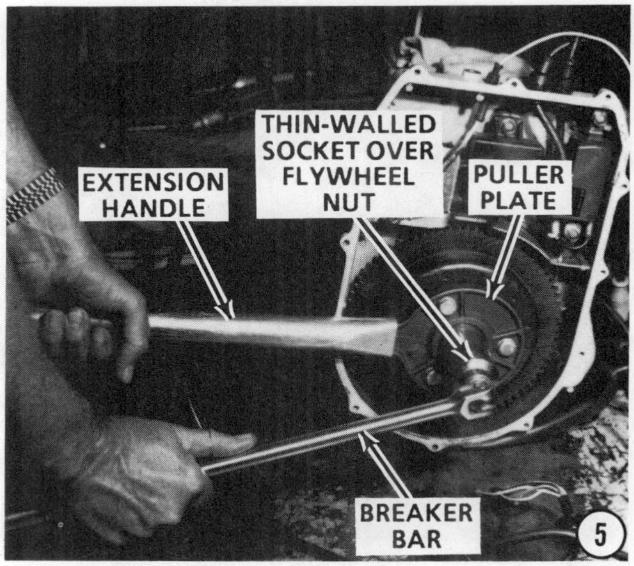

Stator Plate Cover Removal

2- Remove the 12 bolts securing the cover to the stator plate.

3- Lift off the cover. Remove and discard the sealing ring. This type ring cannot be used a second time -- once used, it will never seal properly again.

4- Obtain Puller Plate P/N 290-876-080, Extension Handle P/N 290-000-003, and three M8x25mm bolts. Install the plate onto the flywheel using the three bolts. Tighten the bolts securely.

Install the extension handle onto the puller plate arm.

5- Install a 22mm thin-walled socket over the flywheel nut. The socket **MUST** fit inside the puller plate hole.

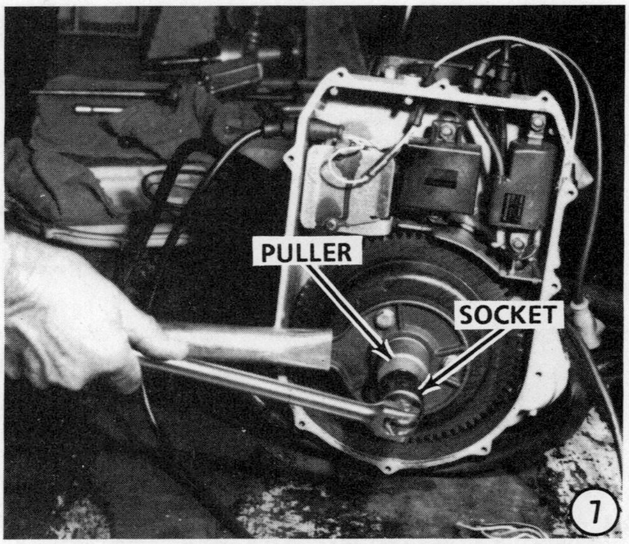

7

If the only available socket will not fit into the puller plate hole, grind down the outside diameter of the socket until it will fit into the hole. **DO NOT** grind on the internal threads of the puller plate.

Place a breaker bar into the socket and rotate it **COUNTERCLOCKWISE** to "break" the flywheel nut free. Remove the nut and washer.

6- Obtain Puller and Puller Bolt P/N 290-876-065. Thread the bolt into the puller

8

until just a couple threads are engaged. Thread the puller into the internal threads of the puller plate. Thread the bolt into the puller until the bolt makes contact with the end of the crankshaft.

7- Obtain the correct size wrench or socket and place it over the puller bolt.

Now, continue to tighten the bolt and at the same time, shock the crankshaft with a gentle to moderate tap on the end of the bolt, with a soft head mallet. This shock will assist in "breaking" the flywheel loose from the crankshaft.

DO NOT "pound" on the bolt. The force of such action would be transmitted to the seals and bearings along the crankshaft and cause damage. **NEVER** strike the flywheel directly, because this type of shock would surely cause the flywheel magnets to loose their magnetism.

8- Pull the flywheel free of the crankshaft. Remove and **SAVE** the Woodruff key from the crankshaft keyway.

SPECIAL WORDS ON WOODRUFF KEYS

Inspect the condition of the Woodruff key and its keyway in the crankshaft. If the key shows any sign of shearing, it is probably caused by jumping waves with the watercraft and the effect will be a loss of power.

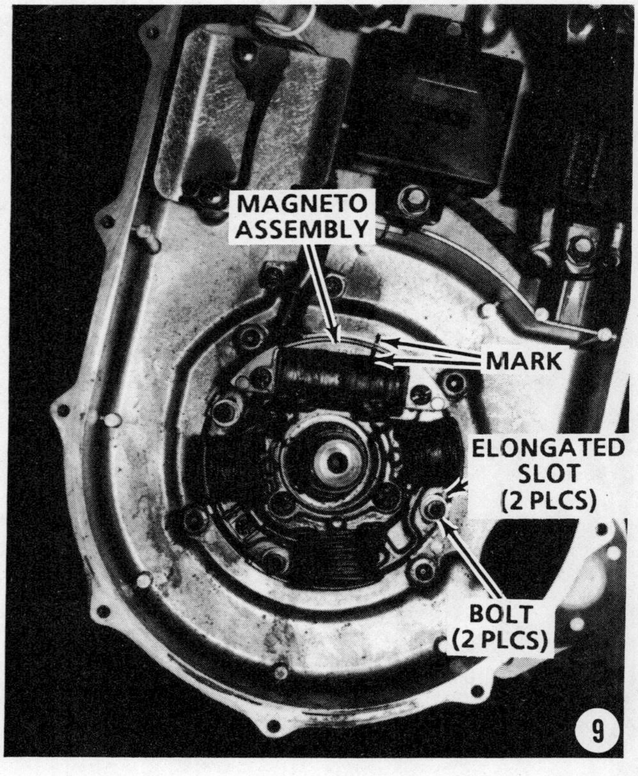

9

When the craft is clear of the water, the engine rpm approaches the limit set by the rpm limiter in an unloaded condition.

When the craft returns to the water surface, the sudden load on the impeller is transferred along the driveline. Speed of the rotating driveshaft and the crankshaft is suddenly reduced while the flywheel continues to spin rapidly, due to its mass. This condition, especially when repeated over a period of time, often shears the Woodruff key and the flywheel becomes repositioned on the crankshaft.

A very small misalignment will result in a change in ignition timing. The craft operator will notice a drastic loss of power. A worse condition would be the flywheel shifting to the degree engine start would not be possible.

Stator Plate and Magneto Assembly Removal

9- The securing screws pass through elongated slots in the magneto assembly. The assembly could be installed as much as 15° off its original position -- drastically affecting the timing characteristics of the engine. Therefore, at the elongated slot in the magneto assembly, make a set of aligned punch marks -- one on the magneto assembly -- the other on the stator plate. These marks will serve as an aid during installation to **ENSURE** correct positioning of the magneto assembly on the stator plate.

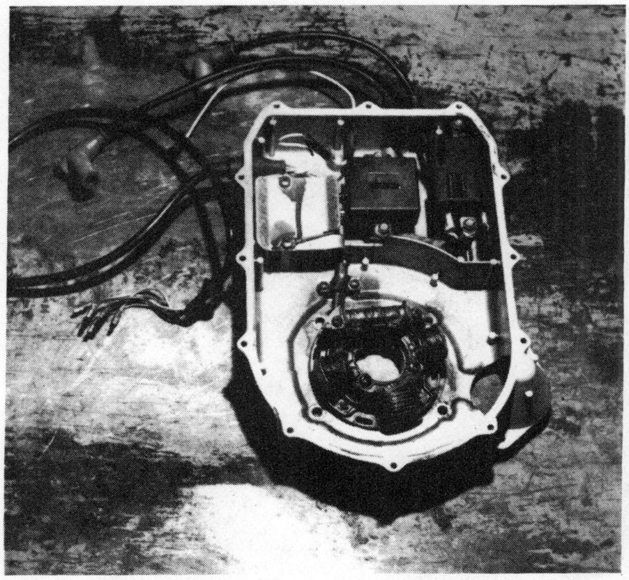

After the stator plate is removed from the engine, all internal and attached components are easily removed and installed.

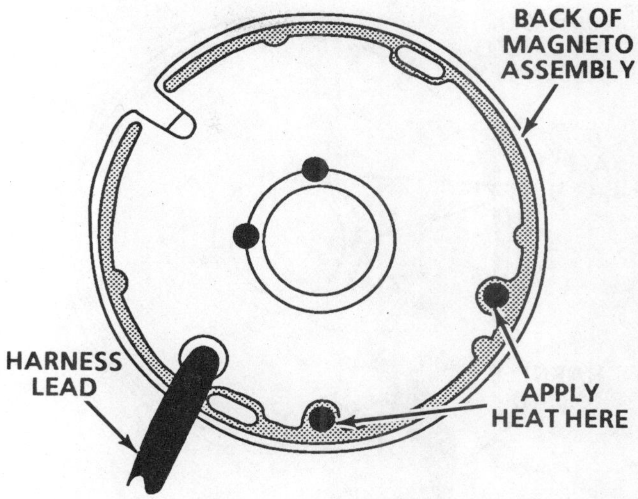

Two holes in the magneto assembly must be heated in order to "break" the Loctite bond and enable removal of a generating coil. This simple line drawing depicts the back side of the magneto assembly with the two holes to be heated identified. Text explains all.

Remove the two screws securing the magneto assembly to the stator plate. These screws must be removed because they are threaded into the crankcase.

Generating Coil Removal and Installation

Lift the magneto assembly off the crankshaft and turn it over. Obtain a blow torch and heat the two screw holes, shown in the accompanying illustration, to break the Loctite bond. **TAKE CARE** not to melt the protective sleeve on the magneto assembly harness lead.

Remove the two securing screws. Snip the Black/White, Black/Red, and Black leads close to the coil. Remove the generating coil.

Remove the shipping nuts from the new generating coil and install the coil over the plate. Apply Loctite 242 onto the threads of the two securing screws. Install and tighten the screws very **SECURELY.** Pass the three leads through the opening in the plate.

Disconnect the connector to the ignition module, but **DO NOT** pull the old leads out

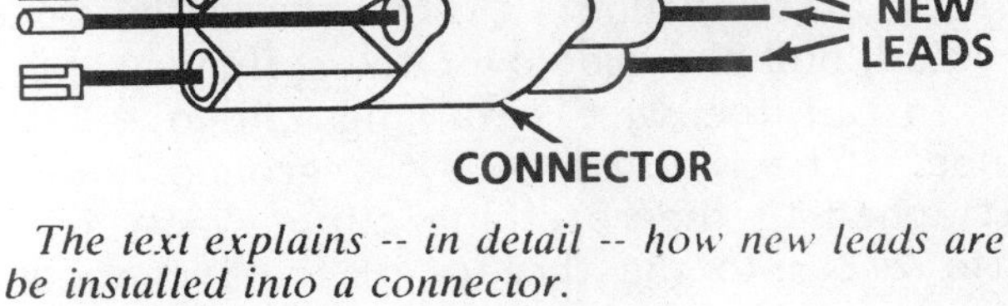

The text explains -- in detail -- how new leads are to be installed into a connector.

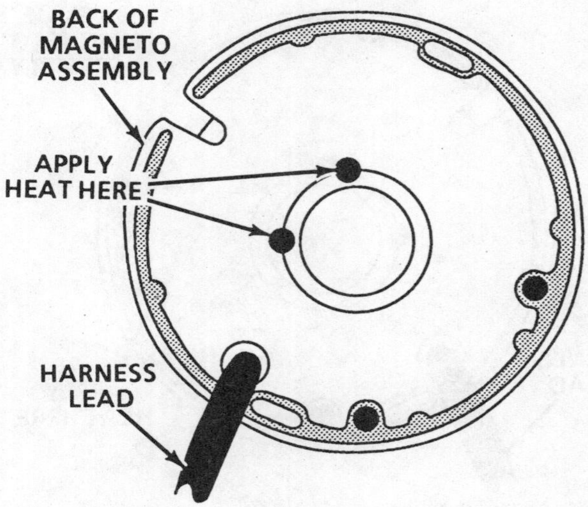

Simple line drawing to depict the back side of the magneto assembly. The two holes to be heated in order to "break" the Loctite bond and remove a battery charging coil, are identified.

of the connector half. Tape the new leads -- matching color-for-color to the old leads, and then pull them through the connector half. Release the new leads from the old leads and connect the new leads matching color-for-color.

Battery Charging Coil Removal and Installation

Use a blow torch and heat the two screw holes, shown in the accompanying illustration to break the Loctite bond. **TAKE CARE** not to melt the protective sleeve on the magneto assembly harness lead. Remove the two securing screws.

Remove the wire retaining clip from the plastic tube. Pull back the White and Black plastic tubes from the spliced connections. Unsolder the connections and remove the three coils.

To install the new coils, beginning by simply placing the three coils over the magneto assembly plate, with the bolt holes aligned. Next, solder the Yellow wire from the new coil to the connector with the White plastic tube. After the soldering is complete and has cooled, pull the tube over the splice. Solder the Yellow/Black wire from the new coil to the connector with the Black plastic tube. After the solder has cooled, pull the tube over the splice.

Install the wire retaining clip over the tube. Check to be sure the ground lead is attached to the retaining clip. Apply Loctite 242 onto the threads of the two securing screws. Install and tighten the screws very **SECURELY.**

CLEANING AND INSPECTING

Inspect the flywheel for cracks or other damage, especially around the inside of the center hub. A damaged flywheel can be **EXTREMELY** dangerous. The forces exerted on the flywheel while it is spinning at high rpm are sufficient to shatter an unbalanced damaged flywheel. Inspect the magnets to see if they are held tightly around the outer circumference of the flywheel rim. Check the surface of the magnets for fractures and grooves.

Grooves around the inner perimeter would indicate a component mounted on the stator plate is not aligned properly or the component is loose. Fractured or damaged magnets cannot be replaced. Both the magnets and flywheel share the same part number, therefore attempting to remove the magnets would be a pointless task. If magnets or the flywheel is defective, the entire flywheel must be replaced.

Check to be sure small metal particles have not become attached to the magnets. Verify each magnet has good magnetism by using a screwdriver or other suitable tool.

Thoroughly clean the inside taper of the flywheel and the taper on the crankshaft to prevent the flywheel from "walking" on the crankshaft during operation.

Test the magneto assembly to verify it is not loose. Attempt to lift each component

*Check the flywheel **CAREFULLY**, to be sure metal particles have not become stuck to the magnets.*

on the plate. There should be little or no evidence of movement.

SPECIAL WORDS ON FLYWHEEL MAGNETS

The outer ends of any magnet are called poles. One end is the north pole and the other end is the south pole. The magnetic field surrounding a magnet is concentrated around these two poles.

Some flywheel magnets are fairly long and curved around the outer perimeter of the underside of the flywheel. Others are short and are mounted around the center hub, depending on the location of the coils mounted on the stator. Magnets are usually installed in pairs with the north pole of one adjacent to the south pole of its neighbor and so on. In this manner continuous magnetic field surrounds the inside of the flywheel.

If a flywheel is accidently dropped, not only could the teeth be damaged, but the impact will weaken the magnetic strength of all the magnets housed in the flywheel.

If one or more of the magnets should break or fracture, two new magnetic poles will be created. A long magnet with two poles will become two short magnets with four poles. The new poles will possess only half the magnetic strength of an original pole. The overall magnetic field will be altered. The new field of the shorter magnets will not extend to cover the area of the flywheel.

Serious consequences apply to a CDI type ignition system in the event of a flywheel magnet fracture. Charging and generating coils are energized by the concentrated magnetic field at the magnet poles. If new poles are suddenly created, the ignition module will receive conflicting signals from the magnets and may even attempt to fire the cylinder twice in one revolution.

All these reasons require the flywheel to be handled with **CARE.**

If there is evidence of oil on the magneto assembly, the oil seal/s at the forward main bearing inside the crankcase may have failed. Refer to Chapter 8, **ENGINE.**

ASSEMBLING AND INSTALLATION

Magneto Assembly

1- Coat the threads of the two attaching Allen head bolts with Loctite 242. Install the magneto assembly onto the stator plate. Align the marks made during disassembling. Install the bolts, washers, and lockwashers. Tighten the bolts alternately and evenly to a torque value of 16 ft lb (22Nm). Check to

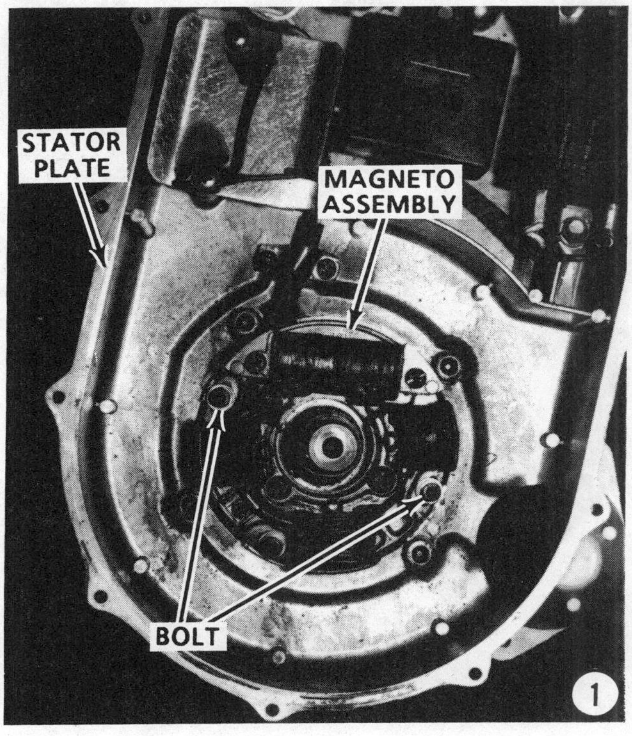

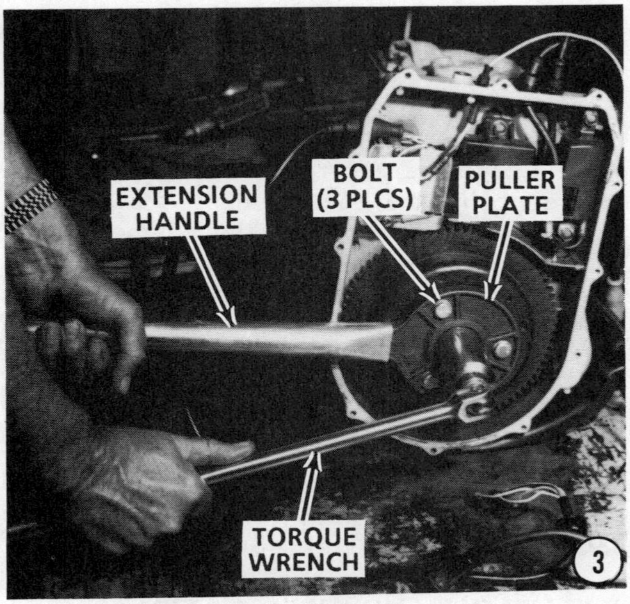

be sure the two marks remained aligned while the bolts were tightened.

Flywheel Installation

2- Place a dab of thick lubricant onto the curved surface of the Woodruff key to hold it in place while the flywheel is being installed. Press the Woodruff key into place in the crankshaft recess. Wipe away any excess lubricant to prevent the flywheel from "walking" during engine operation. Check the flywheel magnets to ensure they are free of any metal particles. Double check the taper in the flywheel hub and the taper on the crankshaft to verify they are clean and free of any oil.

Now, slide the flywheel onto the crankshaft, with the Woodruff key in place. Rotate the flywheel **COUNTERCLOCKWISE** to be sure it does not contact any part of the magneto assembly or wiring on the stator plate.

3- Attach Puller Plate P/N 290-876-065 to the flywheel with the three M8x25mm bolts. Install the extension handle onto the puller arm. Apply Loctite 242 onto the crankshaft taper and threads. Install the flywheel nut and washer. Tighten the nut to a torque value of 77 ft lb (105mm).

4- Apply Loctite 242 to the threads of the twelve stator plate cover securing bolts. Install a new sealing ring into the groove of the stator plate cover.

5- Install the cover and secure it in place with the attaching bolts. Tighten the bolts to a torque value of 35 in lb (4Nm).

6- Attach the high tension leads to the spark plugs.

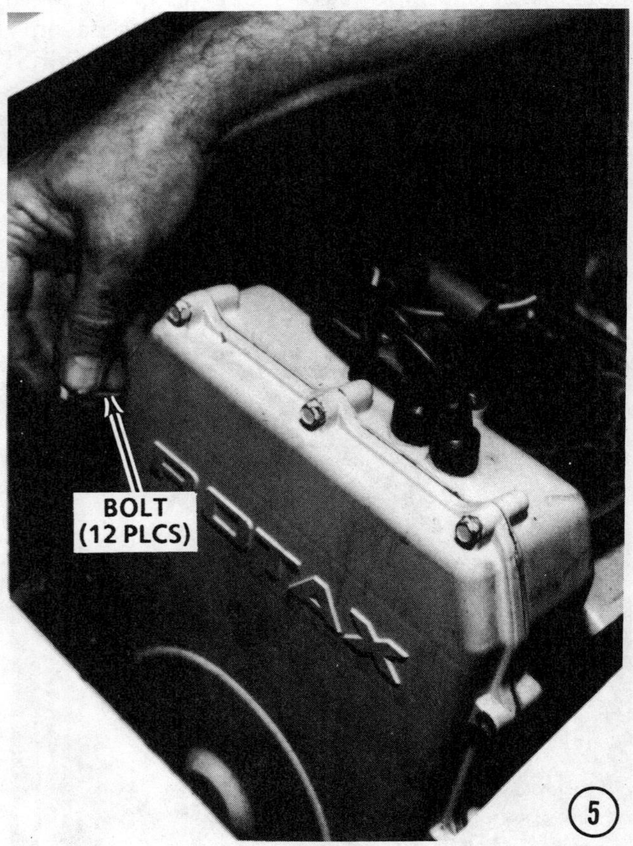

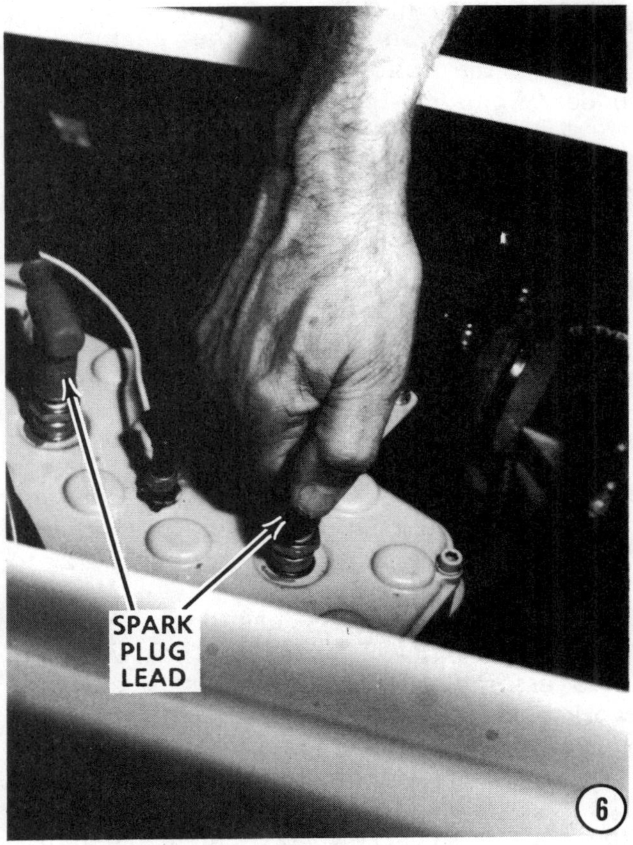

7-7 IGNITION TIMING ADJUSTMENTS

A Nippondenso Electronic Ignition System is installed on all Bombardier watercraft covered in this manual. This system has no moving or sliding parts. Therefore, adjustment is not required. Ignition advance timing is accomplished electronically. Signal output increases proportionately to engine speed. The timing is set at the factory and should never require adjustment **UNLESS** the magneto assembly becomes misaligned or an ignition component has been replaced.

Therefore, the purpose of this section is to give instructions on how to verify the flywheel timing mark position and to give procedures to correctly time a new flywheel which has been purchased and does not have a timing mark.

A new flywheel will always be indexed with the crankshaft, because it is installed over the Woodruff key. **HOWEVER**, the timing mark on the magneto assembly may not match with the characteristics of the new flywheel.

Craft Location for Timing Adjustments

The watercraft **MUST** be moved to a test tank or to a slip in a body of water where it can be restrained from moving with adequate mooring lines.

Determining TDC (Top Dead Center)

Remove the forward spark plug. Install a dial indicator into the spark plug opening and determine TDC of the piston.

If a dial indicator is not available, use a pencil, straw, or equivalent, resting on top of the piston dome. Determine the highest point to which the inserted object rises when the crankshaft is **SLOWLY** rotated.

Timing Marks

Locate the timing marks. One is a line embossed onto the flywheel. The other mark is a notch cast into the stator plate flywheel guard -- just below the ignition module. If necessary, apply "whiteout" or other white substance to the marks to make them more visible. At TDC, the marks will **NOT** align. This is a normal condition -- no problem.

With the piston at TDC -- as explained in the previous paragraphs -- zero the dial indicator gauge. Rotate the flywheel **COUNTERCLOCKWISE** about 1/4 revolution, then rotate the flywheel **CLOCKWISE** until the reading on the dial indicator is 0.086" (2.18mm), BTDC. Now, the mark on the flywheel should align with the mark on the stator plate flywheel guard.

If the marks align -- proceed to the paragraph under the heading "Adjusting Ignition Timing".

If the marks do not align scribe a new mark on the flywheel which does align with the timing mark on the guard. Obliterate the old flywheel mark, leaving only the new mark. Do not rely on the engine having correct timing just because the marks align. The timing **MUST** be verified dynamically as the final check for mark alignment. With the engine in a static condition -- not operating -- shut down -- the marks may be aligned, but the magneto assembly may have changed position on the stator plate. Such a change makes the mark on the flywheel guard meaningless.

The timing mark on the flywheel guard and the timing mark on the flywheel will be compared during the following timing procedure. **THEREFORE**, make any new marks as close together as practical -- extend the embossed mark, if necessary.

Timing Procedure

Connect a timing light to the battery and hook the pickup around the forward spark plug high tension lead. Check to be sure the direction of the embossed arrow on the timing light pickup is **TOWARDS** the spark plug.

Connect a tachometer to the engine.

SPECIAL WORDS ON TACHOMETERS

The manufacturer recommends the use of a Bombardier Electronic Tachometer. Instructions for connecting the tachometer leads are provided with the meter.

However, on a two-stroke engine with CDI type ignition, the tachometer leads are usually connected across the battery charging coil leads -- Yellow and Yellow/Black.

In a test tank or with the craft in a body of water and restrained with adequate mooring lines to prevent the craft from moving, start the engine and allow it to warm to operating temperature, and the rpm rate to stabilize.

CAUTION

Water must circulate through the jet pump -- to and from the engine, anytime the engine is operating. Circulating water will prevent overheating -- which could cause damage to moving engine parts and possible engine seizure.

NEVER, again **NEVER**, operate the engine at high speed with a flush device attached. An engine operating at high speed with such a device attached, would **RUNAWAY** from lack of a load on the impeller shaft, causing extensive damage.

DYNAMIC TIMING CHECK

After the engine has reached normal operating temperature -- with the timing light and the tachometer connected --advance the throttle lever until the engine has reached 6000 rpm. Aim the timing light at the two timing marks. If the mark on the flywheel aligns with the stationary mark, the timing is set correctly. **HOWEVER**, if the mark on the flywheel is not aligned with the stationary mark -- a timing adjustment is required. Shut down the engine and perform the following procedures outlined under "Adjusting Ignition Timing".

Adjusting Ignition Timing

Ignition timing adjustment is performed by altering the position of the magneto assembly in relation to the stator plate. The magneto assembly is secured to the stator plate with two screws passing through elongated slots. The elongated slots permit the magneto assembly to be rotated within a small limit.

Unfortunately, the flywheel must be removed to gain access to the two screws and alter the position of the magneto assembly. See Section 7-6, beginning on Page 7-12, to remove the flywheel.

After the flywheel has been removed: loosen but **DO NOT REMOVE** the two magneto assembly screws. If this is the first work performed since the unit left the factory, these screws are almost impossible to loosen "cold". While attempting to rotate the screwdriver **COUNTERCLOCKWISE**, tap the end of the screwdriver with a hammer to jar the screw loose.

Once the screws have been loosened slightly, very **CAREFULLY** rotate the magneto assembly either clockwise or counterclockwise, as necessary to correct the timing.

Rotate the magneto assembly **CLOCKWISE** to retard spark advance. Rotate the assembly **COUNTERCLOCKWISE** to advance the spark.

SPECIAL WORDS ON SPARK ADVANCE

This particular Nippondenso ignition system allows for a timing decrease in advance as engine speed increases. However, at 6000 rpm, when the marks are aligned, the spark still occurs at BTDC.

When the two marks align, and this process is "trial and error", tighten the two securing screws very **SECURELY**. If these screws should happen to vibrate loose, they could jam between the rotating flywheel magnets and the coils on the magneto assembly -- causing very serious and expensive damage.

Install the flywheel. Refer to Steps 2 thru 6 beginning on Page 7-18.

WORDS FROM EXPERIENCE

The two most important items in establishing and maintaining perfect timing on an engine are the magneto assembly securing screws and the crankshaft Woodruff key.

Magneto Assembly Securing Screws

The magneto assembly is mounted on top of the stator plate and has two elongated slots for the securing screws to pass. If the magneto assembly is allowed to rotate just a small amount because the securing screws have worked loose, the ignition timing may be off as much as 15°. The symptoms of this misalignment are loss of power and difficult starting.

Crankshaft Woodruff Key

The same symptoms apply to a sheared Woodruff key allowing the flywheel -- actually the flywheel magnets -- to become out of position on the crankshaft.

During wave jumping or using a wet ramp, when the craft clears the water, the engine RPM approaches the limit set by the RPM limiter in an unloaded condition.

Once the craft returns to the water surface, the sudden load on the impeller is transferred along the driveline. Speed of the rotating pump shaft and the crankshaft is suddenly reduced while the flywheel continues to spin rapidly due to its mass. This condition, especially when repeated over a period of time, often shears the Woodruff key and the flywheel becomes repositioned on the crankshaft.

A very small misalignment will result in a change in ignition timing. The craft operator will notice a drastic loss of power. A worse condition would be the flywheel shifting to the degree engine start would not be possible.

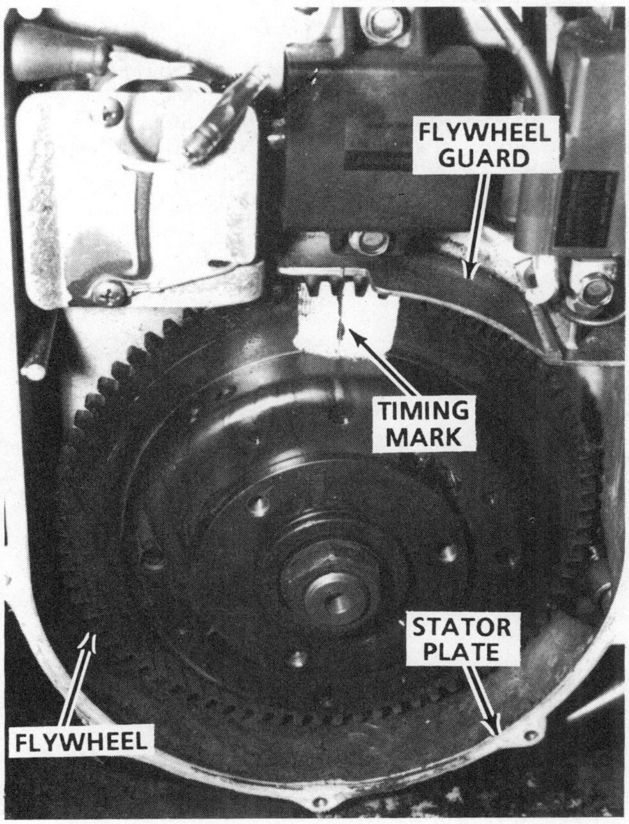

Identification of the flywheel timing mark and the notch on the flywheel guard. These two marks align at 0.086" (2.18mm), BTDC.

*The ignition timing **MUST** be verified dynamically -- with the engine operating -- to match the position of the magneto assembly with the notch on the flywheel guard, as explained in the text.*

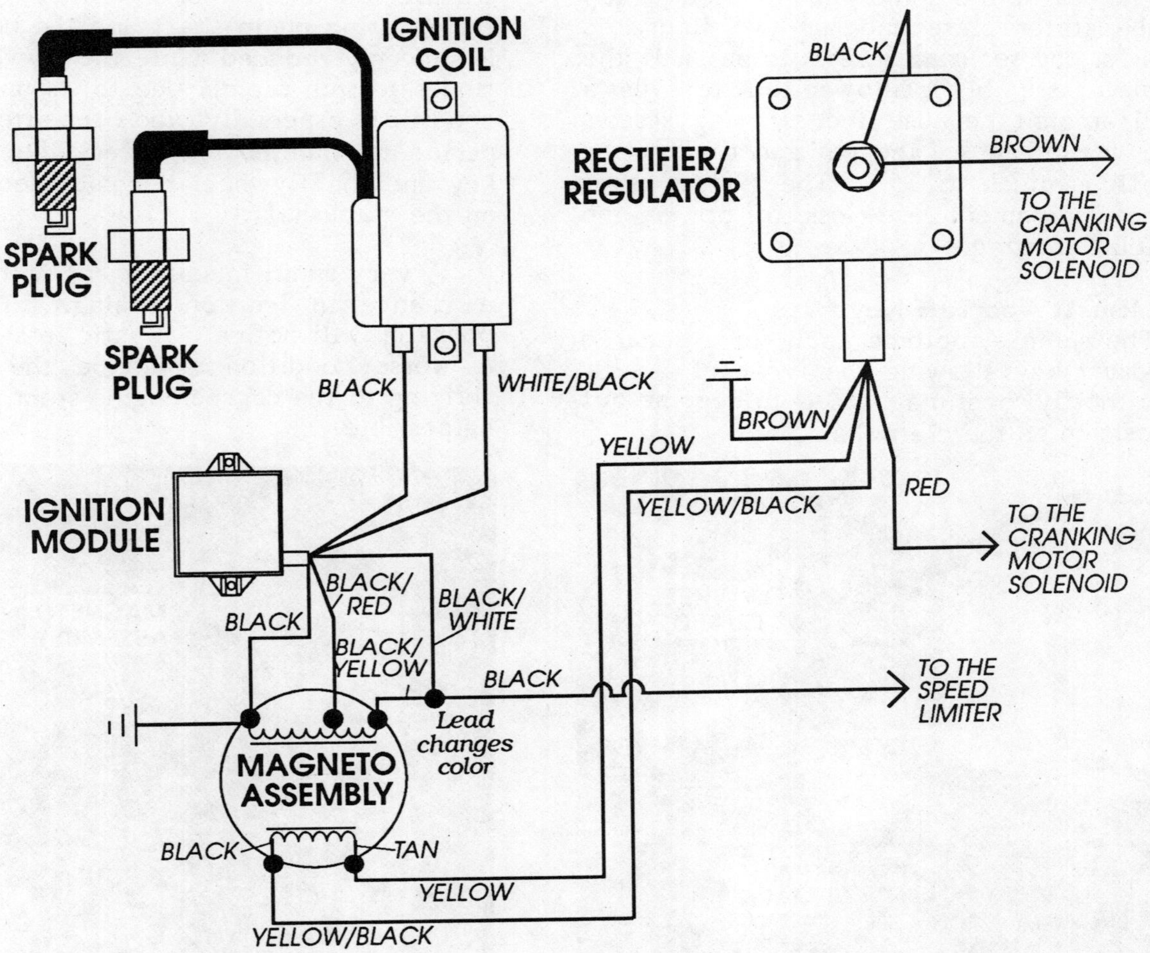

Wire identification and routing for the stator plate, magneto assembly, and rectifier/regulator circuits.

8
ENGINE

8-1 INTRODUCTION AND CHAPTER ORGANIZATION

The only engine used on Bombardier watercraft is the Rotax 587, rotary valve, water-cooled, two-cylinder, two-stroke engine. This chapter is divided into five main working sections as follows:

8-2 Description and Operation -- two-stroke engines.
8-3 Removal and Disassembling
8-4 Cleaning and Inspecting
8-5 Sealants, Adhesives, Lubricants and Fuel Stabilizers
8-6 Assembling and Installation

The carburetion and ignition principles of two-stroke engine operation must be understood in order to perform service work on the marine engine covered in this manual.

REPAIR PROCEDURES

Service and repair procedures will vary slightly between individual models, but the basic instructions are almost identical. Any variations between models are clearly indicated. Special tools may be called out in certain procedures. These tools may be purchased through the local dealer. In certain instances, an alternate tool may be obtained or fabricated.

TORQUE VALUES

All torque values must be met when they are specified. Torque values for various parts of the engine are given in the text.

A torque wrench is essential to correctly assemble the engine. Never attempt to assemble an engine without a torque wrench. Attaching bolts must be tightened to the required torque value in two progressive sequences, following the specified order. On the first sequence, tighten the bolts to half the torque value. On the second sequence, tighten the bolt to the full torque value.

OTHER ENGINE COMPONENTS

Service procedures for the carburetor, oil injection system, fuel pump, cranking motor and other components are presented in separate chapters. See the Table of Contents. Due to the unique design of these engines, a rotary valve is used in place of a reed valve system.

CLEANLINESS

Make a determined effort to keep parts and the work area as clean as possible. Parts **MUST** be cleaned and thoroughly inspected before they are assembled, in-

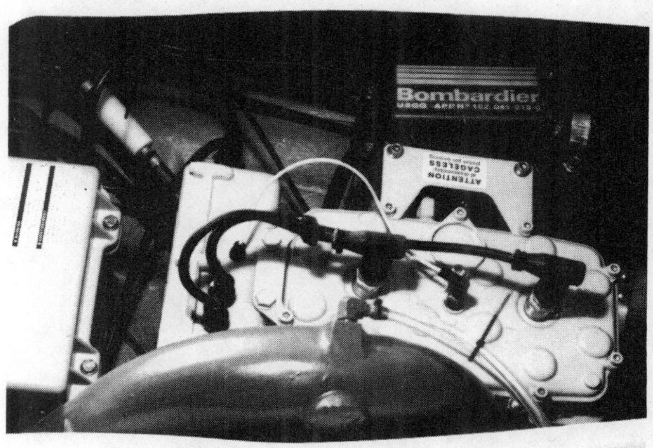

*A photograph taken with an instant-type camera is an excellent means of recording the arrangement of parts, especially electrical connections, **BEFORE** the disassembly work is started.*

stalled, or adjusted. Use proper lubricants, or their equivalent, whenever they are recommended.

8-2 TWO-STROKE ENGINE DESCRIPTION AND OPERATION

Two-stroke engines utilize an arrangement of port openings to admit fuel into the combustion chamber and to purge the exhaust gases after burning has been completed. The ports are located in a precise pattern in order for them to be opened and closed at an exact moment by the piston as it moves up and down in the cylinder bore. The exhaust port is located slightly higher than the fuel intake port. This arrangement opens the exhaust port first as the piston starts downward and therefore, the exhaust phase begins a fraction of a second before the intake phase.

LUBRICATION

A two-stroke engine is lubricated by mixing oil with the fuel. Therefore, various parts are lubricated as the mixture passes through the crankcase and the cylinder. All Bombardier Sea-Doo engines are equipped with an oil injection system which delivers oil directly into the intake manifold.

Oil Injection Pump

An oil injection pump is driven by a rotary valve shaft which meshes with a gear on the crankshaft. The oil injection pump is mounted on the rotary valve housing. The throttle cable from the handle bar splits into two cables. One cable controls engine rpm at the carburetor and the other cable is attached to the oil injection pump. Therefore, the rate of oil injection is controlled by engine rpm.

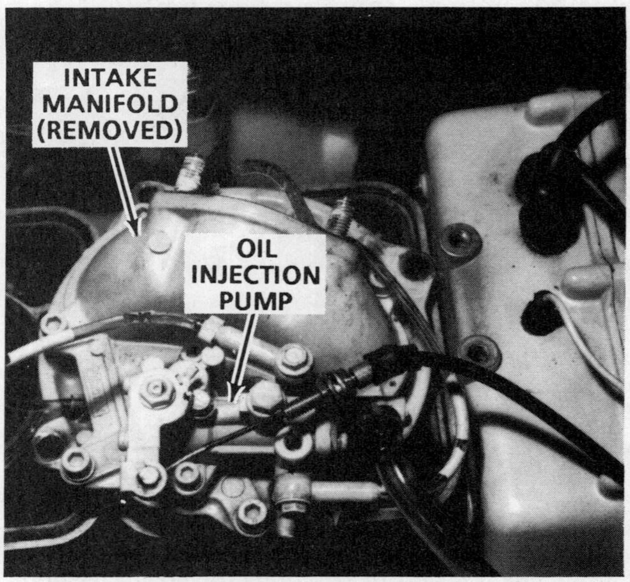

The oil injection pump is mounted on the starboard side of the engine -- on the intake manifold.

Physical Laws

The two-stroke engine is able to function because of two very simple physical laws.

1- Gases will flow from an area of high pressure to an area of less pressure. A tire blowout is an example of this principle. The high pressure air escapes rapidly if the tire is punctured.

2- If a gas is compressed into a smaller area, the pressure increases, and if a gas expands into a larger area, the pressure decreases.

If these two laws are kept in mind, the operation of the two-stroke engine will be more easily understood.

ACTUAL OPERATION TWO-STROKE ENGINE WITH ROTARY VALVE

On a four-stroke engine, the four strokes are: intake, compression, power, and exhaust. On a two-stroke engine: the first stroke shares intake and compression, and the second stroke shares power and exhaust. Beginning with the piston at the start of the upward stroke -- the intake stroke -- the exhaust port is closed by the piston; the rotary valve window is over the intake port; and the oil/air/fuel mixture enters the crankcase/combustion chamber.

As the piston travels up the bore, approaching TDC, the spark plug fires and the compressed mixture is ignited -- when the compression stroke is completed. As the piston passes TDC and begins to move downward -- on the power stroke -- the combustion chamber is filled with burning gases. The exhaust port is uncovered. The gases, which are under tremendous pressure, escape rapidly through the exhaust port. When the piston reaches 52° ATDC, the intake port is completely closed by the rotary valve. The piston completes the exhaust stroke when it reaches BDC (bottom dead center). At this instant, both intake and exhaust ports are closed.

As soon as the piston reaches 132° BTDC, the rotary valve window begins to open the intake port. The intake port remains fully or partially open for 184° (132° plus 52°) of crankshaft rotation -- more than half the two-stroke cycle. The intake port is fully open by the time the piston completes its intake portion of the upward stroke.

The cycle then repeats, as described.

TIMING TWO-STROKE ENGINES

The exact time of spark plug firing depends on engine rpm. At low rpm, the spark is retarded, fires later than when the piston is at or beyond TDC. Engine timing is built into the unit at the factory. At high rpm, the spark is advanced, fires earlier than when the piston is at TDC. No adjustment to this timing isnecessary , unless the magneto assembly has been removed.

8-3 ENGINE SERVICE

Before commencing any work on the engine, an understanding of two-stroke engine operation will be most helpful. Therefore, it would be well worth the time to study the principles of two-stroke engine operation, as outlined briefly in the previous section -- Section 8-2. A Polaroid, or equivalent instant type camera is an extremely useful item, providing the means of accurately recording the arrangement of parts and wire connections before the disassembly work begins. Such a record is most valuable during assembling.

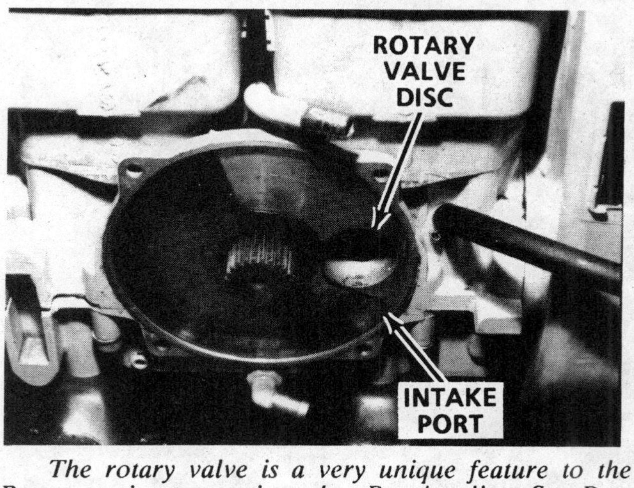

The rotary valve is a very unique feature to the Rotax engine powering the Bombardier Sea-Doo watercraft. A detailed description and explanation of the operating principles of this valve is presented in the text.

PRELIMINARY TASKS

Disconnect the negative and positive leads from the battery. Disconnect the throttle and choke cables from the carburetor.

Disconnect the fuel hoses and remove the carburetor. See Chapter 6 -- **FUEL.**

Remove the driveshaft from the jet pump. See Chapter 10 -- **JET PUMP.**

ENGINE REMOVAL

FIRST, THESE WORDS

The cylinder head, pistons, cylinder bores, oil pump, stator plate, rotary valve disc, and the flywheel may all be serviced while the engine remains in the craft. **HOWEVER,** to service the rotary valve shaft and the crankshaft, the engine must be removed.

The following series of photographs were taken while the engine was removed after all the "hang-on" items were removed. The hull cavity is too cramped an environment to show clearly with photographs the necessary work to be performed. Therefore, depending on the parts requiring service, the engine may remain in place and the unnecessary steps in the following procedures may be skipped.

Briefly stated: perform only the necessary work to return the engine to satisfactory operation. Unnecessary work may include leaving bearings, seals, O-rings, etc. in place, if they are fit for further service.

1- Loosen the clamp on the cooling water hose, and then slide it up the hose. Ease the hose free of the fitting on the jet pump.

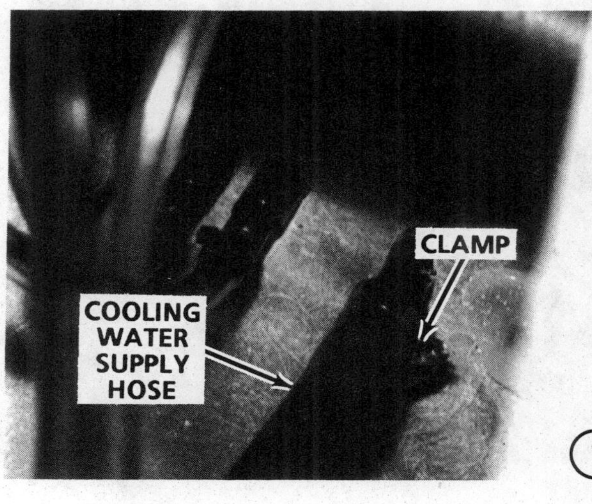

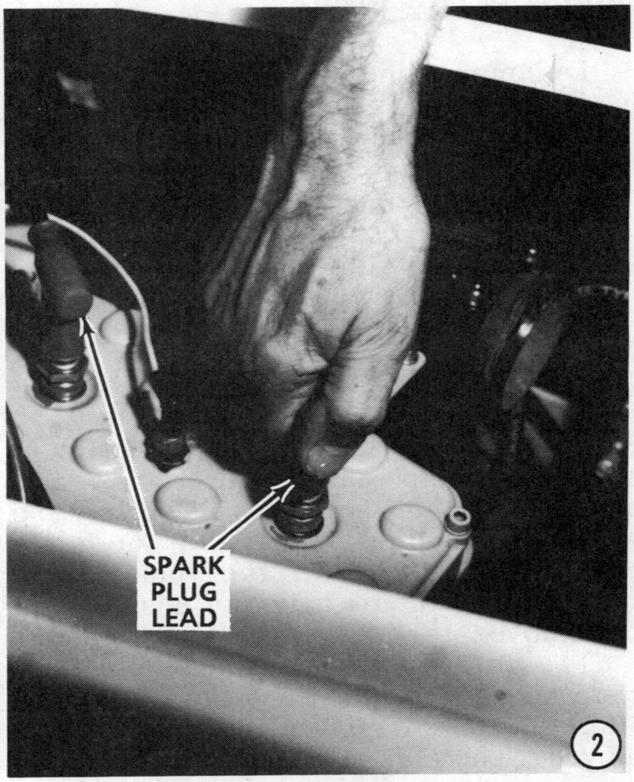

Cylinder Head Removal

2- Remove both spark plug leads.

3- Remove both spark plugs. Remove the lead from the temperature sensor mounted on the cylinder head cover.

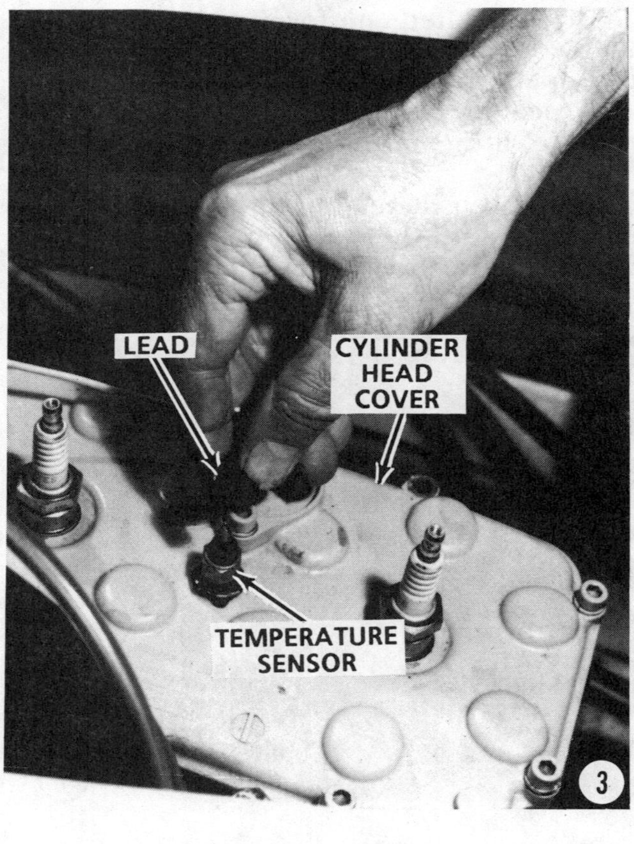

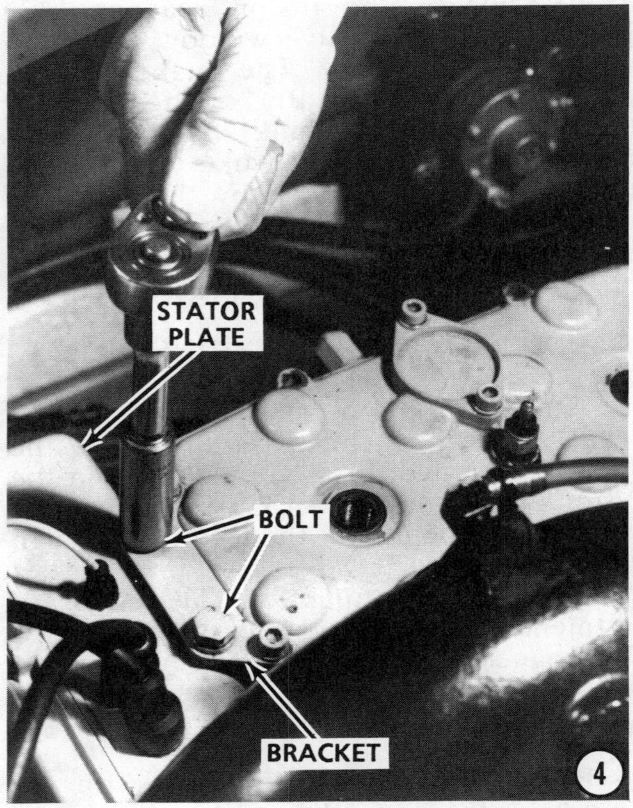

4- Remove the two bolts and washers securing the support bracket to the stator plate.

5- Remove the nine M6 Allen head bolts and washers securing the cylinder head cover to the cylinder head. The two forward bolts also secure the stator plate

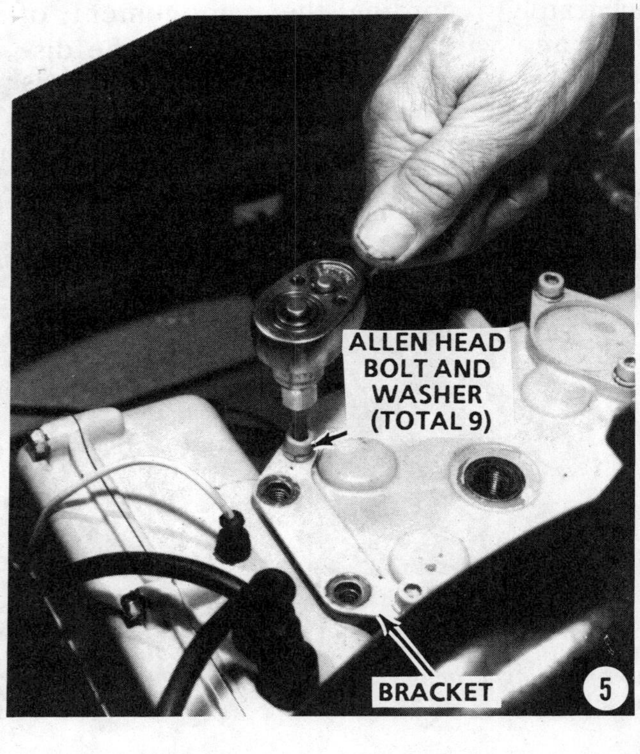

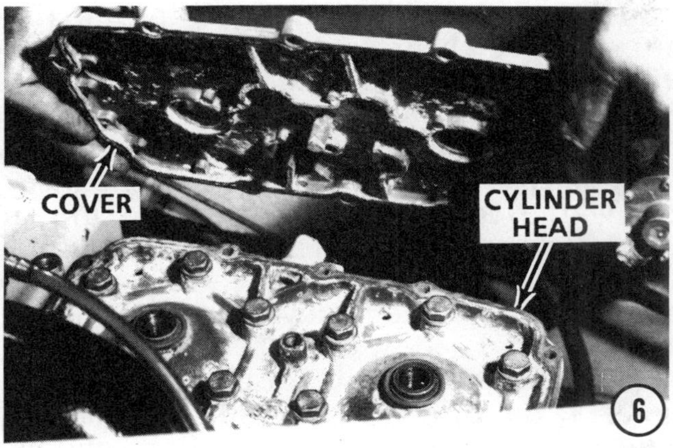

support bracket to the cylinder head cover. Remove the bracket. (Two M6 Allen head bolts were removed when the flame arrestor was removed as part of carburetor removal.)

6- Lift off the cylinder head cover. If the cover seems to be stuck in place, use a soft head mallet and tap alternately and evenly on both sides to jar the cover free. Remove and discard the sealing ring.

7- Remove and discard the single small O-ring in the center of the head and the two larger O-rings around the spark plug openings. Remove the twelve bolts and

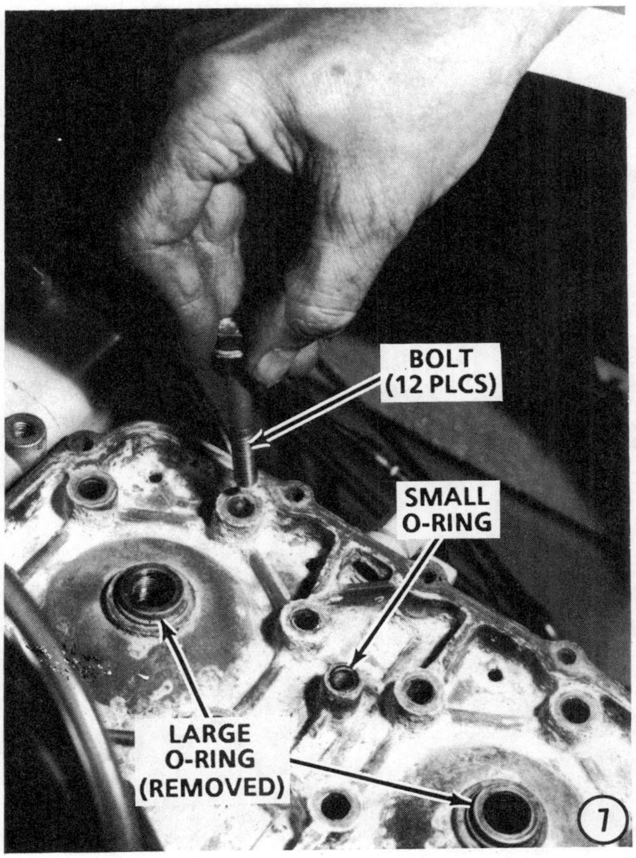

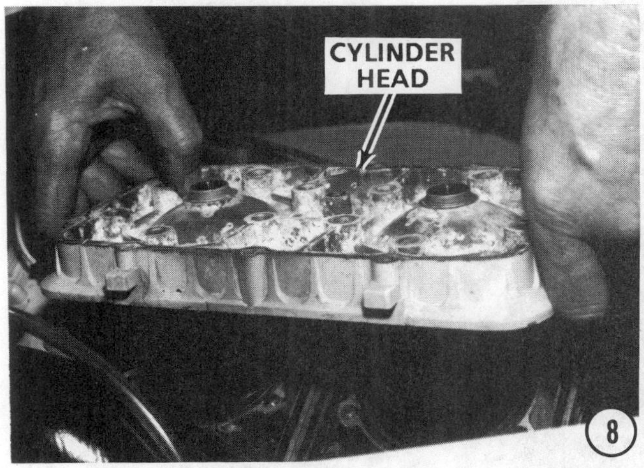

washers securing the cylinder head to the two cylinders.

8- Lift off the cylinder head from the cylinders. If the head is stuck, use a soft head mallet and tap alternately and evenly on both sides to jar the head loose.

9- Remove and discard the four sealing rings, one ring around each cylinder bore and one ring around each water jacket. Remove all traces of sealing compound from the sealing ring groove in the cylinders and the corresponding grooves in the head.

Rotary Valve and Oil Injection Pump Removal

10- Obtain hose pincer tool P/N 295-000-076. Clamp the tool around the oil return hose. If this tool is not available, make arrangements to plug the disconnected hose when it is pulled free of the fitting in the next step.

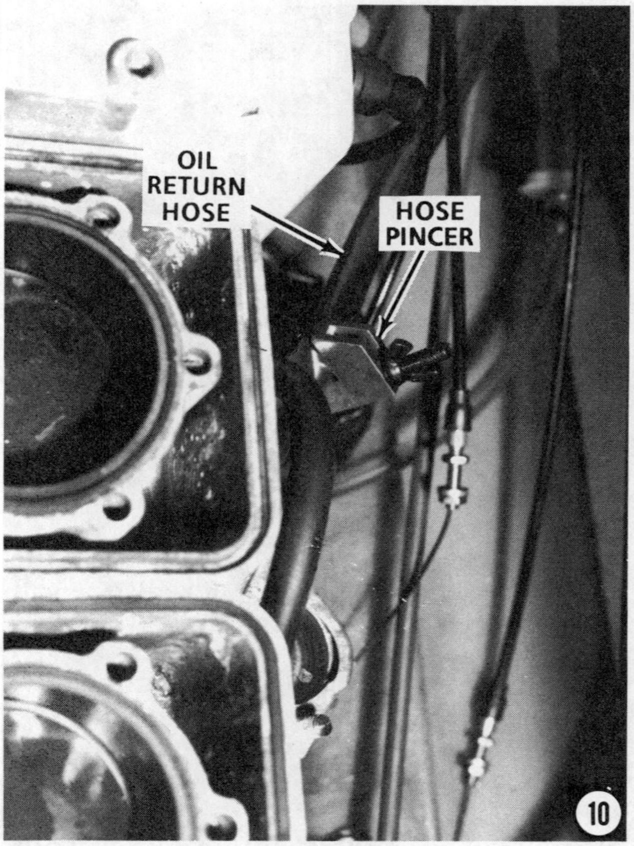

11- Pull the oil return hose free of the fitting above the rotary valve housing. If the hose was not pinched to halt the flow of oil, quickly plug the end of the hose to prevent losing oil into the bilge.

Remove the four bolts and washers securing the rotary valve housing to upper and lower crankcase. The intake manifold and the oil pump are attached to the valve housing and will come away with the housing. Remove and discard the sealing ring.

12- The rotary valve disc is now accessible. If the disc is to be replaced, make a note of the alignment of the disc window edges in relation to the housing, **BEFORE** disturbing the disc. If the disc is suspected of not being properly timed, proceed to Step 29 on Page 8-42 to properly time the disc. If the disc is merely being replaced with a new one, install the new disc in **EXACTLY** the same position as the disc removed. If this is the only service work to be performed, proceed directly to the appropriate area of assembling.

Install a hose pincer tool onto the oil supply hose, and then remove the hose from the fitting at the base of the rotary

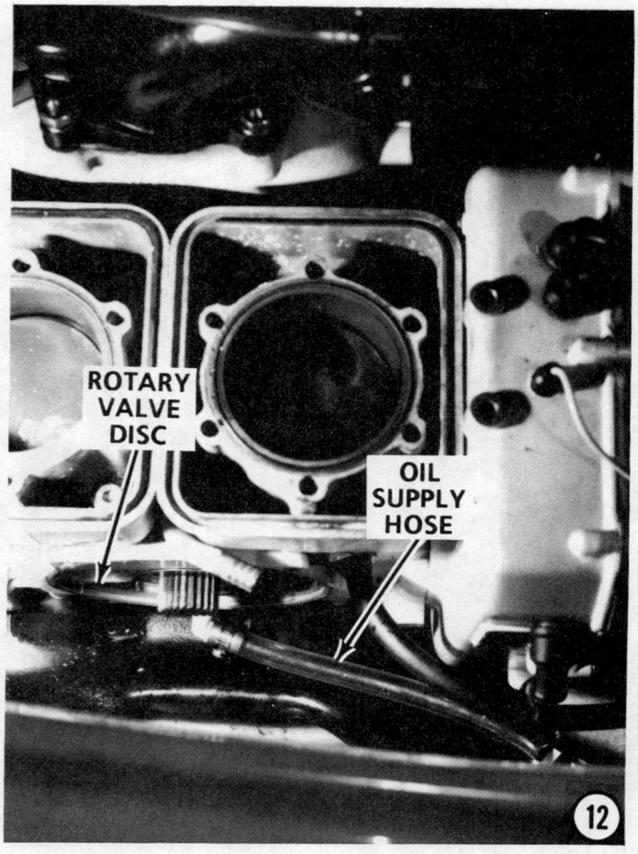

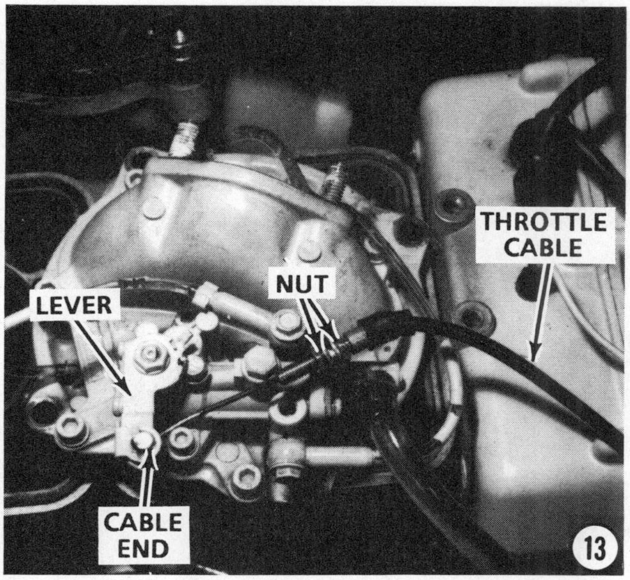

valve housing. If this tool is not available, make adequate preparations to catch oil when the hose is disconnected and to plug the hose immediately.

13- Disconnect the throttle cable from the oil injection pump by disengaging the cable end from the slotted hole in the lever, and loosening the two nuts -- one on each side of the support bracket.

14- Remove the six Allen bolts and the washers securing the intake manifold to the rotary valve housing. The oil injection pump will remain on the manifold.

15- Lift the intake manifold free of the rotary valve housing. **TAKE CARE** not to lose the single dowel pin from one of the mating surfaces. The dowel pin may remain in either surface. Remove and discard the sealing ring.

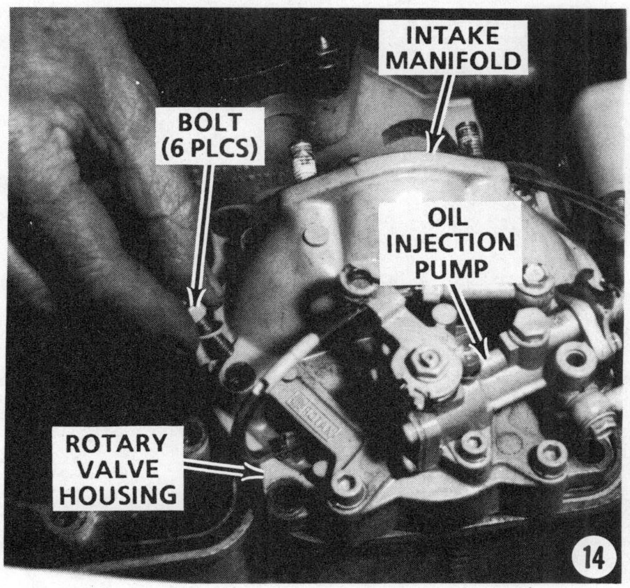

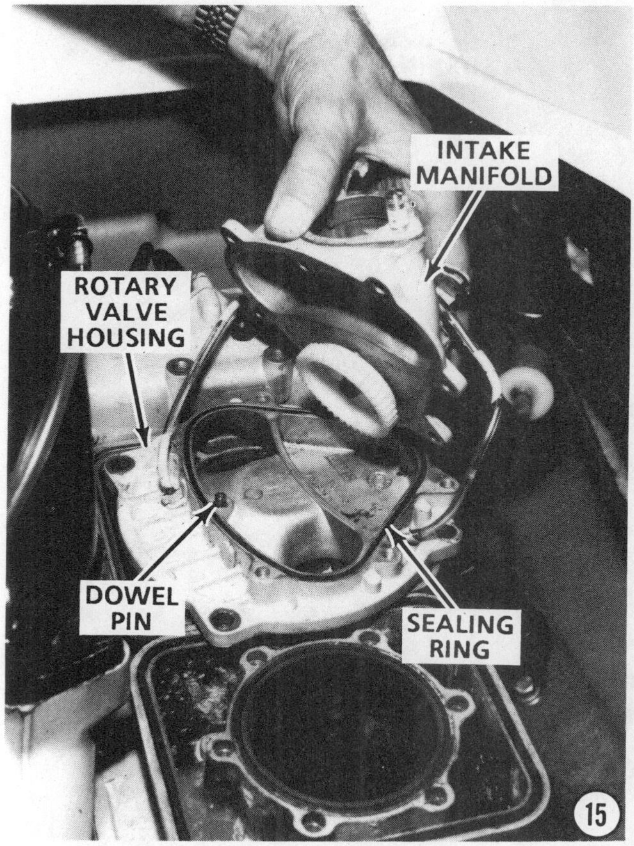

16- Obtain gear holder P/N 290-277-905 and use it to prevent the oil pump gear from rotating while the nut is loosened. If this tool is not available, hold the gear with a shop cloth, as shown. Remove the nut and washer, gear, and another washer. Remove the two mounting bolts. Lift the pump off the intake manifold, and then remove and discard the **O**-ring.

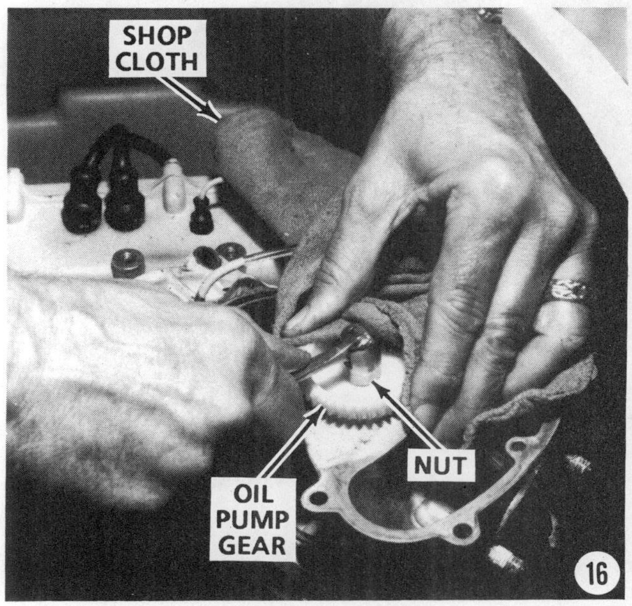

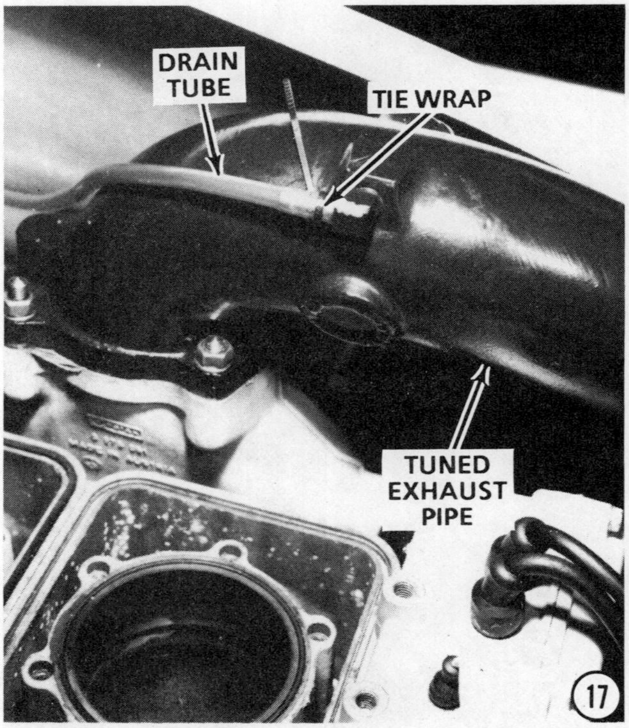

17

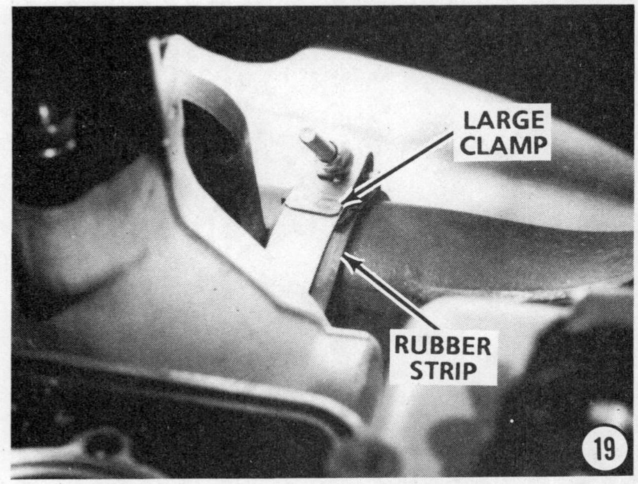

19

GOOD WORDS

No individual parts are available if the pump is not functioning properly. **HOWEVER**, a leaking oil pump can be overhauled with a gasket kit. If the problem cannot be corrected with the gasket kit, the pump **MUST** be replaced.

Exhaust System Removal

17- Snip the tie wrap and remove the drain tube from tuned exhaust pipe.

18- Loosen the clamps on the water injection hose and the inlet hose on the tuned exhaust pipe. Pull the hoses free of the fittings.

19- Loosen the large clamp securing the tuned exhaust pipe to the engine. Remove and **SAVE** the rubber cushioning strip.

20- Loosen the two clamps on the hose connecting the tuned exhaust pipe to the muffler. This disconnect is necessary to provide some free play for the exhaust pipe to be moved forward and removed from the craft at a later time.

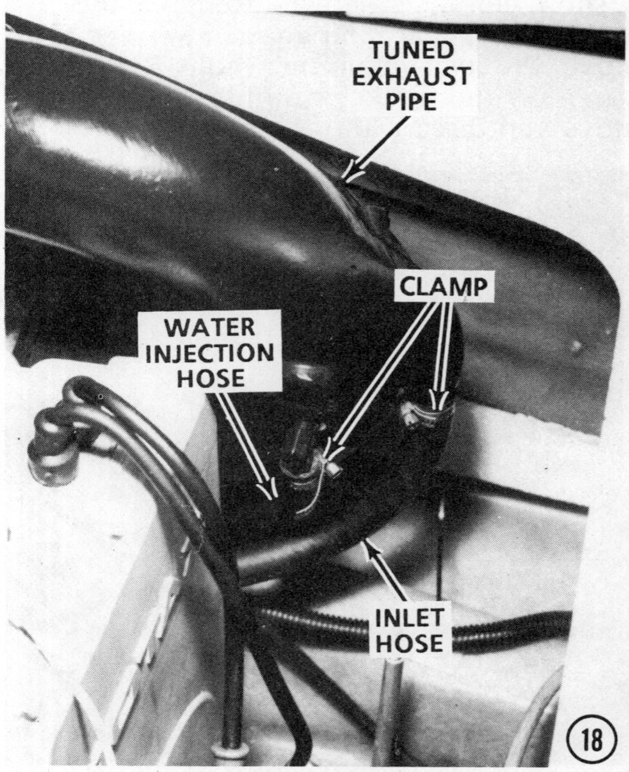

18

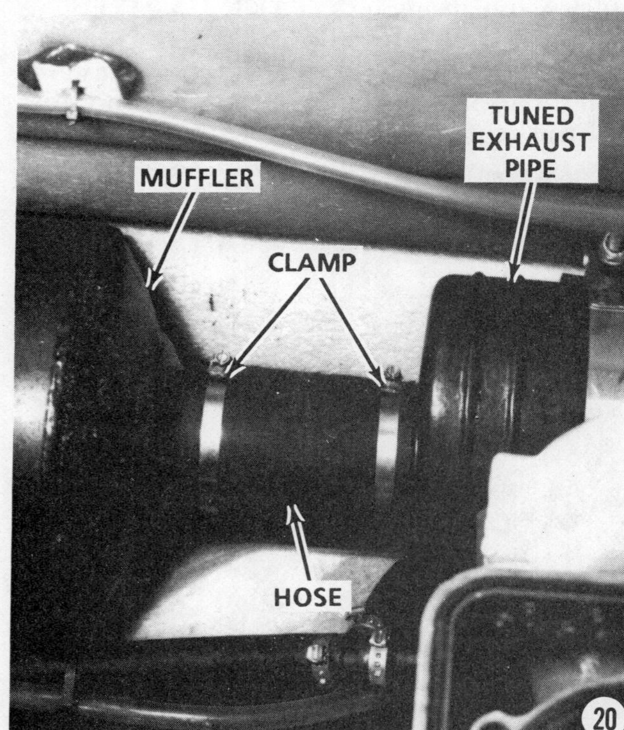

20

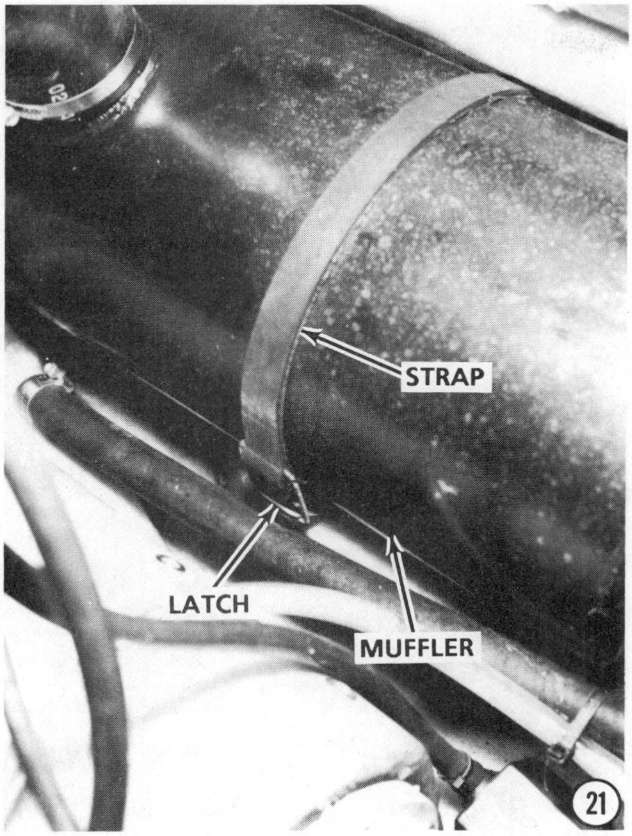

21- Unhook the muffler from the latch on the hull.

22- Remove the four nuts and washers securing the tuned exhaust pipe flange to the exhaust manifold. There is not enough room to lift the tuned exhaust pipe up over the exhaust manifold studs and out of the craft. Therefore, the exhaust pipe must be lifted up and moved aft while the studs are removed.

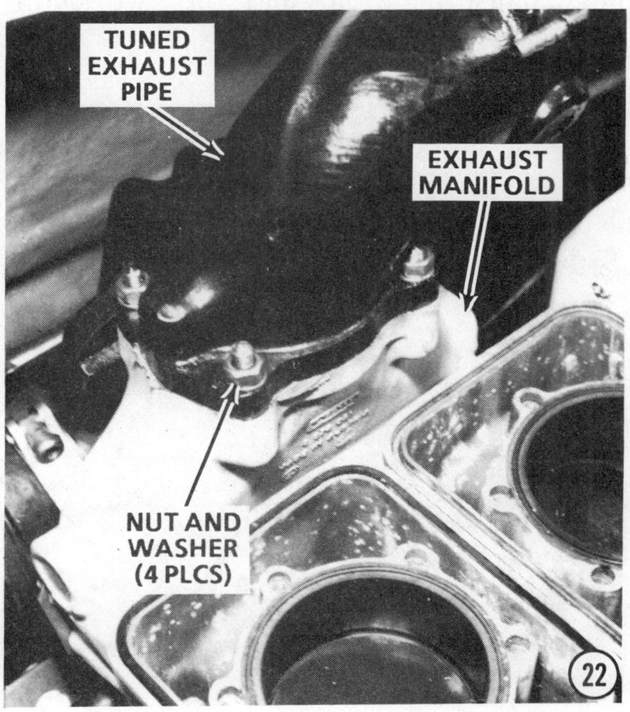

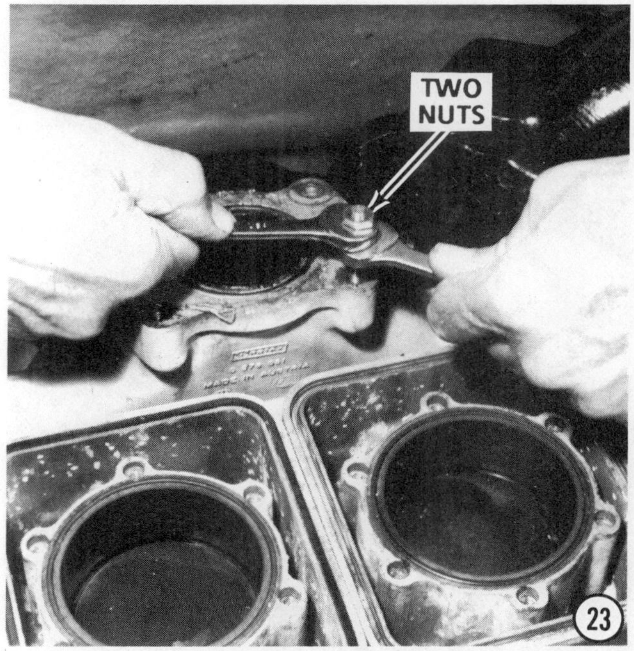

23- Install two of the nuts removed in the previous steps onto one of the studs. Tighten the two nuts against each other.

24- Remove the stud by rotating the lower nut **COUNTERCLOCKWISE**. The upper nut will prevent the lower nut from threading upward on the stud. In this manner the stud may be removed from the exhaust manifold flange. Repeat this procedure and remove the other studs.

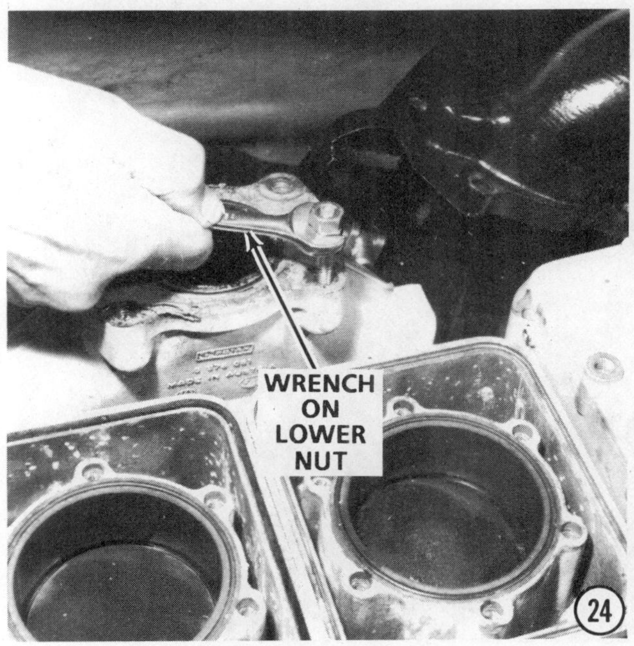

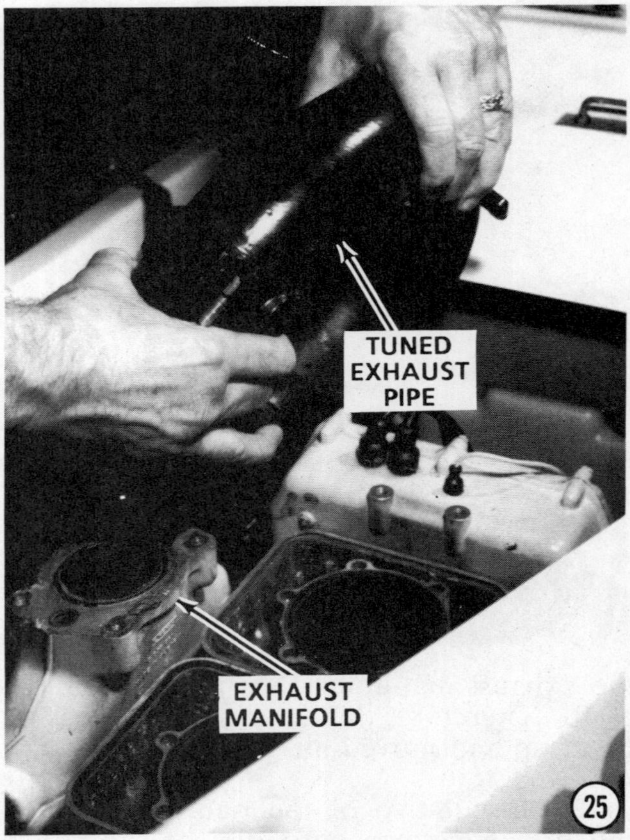

25- Ease the tuned exhaust pipe over the exhaust manifold and clear of the engine compartment.

26- Remove and discard the large exhaust passage O-ring and the two smaller water passage O-rings from the exhaust

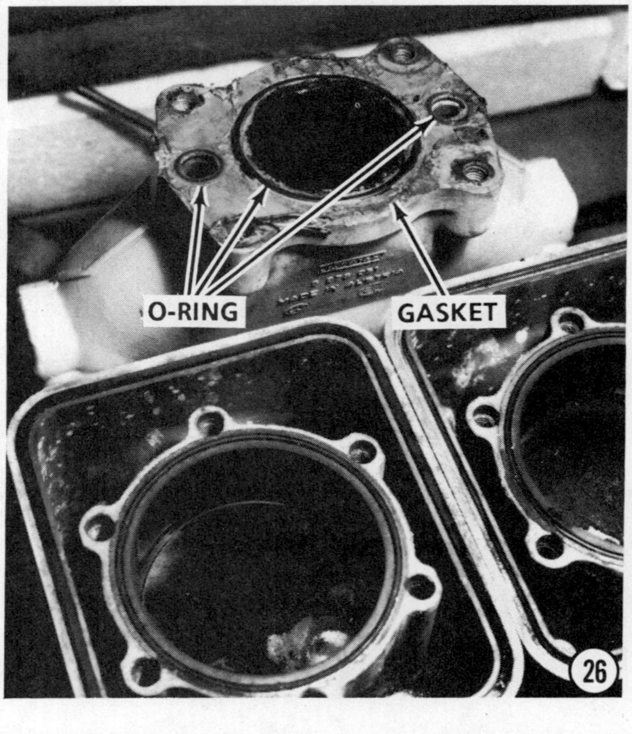

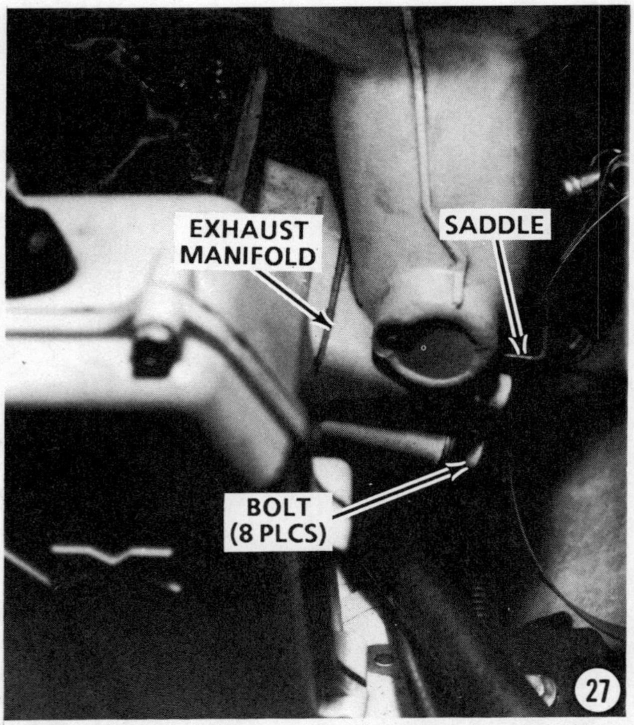

manifold flange. Remove and discard the gasket.

27- Remove the eight Allen head bolts and washers securing the exhaust manifold to both cylinders. The four forward bolts also secure a saddle to the manifold. Remove the saddle, washer, and rubber pad. This arrangement supports the cranking motor support bracket.

28- Insert a long pointed tool between the exhaust manifold and the two cylinders, as shown. Pry the manifold away from the cylinders.

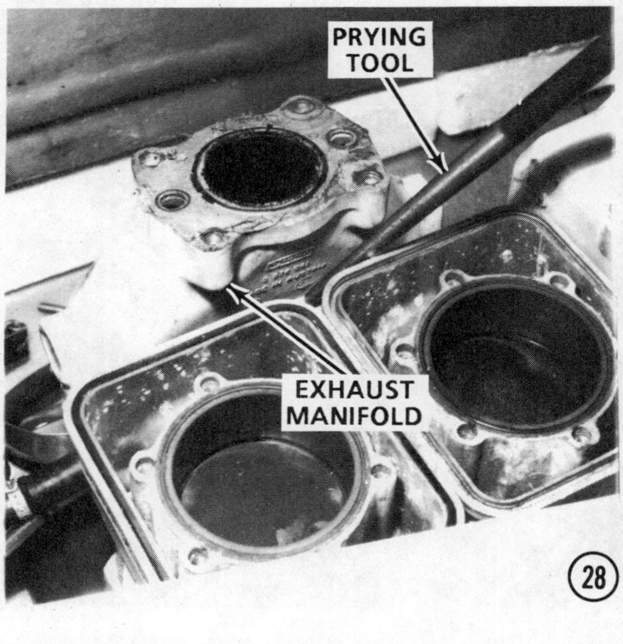

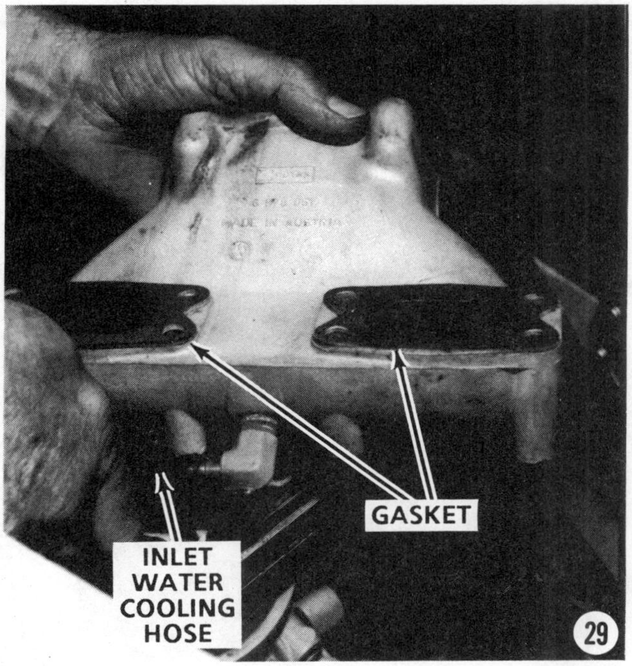

29- Lift out the exhaust manifold from the engine compartment. Remove the inlet cooling hose from the base of the exhaust manifold. Remove and discard the two gaskets. If the gasket material is stuck to the mating surfaces, take time to clean away the old gasket material. An exhaust leak will develop, if new gaskets are installed over traces of old material.

30- Remove the water hose from the outlet fitting on the port side of the engine.

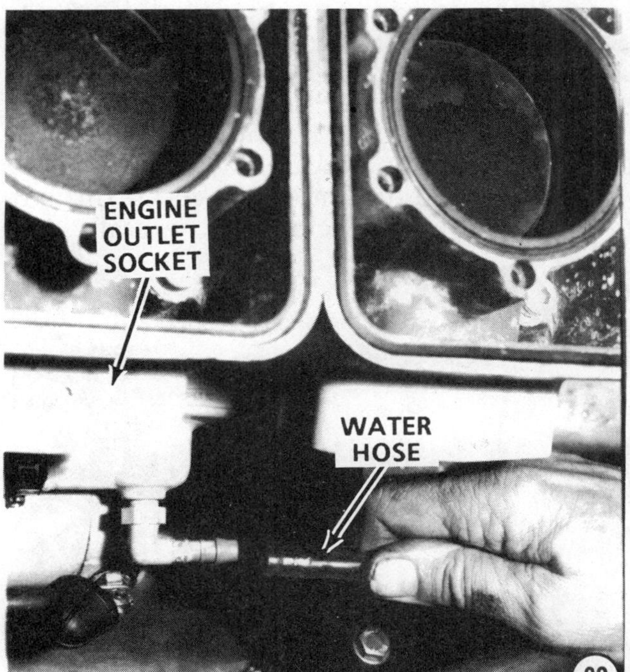

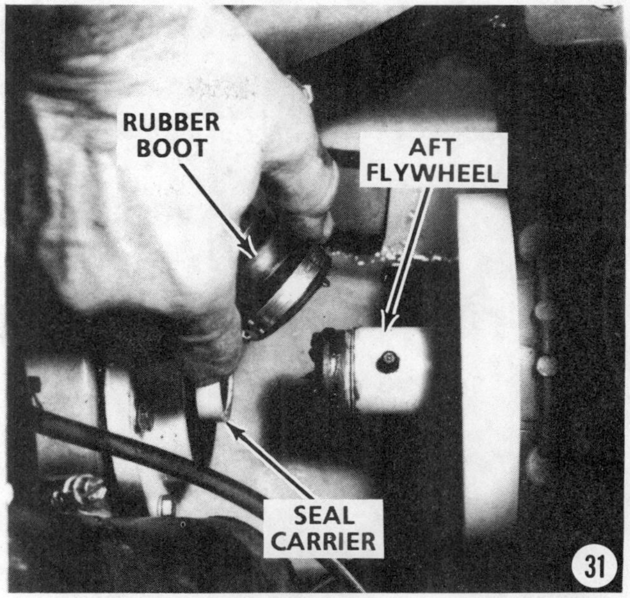

SPECIAL WORDS

The rubber boot mentioned in the next step cannot be removed, unless the driveshaft has been removed from the craft, as instructed at the beginning of this section.

31- Loosen the two clamps and remove the rubber boot between the flywheel and the seal carrier.

Stator Plate Cover Removal

32- Remove the twelve bolts securing the cover to the stator plate.

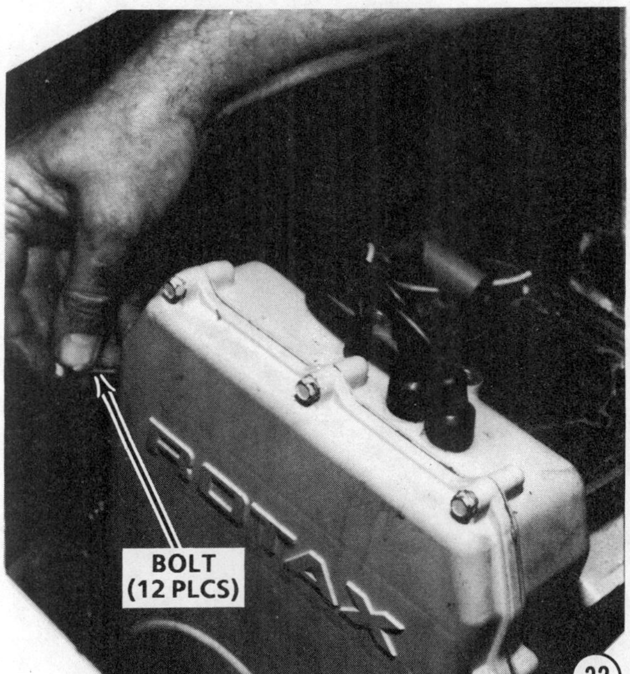

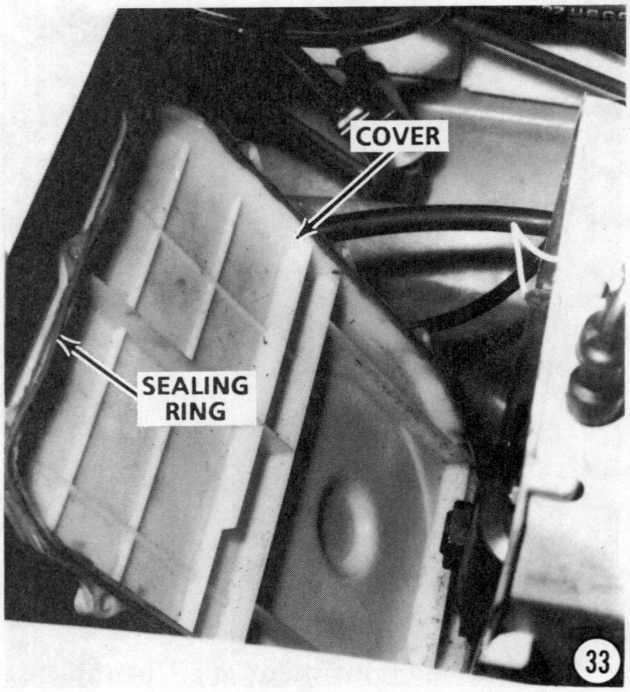

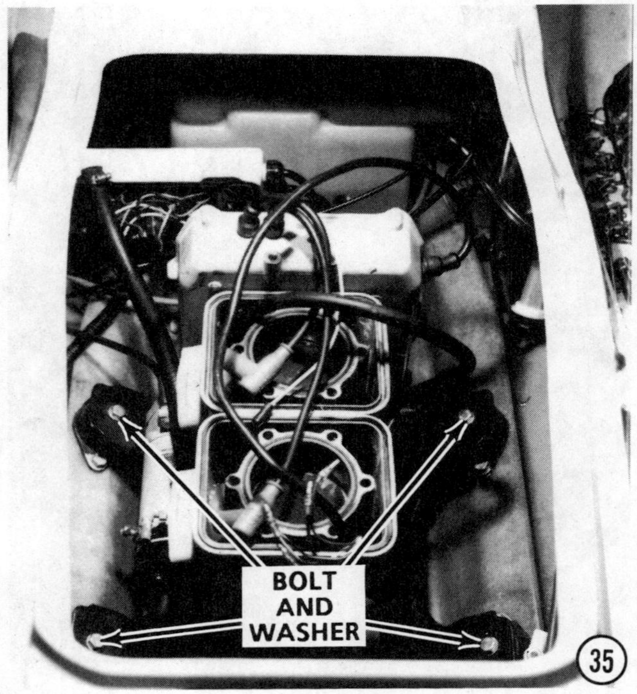

33- Lift off the cover. Remove and discard the sealing ring.

Cranking Motor Disconnect

34- Disconnect the large Red lead from the terminal on the cranking motor. Remove the upper mounting bolt and the large Black ground lead. If this is difficult to accomplish, at this time, the lead may be left in place until the engine is removed from the craft, **PROVIDED** the other end of the Black lead is no longer connected to the battery.

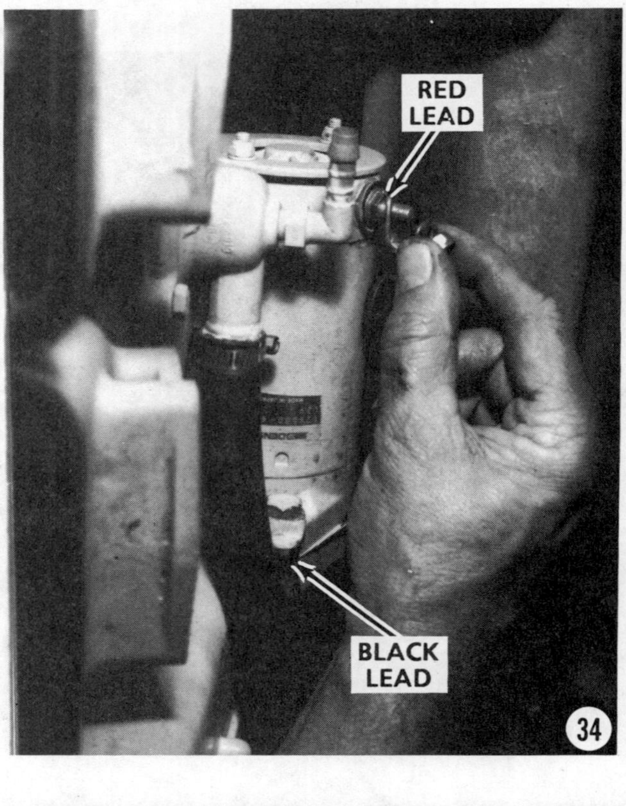

Engine Removal

35- Disconnect the overheat buzzer, tether cord switch, and the starting switch from the engine harness leads. Make one final check to ensure everything attached to the engine has been disconnected and the engine is ready to be lifted from the craft. Remove the four bolts and washers securing the engine and engine support bracket to the engine mounts.

DO NOT disturb the pieces of shim material at the engine mounts.

36- Two adults should have no difficulty lifting the engine clear of the craft. Lift

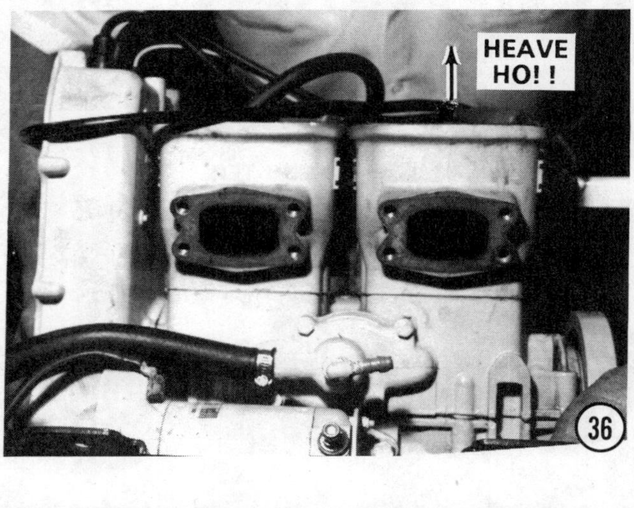

the engine and rotate the cylinders to **PORT** to permit the engine support bracket to clear the edge of the engine compartment. Watch for any hose or electrical leads still attached to the engine. Move the engine to a suitable work station.

37- DO NOT disturb the shim material or the engine mounts, unless a mount is to be replaced. Use the original amounts of shim material at each location, if new mounts are installed. Use some type of commercial cleaner and "square away" the bilge and engine compartment.

38- If there is an excessive amount of water in the bilge, the seals in the seal carrier may have failed and require replacement. Service to the seal carrier is outlined in detail in Chapter 10 -- **JET PUMP**.

ROTARY VALVE BACKLASH CHECK

The rotary valve backlash **MUST** be checked **BEFORE** any further disassembling work is performed.

Grasp the rotary valve shaft and rock it back and forth to determine the degree of movement. Rotate the flywheel to turn the crankshaft through 1/8 revolution. Check the backlash again. Repeat the procedure to check the backlash through one complete revolution of the crankshaft. The backlash **MUST** be the same at all position of the crankshaft.

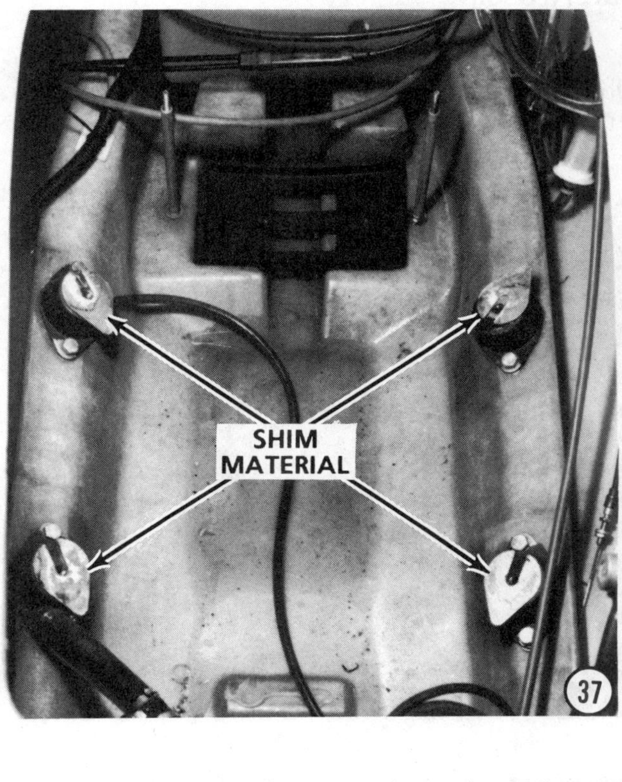

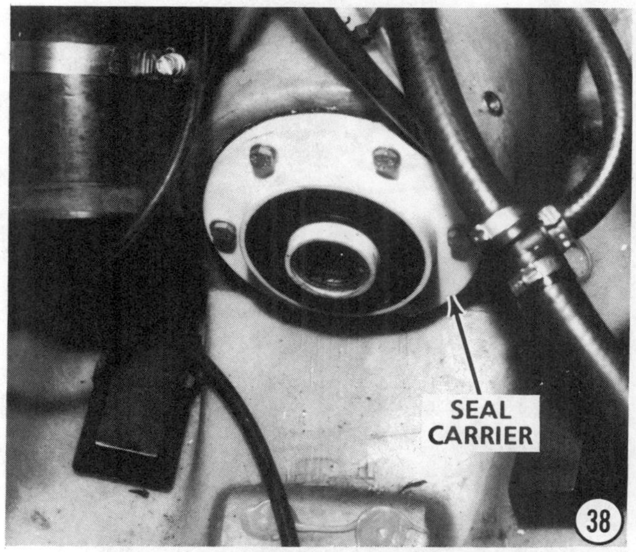

If the backlash is not consistent, the crankshaft and rotary valve shaft must be inspected to determine the cause of the binding:

The gear teeth may be stripped.
The bearing may have failed.
The rotary valve shaft may be bent.
The crankshaft may be bent.

The rotary valve shaft is serviced on Page 8-42, after the crankshaft has been removed.

*The rotary valve backlash check **MUST** be completed **BEFORE** any further disassembling work is performed, as described in the text.*

39- Slide the rotary valve disc from the shaft.

Cranking Motor Removal

40- Remove the two bolts securing the cranking motor to the stator plate -- one may have been removed earlier -- and a third bolt securing the aft end of the motor to the lower crankcase.

41- Pull the cranking motor away from the stator plate. If the motor is stuck, tap the mounting flange with a soft head mallet to jar it loose. Remove and discard the gasket.

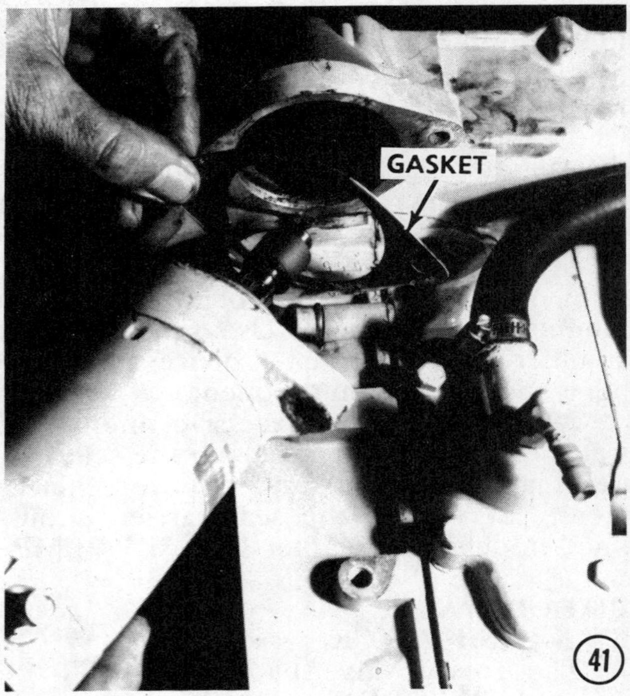

Aft Flywheel Removal

CRITICAL WORDS

NEVER attempt to lock the crankshaft by inserting any type of tool into the crankcase to wedge the crankshaft. This includes attempting to insert a rod through both small ends of the connecting rods to prevent rotation. Extensive damage to the connecting rods and/or crankshaft will surely occur.

42- Obtain Puller Plate P/N 290-876-080, Extension Handle P/N 295-000-003, and three M8x25mm bolts. Install the

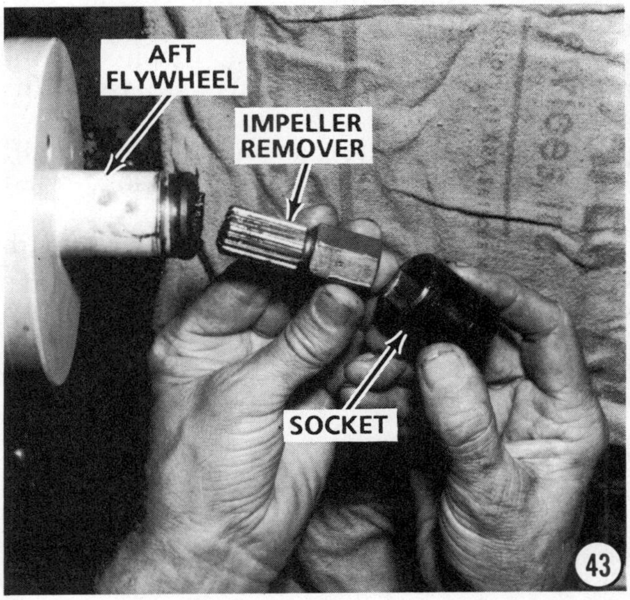

plate onto the flywheel using the three bolts. Tighten the bolts securely. This plate will be used in this step and the next four steps, therefore, leave it in place.

Install the extension handle onto the puller plate arm. Rotate the flywheel **CLOCKWISE** until the handle rests on the work bench, as an aid in loosening the flywheel.

43- Obtain Impeller Remover P/N 295-000-001. Insert the splined end of the tool into the aft flywheel. Use a suitable wrench or socket on the tool, and "break" the aft flywheel loose, by rotating the wrench or socket **COUNTERCLOCKWISE.**

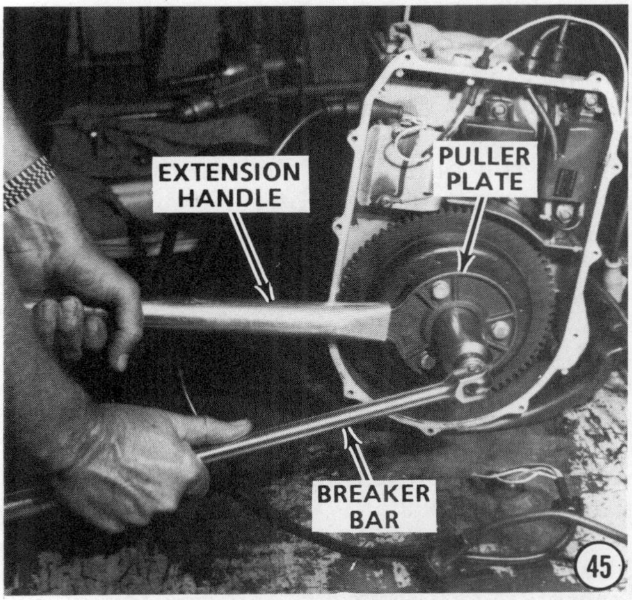

44- Continue to unthread the aft flywheel from the threaded end of the crankshaft. Remove all traces of old sealant from both sets of threads.

Forward Flywheel Removal

45- Install a 22mm thin-walled socket over the flywheel nut. The socket **MUST** fit inside the puller plate hole.

If the only available socket will not fit into the puller plate hole, grind down the outside diameter of the socket until it will fit into the hole. **DO NOT** grind on the internal threads of the puller plate.

Place a breaker bar into the socket and rotate it **COUNTERCLOCKWISE** to

"break" the flywheel nut free. Remove the nut and washer.

46- Obtain Puller and Puller Bolt P/N 290-876-065. Thread the bolt into the puller until just a couple threads are engaged. Thread the puller into the internal threads of the puller plate. Now, Thread the bolt into the puller until the bolt makes contact with the end of the crankshaft.

47- Obtain the correct size wrench or socket and place it over the puller bolt.

Now, continue to tighten the bolt and at the same time, shock the crankshaft with a gentle to moderate tap on the end of the bolt, with a soft head mallet. This shock will assist in "breaking" the flywheel loose from the crankshaft.

DO NOT strike the bolt with a severe blow. The force of such action would be transmitted to the seals and bearings along the crankshaft and cause damage. **NEVER** strike the flywheel directly, because this type of shock would surely cause the flywheel magnets to loose their magnetism.

48- Pull the flywheel free of the crankshaft. Remove and save the Woodruff key from the crankshaft keyway.

SPECIAL WORDS ON WOODRUFF KEYS

Inspect the condition of the Woodruff key and its keyway in the crankshaft. If the key shows any sign of shearing, the cause is probably from jumping waves and the effect will be a loss of power. When the craft is clear of the water, the engine rpm approaches the limit set by the rpm limiter in an unloaded condition.

When the craft returns to the water surface, the sudden load on the impeller is transferred along the driveline. Speed of the rotating driveshaft and the crankshaft is suddenly reduced while the flywheel continues to spin rapidly, due to its mass. This condition, especially when repeated over a period of time, often shears the Woodruff key and the flywheel becomes repositioned on the crankshaft.

A very small misalignment will result in a change in ignition timing. The craft operator will notice a drastic loss of power. A worse condition would be the flywheel shifting to the degree engine start would not be possible.

Stator Plate and Magneto Assembly Removal

49- The securing screws pass through elongated slots in the magneto assembly. The assembly could be installed as much as 15° off its original position -- drastically affecting the timing characteristics of the engine. Therefore, make a set of aligned punch marks -- one on the magneto assembly -- the other on the stator plate. These marks will serve as an aid during installation to **ENSURE** correct positioning of the magneto assembly on the stator plate.

Remove the two screws securing the magneto assembly to the stator plate. These screws must be removed because they are threaded into the crankcase.

50- Remove the four bolts securing the stator plate to the crankcase. Lift off the stator plate with the magneto assembly and wiring harness attached.

Engine Support Bracket Removal

51- Remove the four nuts and washers securing the engine support bracket to the bottom of the lower crankcase. Lift the bracket off the studs.

Cylinder Removal

52- Remove the four Allen head bolts from each cylinder. These bolts are located inside the water jacket.

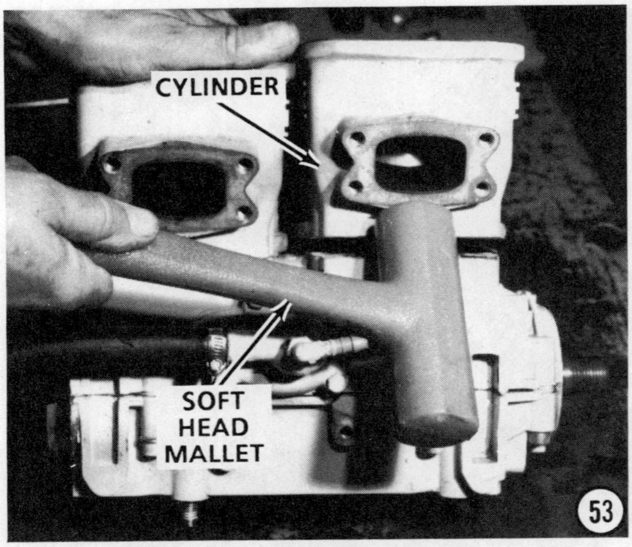

53- Using a soft head mallet, gently tap each cylinder to release it from the upper crankcase.

54- Slowly slide the cylinder up and free of the piston. **TAKE CARE** not to allow the piston to slip out of the cylinder bore and damage the piston skirt or connecting rod on the side of the upper crankcase. Remove the other cylinder in the same manner.

Piston Removal

55- Scribe a mark on the inside of the piston skirt to identify the No. 1 and No. 2 cylinders, before the piston is removed from the connecting rod, as described in the following steps.

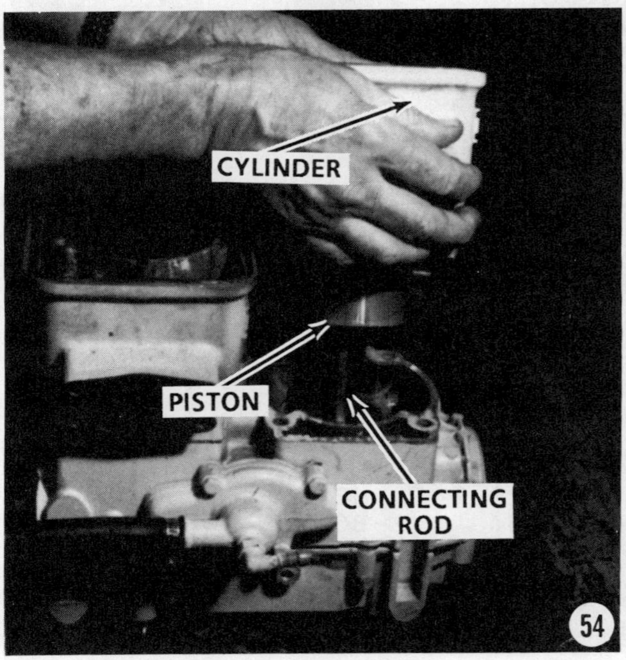

SPECIAL WORDS ON NEEDLE BEARINGS

A decal on the flame arrestor support bracket, warns there is a cageless needle bearing inside the piston. As soon as the piston is separated from the connecting rod, these needle bearings could fall into the crankcase, unless arrangements are made to catch them.

56- Raise the first piston to be disassembled up to the TDC position. Obtain Rubber Pad P/N 295-000-079. Install the

pad around the connecting rod. Temporarily install two of the bolts to prevent the pad from shifting. Stuff a shop cloth into any part of the upper crankcase which the pad does not cover. If the special pad is not available, stuff numerous shop cloths into the upper crankcase. Smooth the tops of the cloths to enable the needles to be found after they fall free.

Some mechanics place a magnet under the piston being serviced to attract the loose needles.

SAFETY WORDS

Snap rings and C-lockrings are made of spring steel and may pop out of the groove with considerable force. Therefore, warn others in the area and **WEAR** eye protection glasses or a shield while removing this type of ring.

57- Insert a small screwdriver or awl into the notch next to the **C**-lockring and pry the ring free. Discard the lockring. These rings stretch and distort during removal. Therefore, they can **NEVER** be used a second time.

SPECIAL WORDS ON PISTON REMOVAL

The piston may be removed using factory special tools or by using readily available shop tools.

If factory tools are available, perform Step 58 -- then proceed to Step 61.

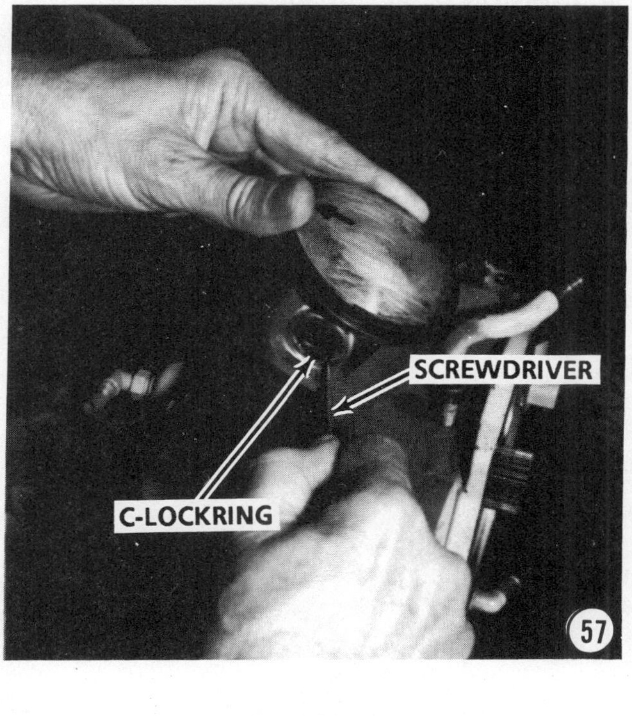

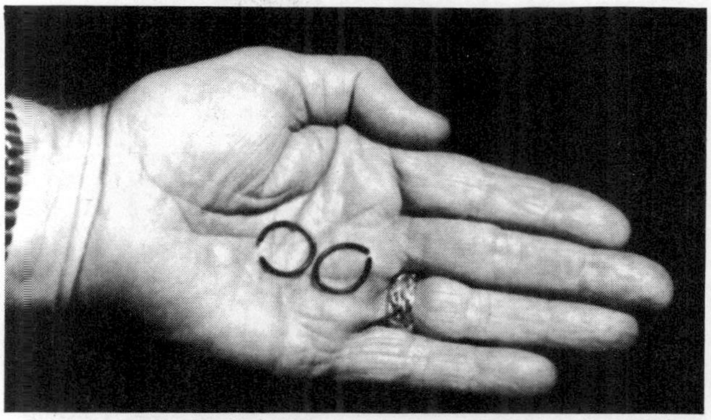

*Piston pin **C**-Lockrings stretch and distort during removal. These rings should **NEVER** be used a second time.*

If shop tools are to be used, perform Step 59 and on.

New needle bearings should be installed in the connecting rods, even though they appear to be in serviceable condition. New bearings will ensure lasting service after the overhaul work is completed. If it is necessary to install the used bearings, keep them separate and identified to **ENSURE** they will be installed back with the same connecting rod from which they were removed.

58- Obtain Piston Pin Puller P/N 290-877-090. Thread the puller handle in as

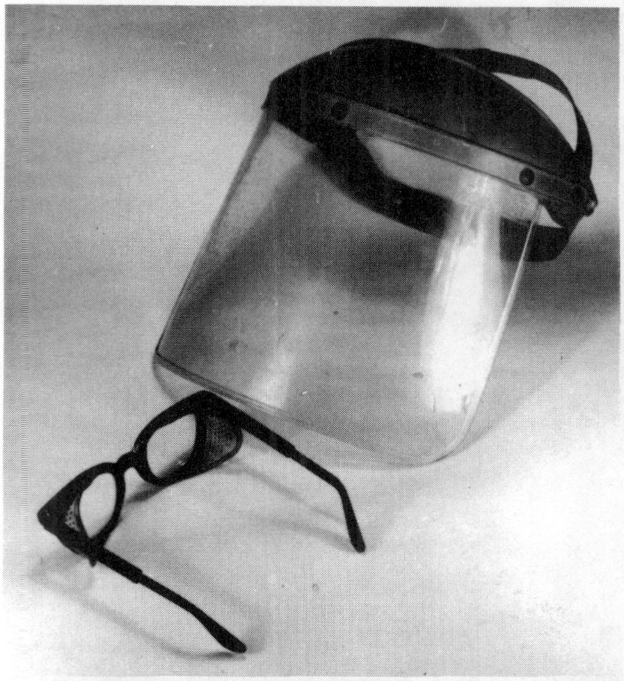

*C-lockrings are under tremendous tension during removal and installation, presenting a potential hazard. As an eye protection measure, safety glasses or a shield should **ALWAYS** be worn during work with such snap rings. Warn others in the area such work is in progress.*

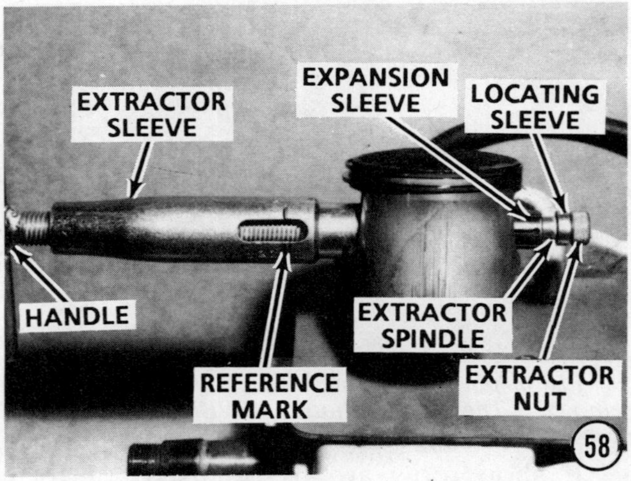

far as possible. Insert the extractor spindle into the piston pin. Slide the expansion sleeve, followed by the locating sleeve onto the spindle. Install the extractor nut onto the spindle threads.

Rotate the handle **CLOCKWISE** to pull the piston pin until the edge of the pin aligns with the reference mark on the extractor sleeve window. Now, rotate the spindle **COUNTERCLOCKWISE** until the extractor nut can be removed. Remove the spindle and extractor sleeve from the piston pin and remove the piston. The expansion sleeve will hold the loose needle bearing and thrust washer in place, preventing the needle bearing from falling apart.

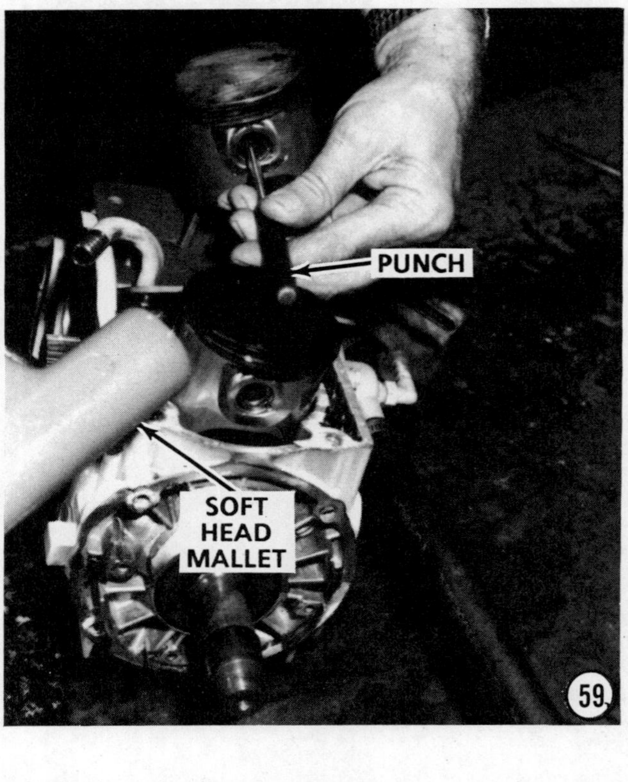

Carefully keep this assembly intact, if the needle bearings are to be used again. Use an extra expansion sleeve to disassemble the other piston using factory tools.

59- If the factory is not available, tap the piston pin free using a narrow punch and soft head mallet. Be sure to tap evenly around the pin, not just on one side or the pin will jam in the bore.

60- Remove the two thrust washers from inside the piston. Recover all 31 needle bearings for each piston, even if they are not to be used again. In other words, all **MUST** be accounted for to ensure none have found their way into the lower crankcase.

GOOD WORDS

Good shop practice dictates to replace each piston ring during an engine overhaul. However, if the rings are to be used again, expand them **ONLY** enough to clear the piston and the groove, because used rings are brittle and break easily.

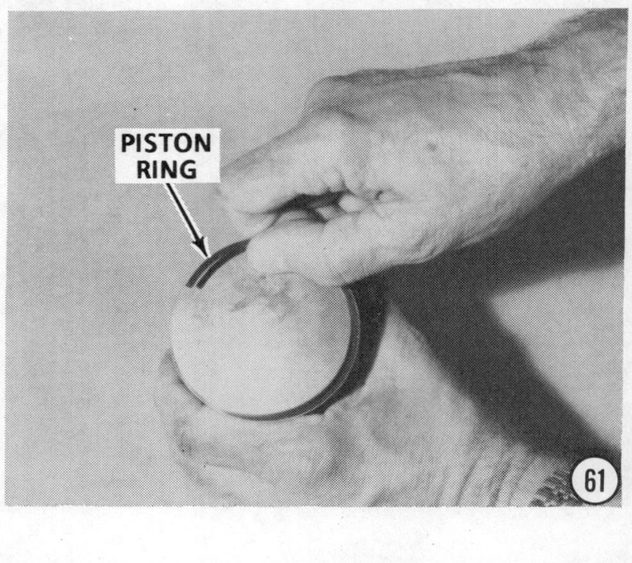

61- Gently spread the top piston ring enough to pry it out and up over the top of the piston. No special tool is required to remove the piston ring. Remove the lower ring in a similar manner. These rings are **EXTREMELY** brittle and **MUST** be handled with care if they are intended for further service.

Crankcase Separation

62- Remove the six hexagon head bolts and eight Allen head bolts securing the two halves of the crankcase together.

63- Separate the two halves of the crankcase by inserting a screwdriver between the lugs specifically provided for this purpose. **DO NOT** pry at any other location. Work around the crankcase split, prying alternately and evenly from side to side.

64- Carefully lift the upper crankcase half up and clear of the lower crankcase. If the two surfaces are stuck together, tap the ends of the crankshaft with a soft head mallet to jar the sealing surfaces loose. Remove all traces of sealant from both mating surfaces.

65- Lift the retaining ring free of the crankshaft worm gear, and then lift the crankshaft out of the lower crankcase.

66- From the forward -- tapered end of the crankshaft -- remove the following in

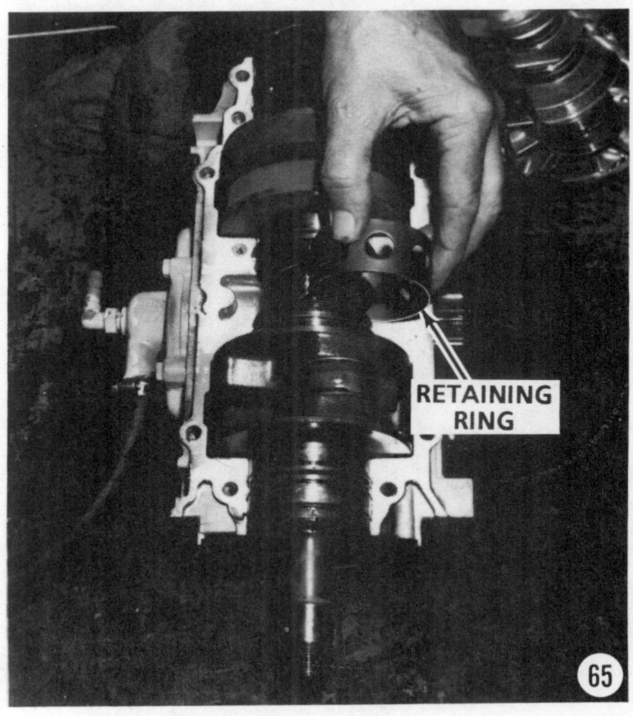

66

The rotary valve disc, attached to the rotary valve shaft, is driven directly off the engine driveshaft through a gear arrangement.

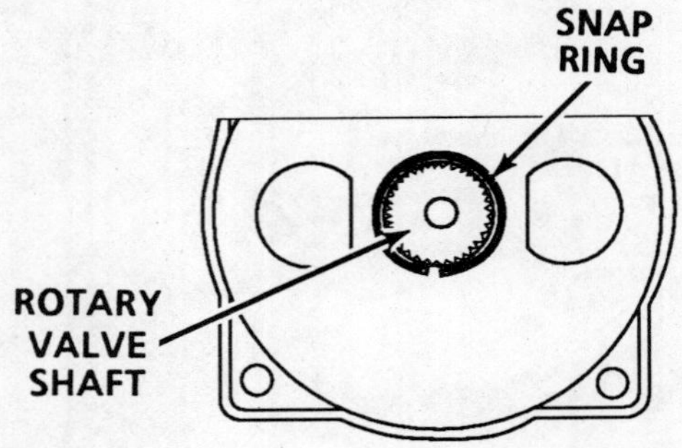

67

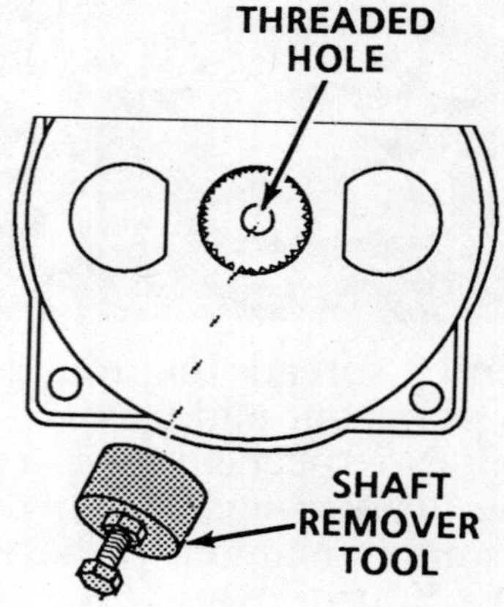

68

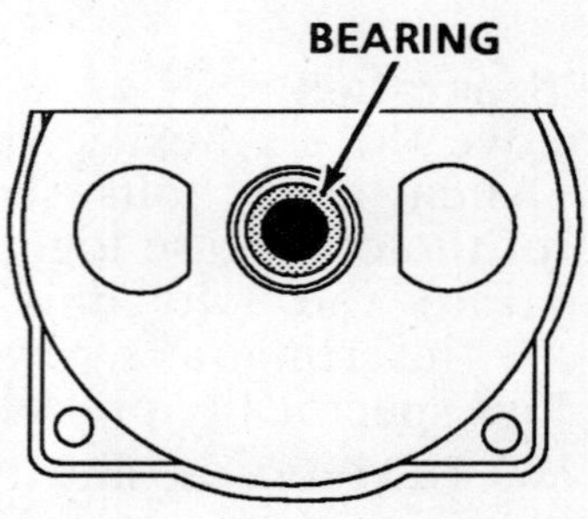

69

order: oil seal, ball bearing with an **O**-ring, spacer shim material, another ball bearing with an **O**-ring, and finally a "distance" ring. Keep these items in order on the work bench to **ENSURE** they will be installed back on the crankshaft in the proper order.

The ball bearings may or may not be a slide fit. If not, they can be pressed free with an arbor press, if they are to be replaced.

Remove the following in order from the aft end of the crankshaft: retaining shim material, oil seal, two ball bearings with **O**-rings, and a "distance" ring.

The connecting rods and rotary valve worm gear are an integral part of the crankshaft and cannot be removed. If they are no longer fit for service, the entire crankshaft **MUST** be replaced.

ROTARY VALVE SERVICE

SAFETY WORDS

Snap rings and C-lockrings are made of spring steel and may pop out of the groove with considerable force. Therefore, warn others in the area and **WEAR** eye protection glasses or a shield while removing this type ring.

67- Use a pair of external snap ring pliers and **CAREFULLY** remove the snap ring from the groove in the rotary valve shaft.

68- Obtain Rotary Valve Shaft Remover P/N 290-876-487. Slide the tool over the shaft and thread the bolt on the tool into the small threaded hole in the end of the shaft. Hold the nut welded onto the tool with the proper size wrench, and then rotate the bolt **CLOCKWISE** with another wrench until the shaft is withdrawn from the lower crankcase.

69- If the rotary valve shaft bearing on the port side of the lower crankcase need to be replaced, obtain a slide hammer with puller jaws attachment. Insert the puller jaws through the opening in the starboard side lower crankcase and hook the jaws around the back of the bearing. Pull the bearing free. **DO NOT** disturb the plug behind the bearing in the port side lower crankcase.

A quick crankshaft alignment check can be made by simply inserting a finger through each connecting rod, lifting the crankshaft to clear the work surface, and then attempting to "rotate" the crankshaft while it is suspended. If the crankshaft has the slightest amount of "twist", it will rotate in an out-of-round motion.

8-4 CLEANING AND INSPECTING

The success of the overhaul work is largely dependent on how well the cleaning and inspecting tasks are completed. If some parts are not thoroughly cleaned, or if an unsatisfactory unit is allowed to be returned to service through negligent inspection, the time and expense involved in the work will not be justified with peak engine performance and long operating life. Therefore, the procedures in the following sections should be followed closely and the necessary work performed with patience and attention to detail.

Crankshaft Service

Clean the crankshaft with solvent and wipe it dry with a lint free cloth. Inspect the connecting rod for cracks, or scratches. Inspect the crankshaft oil seal surface for nicks, sharp edges or burrs which might damage the oil seal during installation or might cause premature seal wear. **ALWAYS** handle the crankshaft carefully to avoid damaging the highly finished bearing surfaces. Blow out all oil passages with compressed air. The oil passageway leads from the rod to the main bearing journal. **TAKE CARE** not to blow dirt into the main bearing journal bore.

A quick check for crankshaft alignment can be made by inserting finger into each connecting rod and attempting to "rotate" the crankshaft in the air. A twisted crankshaft will twist and turn in an out-of-round motion.

Inspect the threads at both ends for signs of abnormal wear.

The connecting rod axial "play" is checked at the piston end to determine the amount of wear at the crankshaft end of the connecting rod. Checking the "play" is described in the following paragraph.

Set up the crankshaft in the **"V"** blocks. Set up a dial indicator to make contact with the flat surface of the piston end of the rod. Now, hold the crankshaft steady in the **"V"** blocks and at the same time, rock the piston end of the rod along the same axis as the crankshaft. If the dial indicator needle moves through more then 0.047" (1.2mm), the "play" is considered excessive. The connecting rods are an integral part of the crankshaft and **CANNOT** be removed. If they are no longer fit for service, the entire crankshaft must be replaced.

Inspect the crankshaft oil seal surfaces to be sure they are not grooved, pitted, or scratched. Replace the crankshaft if it is severely damaged or worn. Check all crankshaft bearing surfaces for rust, water marks, chatter marks, uneven wear or overheating. Clean the crankshaft surfaces with 320-grit carborundum cloth. **NEVER** spin-dry a crankshaft ball bearing with compressed air.

Clean the crankshaft and crankshaft ball bearing with solvent. Dry the parts, but not the ball bearing, with compressed air. Check the crankshaft surfaces a second time. Replace the crankshaft if the surfaces cannot be cleaned properly for satisfactory service. If the crankshaft is to be installed for service, lubricate the surfaces with light oil. **DO NOT** lubricate the crankshaft ball bearing, at this time.

After the crankshaft has been cleaned, grasp the outer race of the crankshaft ball bearing installed on each end of the crankshaft, and attempt to work the race back-and-forth. There should not be excessive "play". A very slight amount of side "play" is acceptable because there is only about 0.001" (0.025mm) clearance in the bearing.

Lubricate the ball bearing with light oil. Check the action of the bearing by rotating the outer bearing race. The bearing should have a smooth action and no rust stains. If the ball bearing sounds or feels rough or catches, the bearing should be removed and discarded.

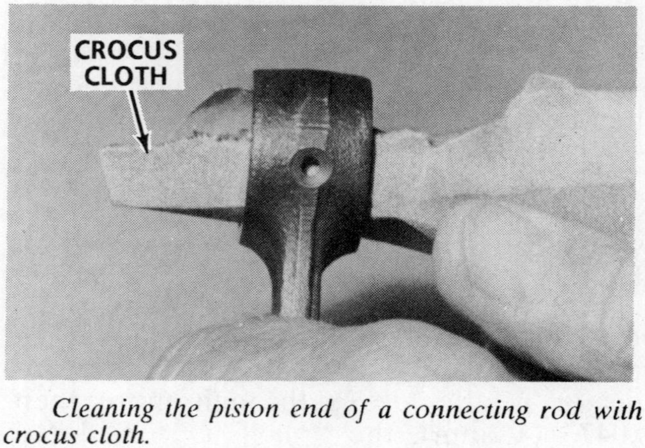

Cleaning the piston end of a connecting rod with crocus cloth.

Connecting Rod Service

Inspect the connecting rod bearings for rust or signs of bearing failure. **NEVER** intermix new and used bearings. If even one bearing in a set needs to be replaced, all bearings at that location **MUST** be replaced.

Clean the inside diameter of the piston pin end of the connecting rod with crocus cloth.

Clean the connecting rod **ONLY** enough to remove marks. **DO NOT** continue once the marks have disappeared.

Assemble the piston end of the connecting rod with loose needle bearings. Insert the piston pin and check for vertical "play". The piston pin should have **NO** noticeable vertical "play".

If the pin is loose or there is vertical "play", check for and replace the worn part/s.

Inspect the piston pin and matching rod end for signs of heat discoloration. Overheating is identified as a bluish bearing surface color and is caused by inadequate lubrication or by operating the engine at excessive high rpm.

Piston Service

Inspect each piston for evidence of scoring, cranks, metal damage, cracked piston pin boss, or worn pin boss. Be especially critical during inspection if the craft and engine has been submerged. If the piston pin is bent, the piston and the pin **MUST** be replaced as a set for two reasons. First, a bent pin will damage the boss when it is removed. Secondly, a piston pin is not sold as a separate item.

Checking the piston end of a connecting rod for vertical free "play" using a piston pin.

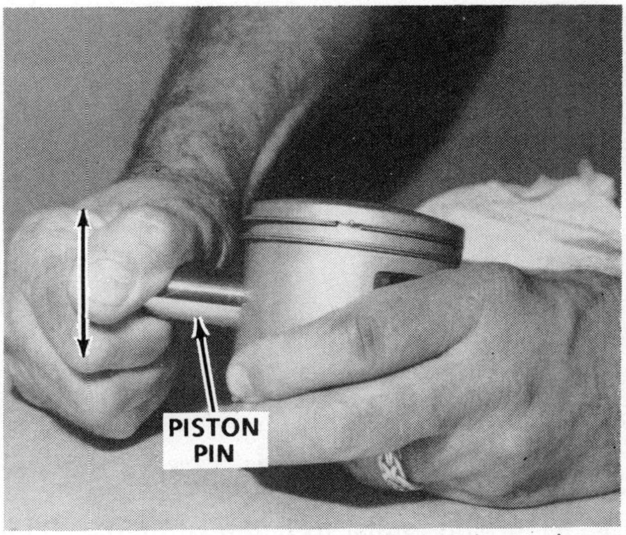

Checking free "play" between the piston pin and the piston boss. There should be NO "play".

Locating pins prevent the rings from rotating and catching on a port in the cylinder wall.

Check the piston ring grooves for wear, burns, distortion, or loose locating pins. During an overhaul, the rings should be replaced to **ENSURE** lasting repair and proper engine performance after the work is completed. Clean the piston dome, ring grooves and the piston skirt. Clean carbon deposits from the ring grooves using the recessed end of a broken piston ring.

NEVER use a rectangular ring to clean the groove for a tapered ring, or use a tapered ring to clean the groove for a rectangular ring.

NEVER use an automotive type ring groove cleaner, because such a tool may loosen the piston ring locating pins.

Clean carbon deposits from the top of the piston using a soft wire brush, carbon removal solution or by sand blasting. If a wire brush is used, **TAKE CARE** not to burr or round machined edges. Clean the piston skirt with crocus cloth.

Install the piston pin through the first boss only. Check for vertical free "play". There should be **NO** vertical free "play". The presence of "play" is an indication the piston boss is worn. The piston is manufactured from a softer material than the piston pin. Therefore, the piston boss will wear more quickly than the pin.

Excessive piston skirt wear **CANNOT** be visually detected. Therefore, good shop practice dictates the piston skirt diameter be measured with a micrometer.

Piston diameters are measured at a definite distance up from the bottom of the skirt at right angles to the piston pin axis.

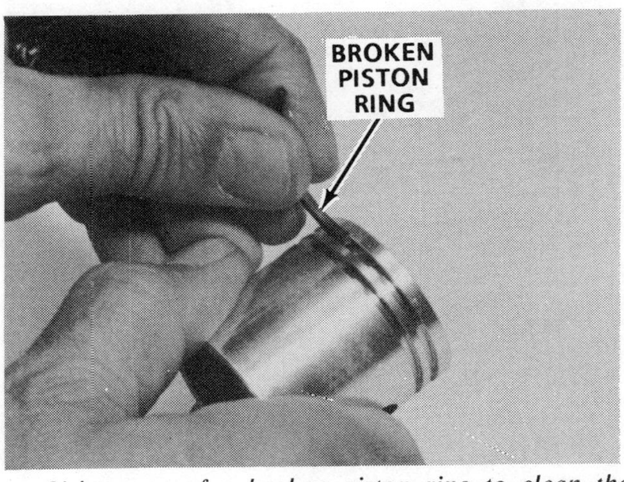

*Using part of a broken piston ring to clean the ring groove. Exercise **CARE** not to disturb the locating pin in each groove.*

It is believed, this crown seized with the cylinder wall when the unit was operated at high rpm and the timing was not adjusted properly. At the same instant, the rod apparently pulled the lower part of the piston downward, severing it from the crown.

The pitted damage to this piston crown was probably caused by a broken piston ring working its way into the combustion chamber. The little "hills" then became "hot" spots on the crown, contributing to "dieseling" after the engine was shut down.

Size limits are as follows:

Std. - New	2.990"	(75.94mm)
1st Oversize	3.000"	(76.19mm)
2nd Oversize	3.020"	(76.44mm)

Ring End Gap

Before the piston rings are installed onto the piston, the ring end gap clearance

The rings on this piston were broken, possibly during installation, and then caused extensive damage to the piston. Ring parts found their way into the combustion chamber and caused damage to the piston crown.

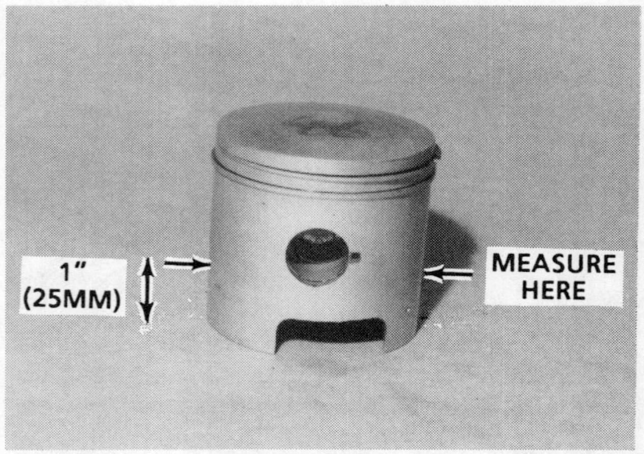

Measure the piston diameter at the specified distance above the bottom of the piston skirt.

for each ring must be determined. The purpose of the piston rings is to prevent the blowby of gases in the combustion chamber. This cannot be achieved unless the correct oil film thickness is left on the cylinder wall. This thin coating of oil acts as a seal between the cylinder wall and the face of the piston ring. An excessive end gap will allow blowby and the cylinder will lose compression. An inadequate end gap will scrape too much oil from the cylinder wall and limit lubrication. Lack of adequate lubrication will cause excessive heat and wear.

IDEALLY, the ring end gap measurement should be taken **AFTER** the cylinder bore has been measured for wear and taper **AND** after any corrective work, such as boring or honing, has been completed.

The rings on this piston became stuck due to lack of adequate lubrication, incorrect timing, or overheating.

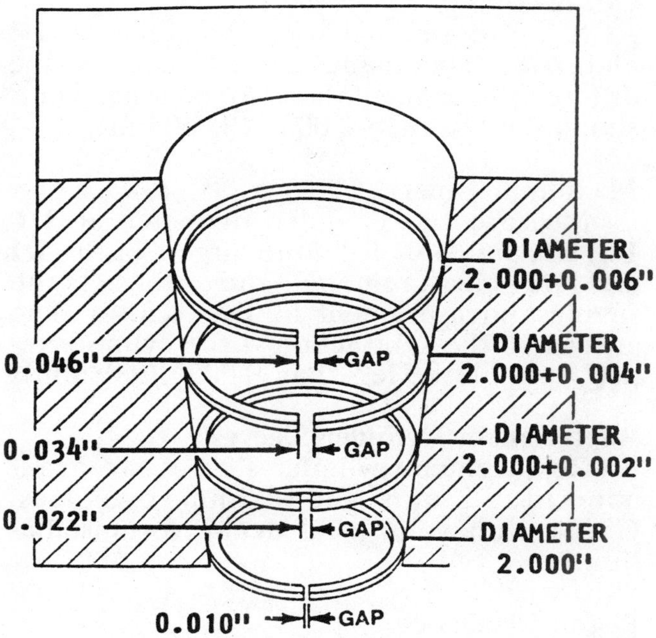

The cylinder taper drastically affects ring and end gap, as shown in this cross-section line drawing.

If the ring end gap is measured with a taper to the cylinder wall, the diameter at the lower limit of the ring travel will be smaller than the diameter at the top of the cylinder.

If the ring is fitted to the upper part of a cylinder with a taper, the ring end gap will not be great enough at the lower limit of ring travel. Such a condition could result in a broken ring and/or damage to the cylinder wall and/or damage to the piston and/or damage to the cylinder head.

If the cylinder is to be only honed, not bored, **OR** if only cleaned, not honed, the ring end gap should be measured at the lower limit of ring travel.

To perform the end gap measurement, the piston crown is used to insert the ring into the cylinder just above the ports, as indicated in the accompanying illustration.

Be sure the ring is parallel to the surface of the block, then measure the ring end gap with a feeler gauge.

If the end gap is greater than the amount listed, replace the entire ring set.

If the end gap is less than 0.010-0.016" (0.25-0.40mm) for both top and bottom rings, carefully file the ends of the ring -- just a little at a time -- until the correct end gap is obtained. If the gap is greater than 0.39" (1.00mm), the ring must be replaced. Remember, if one ring requires replacement, the set **MUST** be replaced.

Inspect the piston ring locating pins to be sure they are tight. There is one locating pin in each ring groove. If the locating pins are loose, the piston **MUST** be replaced.

Ring-To-Groove Clearance

This clearance only applies to the lower rectangular piston ring. Measure the clearance between the bottom of the ring and the groove. The clearance must not exceed 0.008" (0.2mm). If the clearance is greater than specified, the piston **MUST** be replaced.

The piston is manufactured from softer material than the ring. Therefore, the wear will occur at the piston rather than the ring. As a general rule of good shop practice, if the piston is replaced, a new set of rings are installed on the new piston.

Oversized Pistons and Rings

Scored cylinder blocks can be saved for further service by reboring and installing oversize pistons and piston rings. **HOWEVER**, if the scoring is over 0.0075" (0.13mm) deep, the block cannot be effectively rebored for continued use.

Check with the local dealer for oversize piston availability.

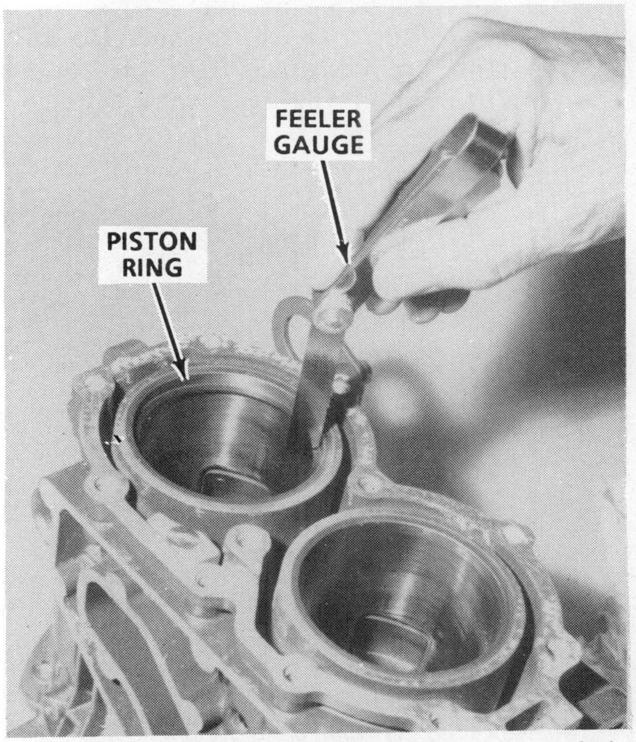

Measuring the piston ring end gap with a feeler gauge, as described in the text.

If oversize pistons are not readily available, the local marine shop may have the facilities to "knurl" the piston, making it larger.

Cylinder Block Service

Inspect the cylinder block and cylinder bores for cracks or other damage. Remove carbon with a fine wire brush on a shaft attached to an electric drill or use a carbon remover solution.

Use an inside micrometer or telescopic gauge and micrometer to check the cylinders for wear. Check the bore for out-of-round and/or oversize bore. If the bore is tapered, out-of-round, or worn more than the wear limit specified by the manufacturer, the cylinder should be rebored -- provided oversize pistons and rings are available.

Check with the local dealer **PRIOR** to reboring. If oversize pistons and matching ring sets are not available, the block **MUST** be replaced or re-sleeved.

Four inside cylinder bore measurements must be taken for each cylinder to determine an out-of-round condition: the maximum taper, and the maximum bore diameter.

In the accompanying illustration, measurements **D1** and **D2** are diameters measured at the top of the cylinder at right angles to each other. Measurements **D3** and **D4** are diameters measured from the top of the cylinder at right angles to each other.

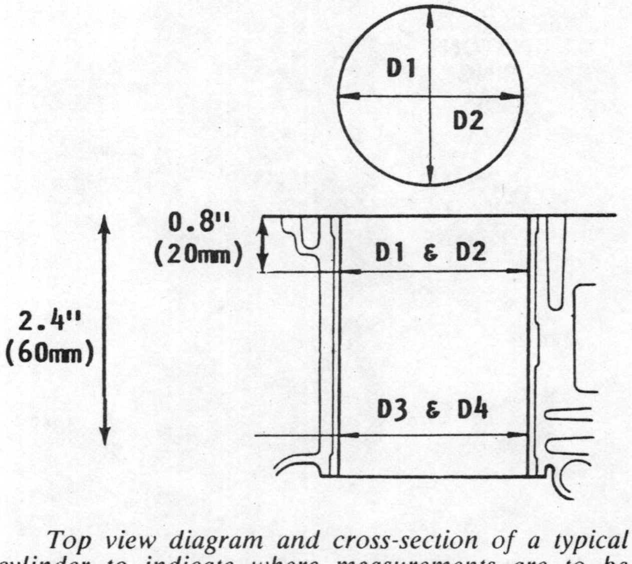

Top view diagram and cross-section of a typical cylinder to indicate where measurements are to be taken for wear limit, taper, and out-of-round limits.

Out-of-Round

Measure the cylinder diameter at **D1** and **D2.** The manufacturer requires the difference between the two measurements should be less than 0.002" (0.050mm).

Maximum Taper

Measure the cylinder diameter at **D1, D2, D3,** and **D4.** Take the largest of the **D1** and **D2** measurements and subtract the smallest measurement at **D3** or **D4.** The answer to the subtraction -- the cylinder taper -- should be less than 0.003" (0.08mm).

Replacement Cylinder Sleeves

Replacement cylinder sleeves are available in 0.25mm and 0.5mm oversizes. Check with your local dealer for installation.

Piston Clearance

Piston clearance is the difference between a maximum piston diameter and a minimum cylinder bore diameter. If this clearance is excessive, the engine will de-

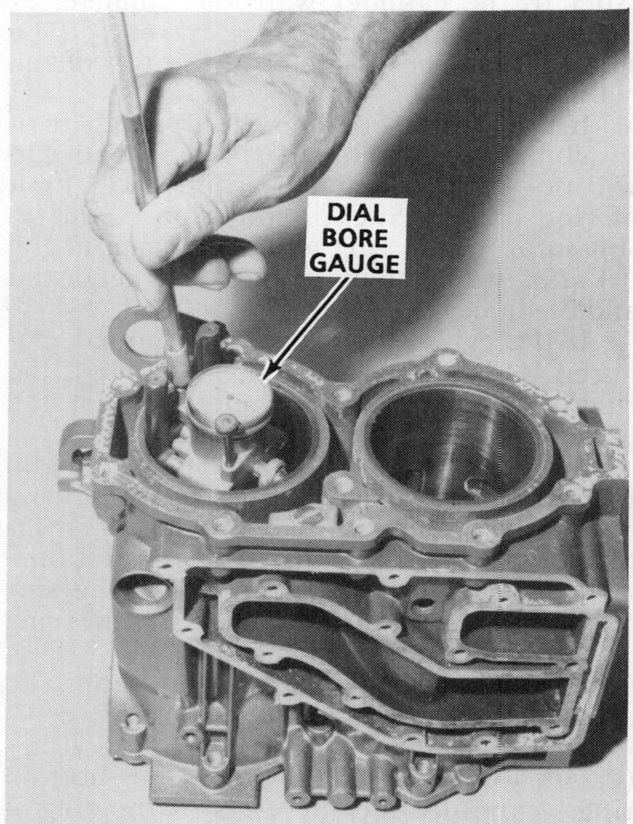

Checking the cylinder for taper using an inside micrometer. One measurement should be taken near the top and another near the bottom of the cylinder. The difference between the two measurements is the amount of taper.

velop the same symptoms as for excessive cylinder bore wear -- namely loss of ring sealing ability, loss of power, increased engine noise, unnecessary vibration, and excessive oil consumption.

Maximum piston diameter was described earlier in this section. Minimum cylinder bore diameter is usually determined by measurement **D3** or **D4,** also described earlier in this section.

If the piston clearance exceeds 0.008" (0.2mm), either the piston or the cylinder block **MUST** be replaced.

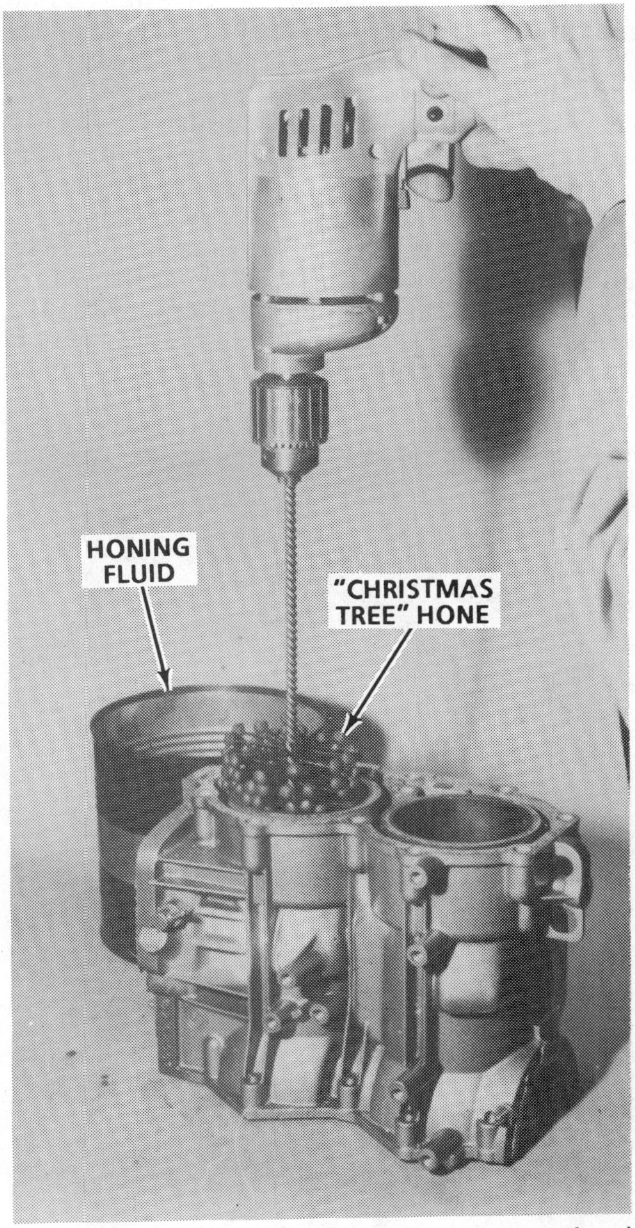

Refinishing the cylinder wall using an electric drill and "Christmas tree" hone. ***ALWAYS*** *keep the tool moving in long even strokes the entire cylinder depth. Use a continuous* ***LIBERAL*** *amount of honing fluid.*

HONING CYLINDER WALLS

Hone the cylinder walls lightly to seat the new piston rings, as outlined in this section. If the cylinders have been scored, but are not out-of-round, or the bore is rough, clean the surface of the cylinder with a cylinder hone as described in the following paragraphs.

SPECIAL WORDS

If overheating has occurred, check and resurface the spark plug end of the cylinder block, if necessary. This can be accomplished with 240-grit sandpaper and a small flat block of wood.

To ensure satisfactory engine performance and long life following the overhaul work, the honing work should be performed with patience, skill, and in the following sequence.

a- Follow the hone manufacturer's recommendations for use of the hone and for cleaning and lubricating during the honing operation. A "Christmas Tree" hone may also be used.

b- Pump a continuous flow of honing oil into the work area. If pumping is not practical, use an oil can. Apply the oil generously and frequently on both the stones and the work surface.

c- Begin the stroking at the smallest diameter. Maintain a firm stone pressure against the cylinder wall to assure fast stock removal and accurate results.

The walls of this cylinder were damaged beyond repair when a piston ring broke and worked its way into the combustion chamber.

d- Expand the stones as necessary to compensate for stock removal and stone wear. The best cross-hatch pattern is obtained using a stroke rate of 30 complete cycles per minute. Again, use the honing oil generously.

e- Hone the cylinder walls **ONLY** enough to de-glaze the walls.

f- After the honing operation has been completed, clean the cylinder bores with hot water and detergent. Scrub the walls with a stiff bristle brush and rinse thoroughly with hot water. The cylinders **MUST** be thoroughly cleaned to prevent any abrasive material from remaining in the cylinder bore. Such material will cause rapid wear of new piston rings, the cylinder bore, and the bearings.

g- After cleaning, swab the bores several times with engine oil and a clean cloth, and then wipe them dry with a clean dry cloth. **NEVER** use kerosene or gasoline to clean the cylinders.

h- Clean the remainder of the cylinder block to remove any excess material spread during the honing operation.

SPECIAL CRANKCASE WORDS

The upper and lower crankcase halves are matched at the factory. They are not interchangeable and are not available as separate items -- only as a matched set.

This cylinder head was removed from the same engine as the pitted piston shown on Page 8-26. In order to return this head to a serviceable condition, the combustion chamber surface must be refinished to a smooth surface using a stone. If left in its present condition, the sharp corners of the gouges will become "hot" spots resulting in pre-ignition and possible "dieseling" after the engine is shut down.

Rotary Valve

Inspect the teeth on the rotary valve shaft gear. Any wear pattern -- there probably will be one -- should be even on all teeth.

Check for the presence of brass filings in the bottom of the lower crankcase. Install the bearing (if it was removed), and the rotary valve shaft into the lower crankshaft. Make a setup with a dial indicator with the probe making contact on the shaft, as shown in the accompanying illustration. Now, rotate the shaft and measure the maximum deflection of the shaft. The maximum deflection should not exceed 0.003" (0.08mm). If a greater deflection is indicated, replace the shaft.

Hold the rotary valve housing -- without its sealing ring in place -- against the lower crankcase. Measure the clearance between the housing and the crankcase, with a feeler gauge. The measured clearance should be 0.012-0.020" (0.3-0.5mm). Take a clearance measurement with the upper crankcase. If the measured clearance is not within the limits given, replace the rotary valve housing.

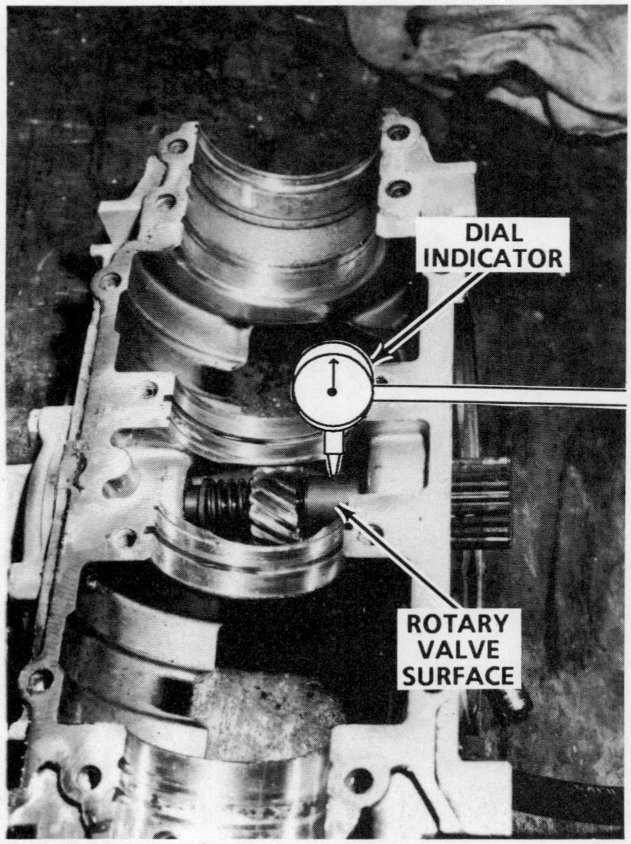

Checking the rotary valve shaft deflection, as explained in the text.

8-5 SEALANTS, ADHESIVES, LUBRICANTS, AND FUEL STABILIZERS

It is common practice for the larger manufacturers of personal watercraft to market their own line of products for use on their craft. Bombardier chemical engineers have developed such a line available for use with Bombardier personal watercraft and snowmobiles. Throughout this manual, the authors recommend the application of the manufacturer's products as a first choice. All products listed are alternatives of equal value, and may be used with confidence, if the manufacturer's line is not available.

Sealants and Adhesives

Many sealants and adhesives are recommended and they are **NOT** interchangeable. Each is designed to perform under a different set of conditions. Follow the directions on the package for cleaning and preparing surfaces. Sealants and adhesives **MUST** be applied **ONLY** to clean dry surfaces. Apply sparingly -- excessive amounts may block oil passageways and cause serious damage.

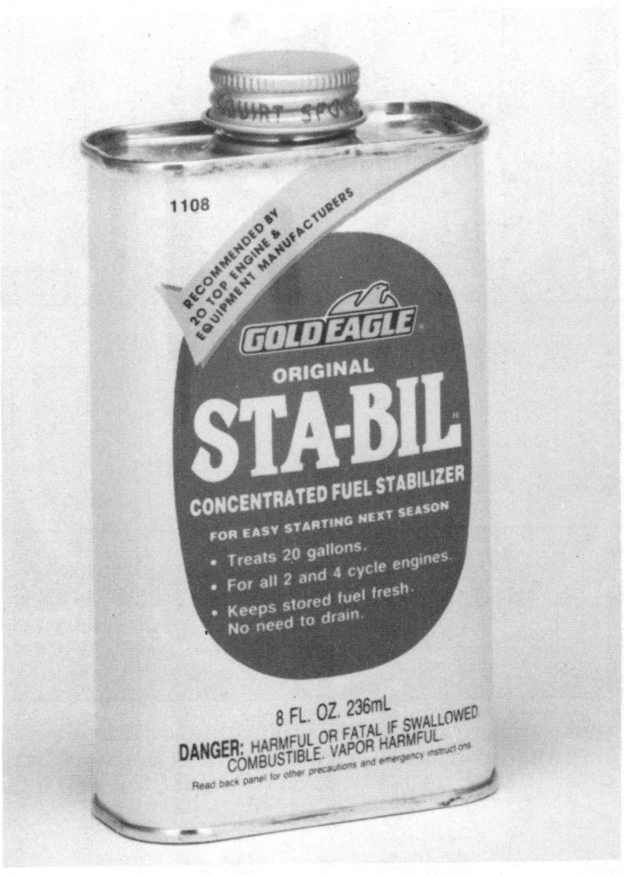

STA-BIL Concentrated Fuel Stabilizer -- a very effective additive -- may be used to prevent fuel from "souring" for up to twelve full months.

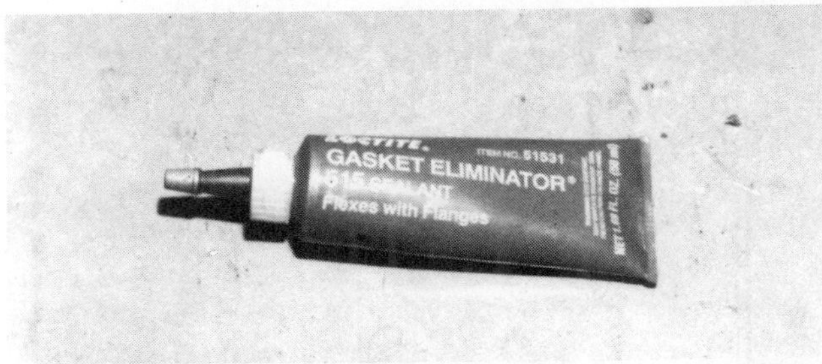

The manufacturer recommends, and the authors instruct, the use of numerous Loctite products during engine assembly, in fact, throughout this manual.

The accompanying table lists all the required products to lubricate and correctly seal any work or maintenance performed on the craft.

Fuel Stabilizer

"Sta-Bil" Fuel Conditioner and Stabilizer is recommended during engine operation and during the storage period. This fluid absorbs water in the fuel system and protects against corrosion.

If used during operation, this fuel additive will prevent the formation of gum and varnish deposits and greatly extend the period between required carburetor overhauls.

When added to the fuel during storage, the additive will prevent the fuel from "souring" for up to twelve full months.

Bombardier chemical engineers have developed their own line of lubricants and sealants for use with Bombardier watercraft and snowmobiles.

FACTORY RECOMMENDED LUBRICANTS AND ADHESIVES

PRODUCT	PART NUMBER	USE
Sea-Doo Lube	293-600-006	Spark plug hole threads
Sea-Doo Injection Oil	293-600-005	Oil injection system and rotary valve face.
Sea-Doo Jet pump Synthetic Oil	293-600-011	Jet pump end cap reservoir
Sea-Doo PL-32 Sealant	293-800-001	Outside diameter of all oil seals
Sea-Doo Bearing Grease	293-550-005	Jet Pump thrust bearing and washer
Loctite 157 Stripper	295-000-040	Anywhere Loctite 515 Gasket Eliminator was used.
Loctite 242 (blue)	293-800-002	Cranking motor, rock grate, steering nozzle, engine support, engine mount, exhaust manifold, flame arrestor, rotary valve, and stator plate mounting bolt threads. Also crankshaft taper and all electrical component mounting bolts.
Loctite 414 Superbond	N/A	Drive shaft plug
Loctite 515 Gasket Eliminator	293-800-007	Cylinder head, crankcase, and nozzle-to-impeller housing mating surfaces.
Loctite 517 or 587 Superflex Ultra Blue	293-800-016	Impeller housing mounting holes, ride shoe perimeter, and all oil seal lips.
Loctite 592	N/A	Exhaust outlet water fittings
Loctite PST 567	293-800-013	All other water fittings, and crankcase drain plug threads.
Loctite 648 (green)	290-899-788	Distance sleeve on rotary valve shaft
Loctite 767 Anti-seize	293-550-001	Crankshaft/flywheel threads, drive shaft and impeller shaft threads, and all bearing race exteriors.
Loctite Primer N	293-600-012	Internal aluminum threads
Loctite U Blue	293-800-006	All exposed cable threads
Silicone 732 RTV	N/A	Electrical harness grommets, hull screw threads, and wear ring lip.
Armour All ®	N/A	Vinyl seat

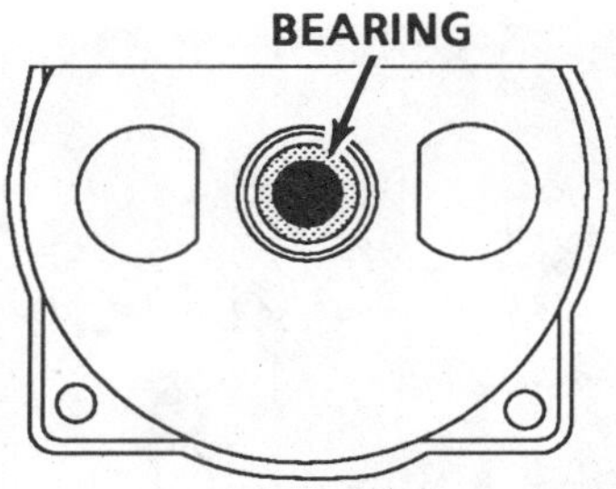

1

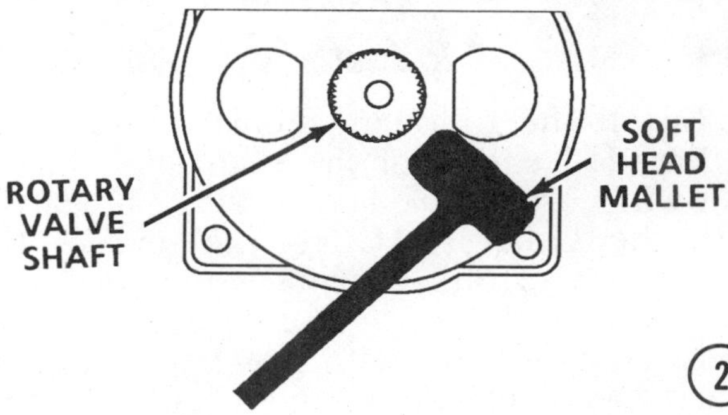

2

8-6 ASSEMBLING AND INSTALLATION

PLAN AHEAD

Apply a light coating of anti-sieze compound to the inner races of all four crankshaft bearings. Pack the lips of the two oil seals with Lithium lubricant. Use the same lubricant on the two O-rings. Immerse the

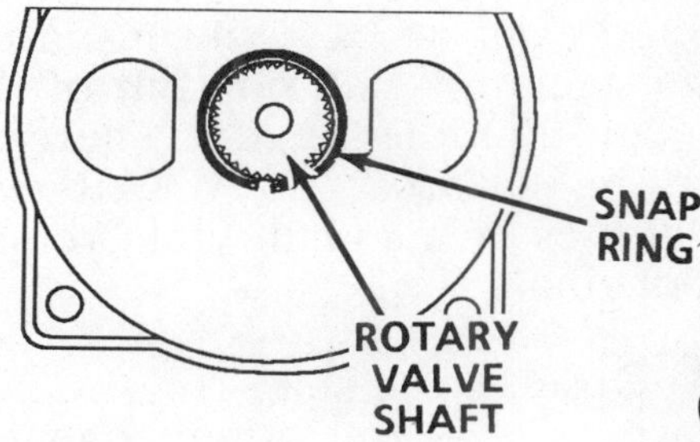

3

four bearings, with the O-rings installed, into a container of oil heated to 167°F (75°C). Heating the bearings in the hot oil will expand them ever so slightly and ease the task of installation. Keep the bearings in the heated oil until the work has progressed to the point the bearings are to be installed in Step 5.

Rotary Valve Shaft Installation

1- Apply water resistant lubricant to the rotary valve shaft ball bearing. Install the bearing with "open" side facing the plug in the crankcase and the "shielded" toward the center of the crankcase. Tap the bearing into place with a suitable driver.

2- Install the rotary valve shaft into the crankcase. Use a soft head mallet and gently tap the shaft into place until it seats.

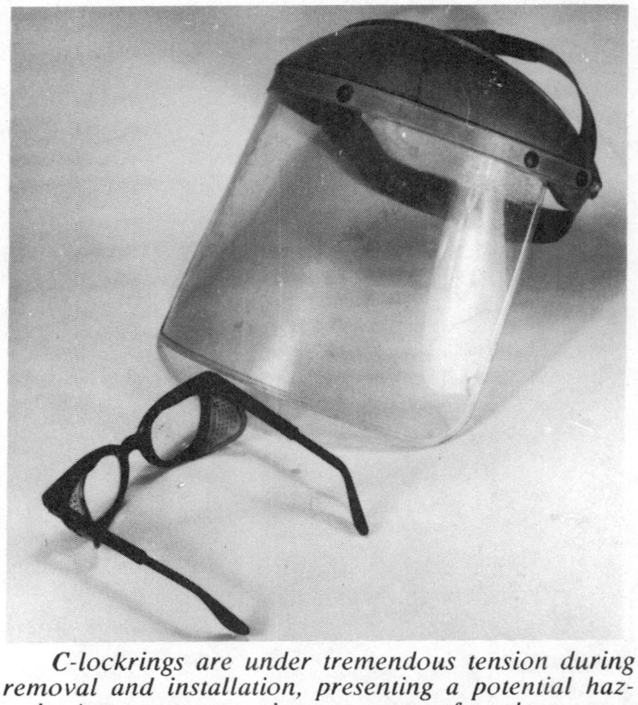

*C-lockrings are under tremendous tension during removal and installation, presenting a potential hazard. As an eye protection measure, safety glasses or a shield should **ALWAYS** be worn during work with such snap rings. Warn others in the area such work is in progress.*

4

SAFETY WORDS

Snap rings and C-lockrings are made of spring steel and may pop out of the groove with considerable force. Therefore, warn others in the area and **WEAR** eye protection glasses or a shield while removing this type of ring.

3- Observe the snap ring. One face has a sharp edge -- the other edge is smooth. Using a pair of snap ring pliers, install the ring, into the groove in the rotary valve shaft with the sharp edge facing **OUTWARD.**

4- Rotate the shaft through a complete revolution to **ENSURE** there is no evidence of binding.

Crankshaft Assembling And Installation

5- Install the following items onto the forward -- tapered -- end of the crankshaft in the following order: the distance ring, a bearing with the **O**-ring on the side closest to the tapered end of the crankshaft, a spacer shim material piece, a second bearing with the **O**-ring on the side closest to the center of the crankshaft, and finally an oil seal with the lip **FACING** the bearing just installed.

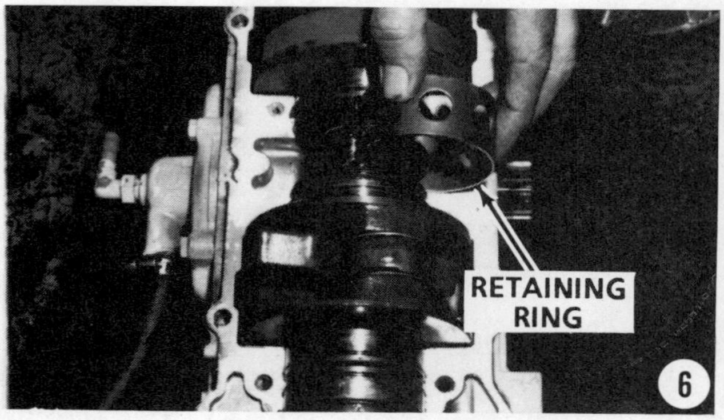

Install the following items onto the aft -- threaded end -- of the crankshaft in the following order: a distance ring, a bearing with the **O**-ring **FACING** the end of the crankshaft, second bearing with the **O**-ring on the side closest to the bearing just installed, an oil seal with the lip facing the bearing just installed, and finally the retainer shim material piece.

The two sets of **O**-rings should be as close to each, as possible.

The outer bearing on the aft end of the crankshaft **MUST** be spaced a specified distance from the other bearing. Obtain Distance Gauge P/N 290-876-826 and insert it between the two bearings prior to installation of the crankshaft. Remove this gauge after the crankshaft is seated.

Position the assembled crankshaft into the lower crankcase with the worm gear on the crankshaft meshing with the gear on the rotary valve shaft.

6- Install the retaining ring over the worm gear on the crankshaft.

Crankcase Assembling

7- Apply a light coating of Loctite Primer N onto the mating surfaces of the

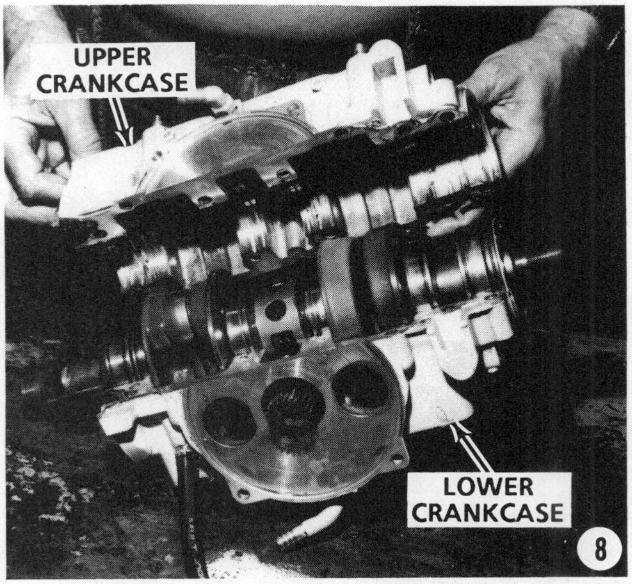

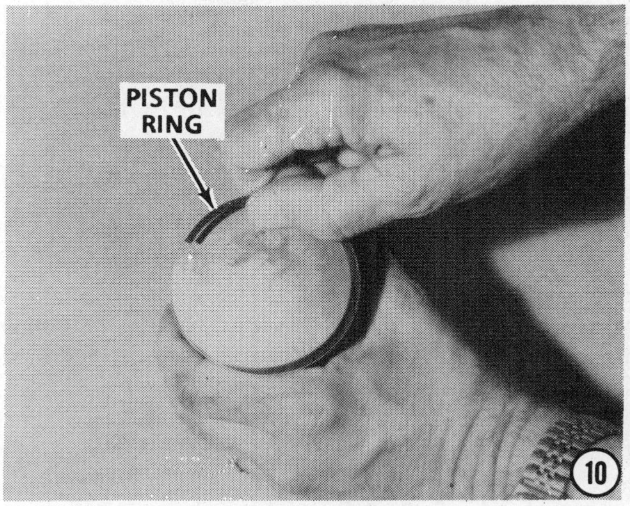

two crankcase halves, working around the bolt holes. Allow about twenty minutes for the primer to dry. Apply a coating of Loctite 515 gasket eliminator on top of the primer. Wipe away any excess Loctite.

8- Bring the two halves of the crankcase together, with the bolt holes aligned as accurately as possible.

9- Apply a coating of Loctite 242 to the threads of the six hex bolts and the eight Allen head bolts. Install all the bolts and tighten them just **HAND TIGHT**, at this time.

Rotate the crankshaft by hand to verify it turns freely with no evidence of binding. If any binding is felt, the crankcase must be opened and the source of the binding discovered quickly -- before the Loctite sets.

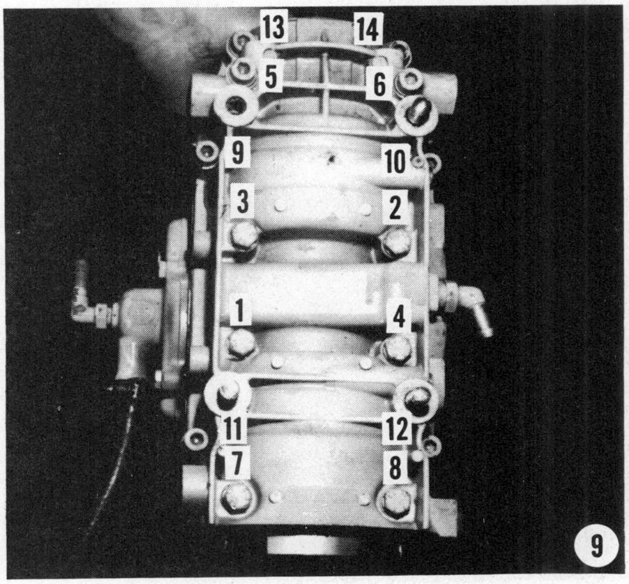

Once the determination is made the crankshaft is rotating freely, tighten the six hex head bolts to a torque value of 16 ft lb (22Nm) and the eight Allen head bolts to a torque value of 80 in lb (9Nm), all in the sequence indicated in the accompanying illustration.

Piston Installation

10- No special tool is necessary to install the piston rings. **HOWEVER**, take care to spread the ring only enough to clear the top of the piston. The rings are **EXTREMELY** brittle and will snap if spread beyond their limit.

Install a new set of piston rings onto each piston, with the rectangular ring in the lower groove. Align the ring end gap over the locating pin in each piston groove.

The purpose of the pin in the groove is to prevent the ring from rotating in the

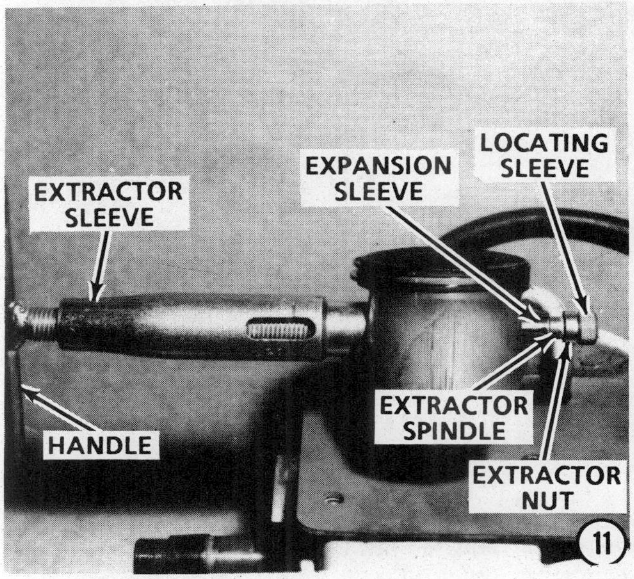

ring groove. The ring end gap must travel up and down a portion of the cylinder wall where there is no port opening. If the ring were free to rotate, an end might catch on the edge of a port opening and break.

SPECIAL PISTON INSTALLATION WORDS

Observe the intake hole in the side of the piston. The piston **MUST** be installed onto the connecting rod with the intake hole on the same side as the rotary valve housing -- the starboard side.

Two methods for installing pistons are presented. Perform Step 11, if the factory special tool is available. Perform Steps 12, 13, and 14, if the special tool is not available.

First, position the rubber mat or shop cloths over the crankcase to prevent needle bearings from being lost.

11- Dip the expansion sleeve, with needle bearings inside into a container of two-stroke engine oil. Insert the expansion sleeve between the piston pin bosses inside the piston. Place a thrust washer on each side of the sleeve. Insert the piston pin into the piston until it is flush with the inner edge of the piston pin boss. Insert the extractor spindle into the pin and install the locating sleeve and the extractor nut.

Rotate the handle **COUNTERCLOCKWISE** to pull the piston pin into the thrust washer, the needle bearings, the second thrust washer, and the piston pin boss. The pin will drive out the expansion sleeve as the pin is pushed in.

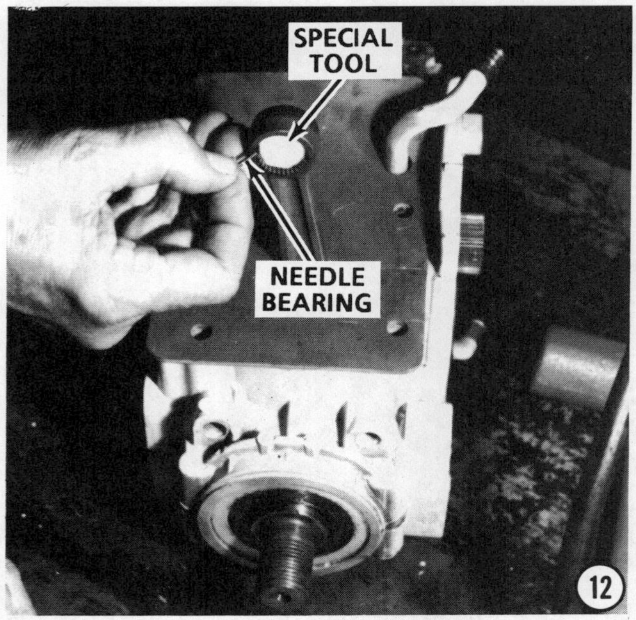

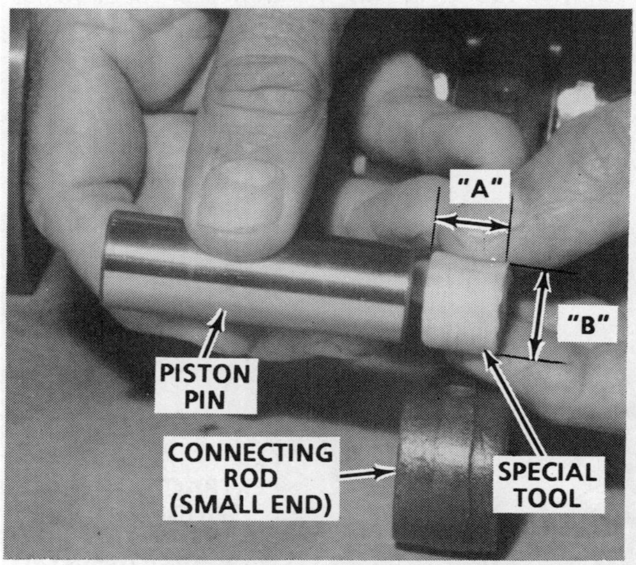

Length "A" must be just a "whisker" smaller than the length of the connecting rod small end. Diameter "B" must be just a "whisker" smaller than the diameter of the piston pin.

Center the pin in the boss. The **C**-lockring grooves **MUST** be visible on both sides of the piston. Disassemble and remove the installation tools.

HOMEMADE TOOL SPECIAL WORDS

A tool must be made to keep the needle bearings in place and prevent them from falling from the perimeter of the small end of the connecting rod, while the piston pin is installed.

Find a bottle cap with straight sides which has the following dimensions: length of cap must be a "whisker" smaller than the length of the connecting rod hole. The diameter of the cap must be a "whisker" smaller than the diameter of the piston pin, as shown in the accompanying illustration.

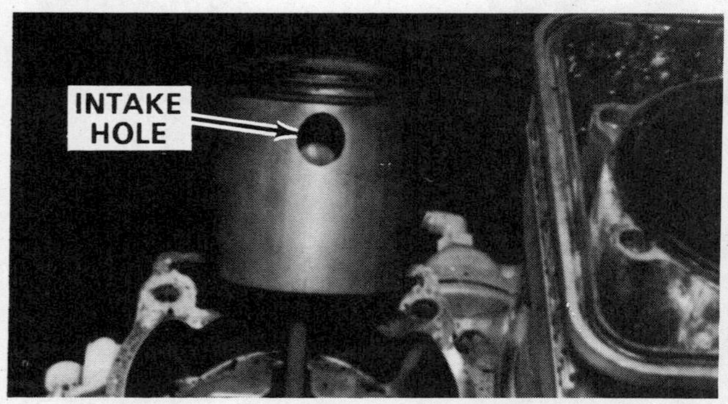

*The piston **MUST** be installed with the intake hole on the starboard side.*

Tape, wound around the cap, can help to increase the diameter of the cap and make it the desired size.

12- Select a set of 31 needle bearings. Cover the inner circumference of the connecting rod with two-stroke engine oil. Begin to position the needle bearings -- one by one -- around the circumference. After about half the needle bearings are in place, slide the special homemade tool inside the semicircle of needle bearings. Continue inserting the needle bearings into the rod end until all 31 are in place.

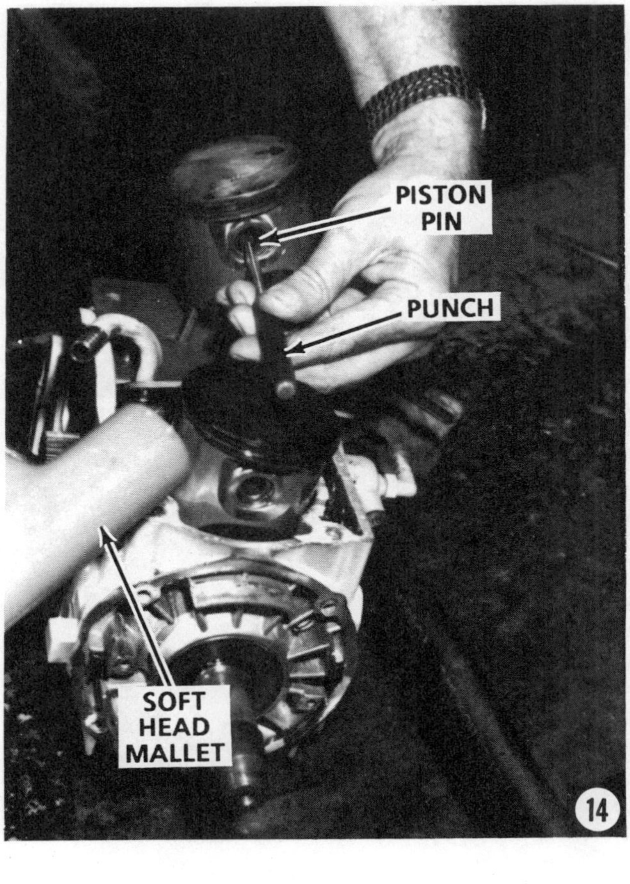

13- Insert the piston pin through the piston just enough to hang a thrust washer on the end of the pin.

14- Install the piston over the end of the connecting rod. **TAKE EXTRA CARE** not to disturb the needle bearings and thrust washer. Check to be sure the intake hole in the piston is on the same side as the rotary valve housing -- the starboard side.

Insert the second thrust washer between the connecting rod end and the other boss. Slide the piston pin through the connecting rod end, thrust washer, and piston pin boss. As the piston pin slides through these parts, the pin will push the special homemade tool ahead of it. Center the piston pin by tapping it with a small punch and soft head mallet, until it is centered in the piston, with both **C**-lockring grooves visible at both ends.

SAFETY WORDS

Snap rings and **C**-lockrings are made of spring steel and may pop out of the groove with considerable force. Therefore, warn

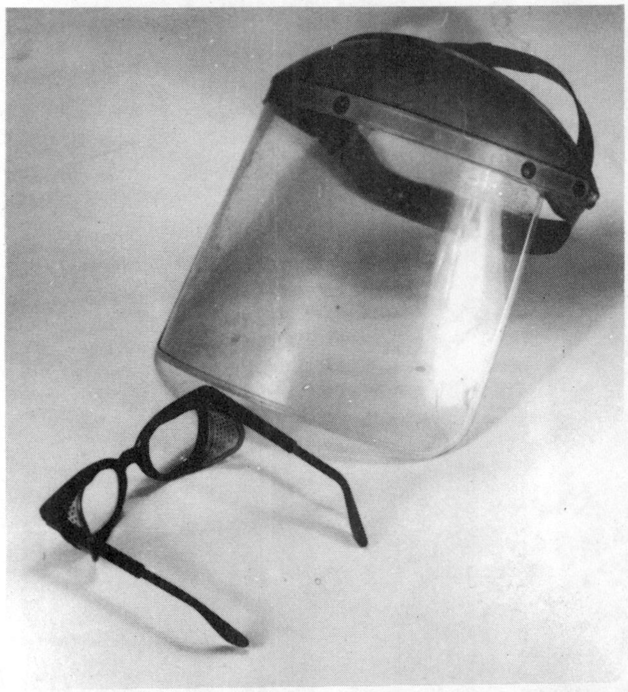

*C-lockrings are under tremendous tension during removal and installation, presenting a potential hazard. As an eye protection measure, safety glasses or a shield should **ALWAYS** be worn during work with such snap rings. Warn others in the area such work is in progress.*

others in the area and **WEAR** eye protection glasses or a shield while removing this type of ring.

15- Install a **C**-lockring in the groove at each end of the piston pin. The lockrings may be worked in **CAREFULLY,** by hand. Place the end gap of the **C**-locking ring at the bottom -- the 6 o'clock position. Remove the rubber pad and shop cloths.

Cylinder Installation

16- Place a new cylinder gasket over the piston and onto the top surface of the upper crankcase. Coat the crown and upper portion of the piston sides and the rings with two-stroke engine oil. Check to be sure the ring end gap straddles the piston locating pin in each groove.

Any one of three methods may effectively be used to install the pistons into the cylinder.

a- Obtain Ring Compressor P/N 290-876-972. **UNFORTUNATELY**, this special tool cannot be used to install oversize pistons. Slide the compressor over the piston, with the tapered end going on first. Hold the piston upright,

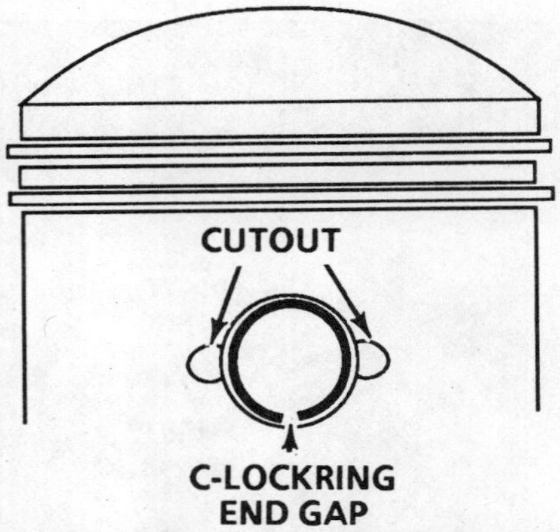

The C-lockring end gap must be positioned away from the two cutout in the piston, as indicated in this line drawing.

and slowly lower a cylinder down over the piston crown and rings, with the cylinder exhaust port on the opposite side to the rotary valve housing.

b- Obtain a large hose clamp and secure it over the piston rings to compress them. Hold the piston upright, and slowly lower a cylinder down over the piston crown and rings, with the cylinder exhaust port on the opposite side to the rotary valve housing.

c- Obtain the services of an assistant to compress the rings while the cylinder is lowered. Hold the piston upright, and

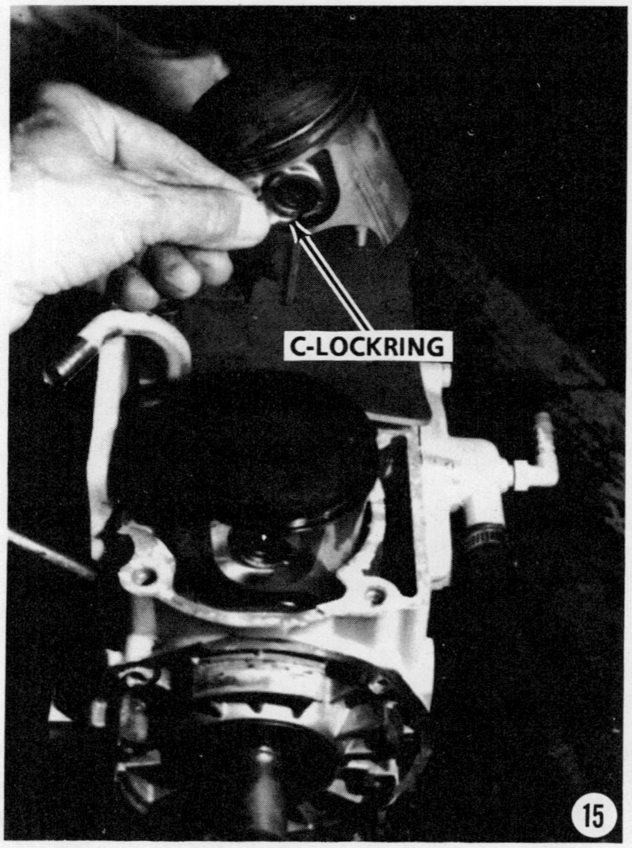

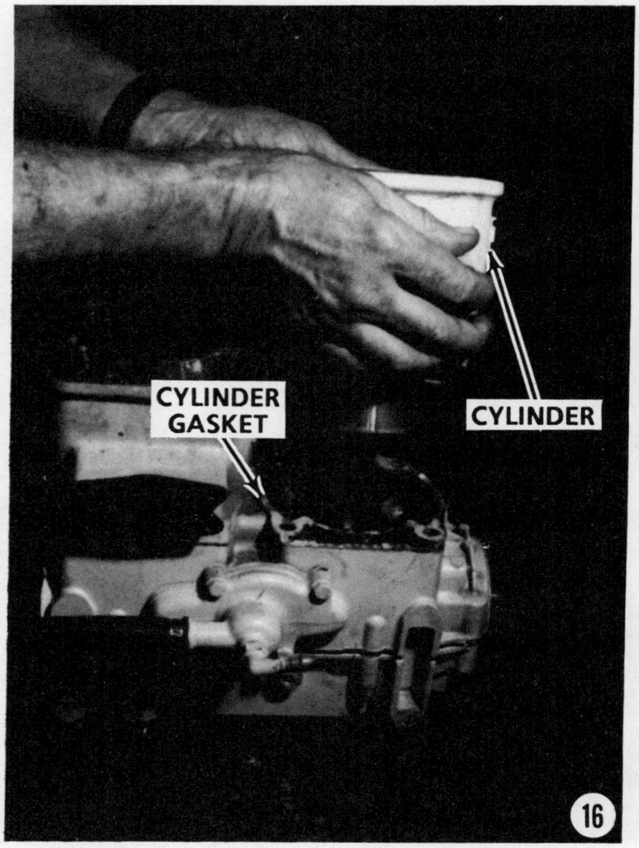

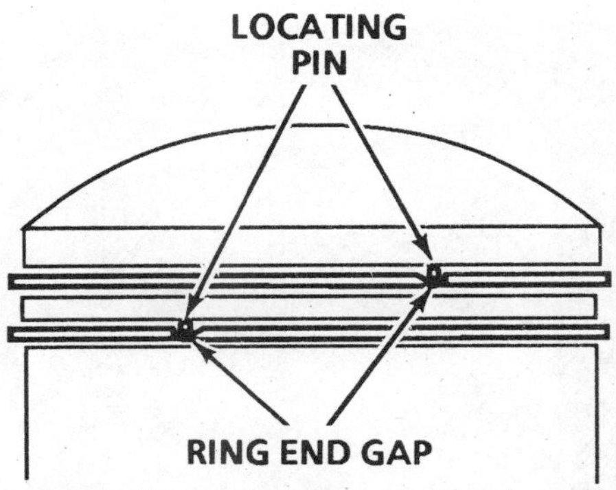

Align the ring end gap over the locating pin in each piston groove to prevent the ring from rotating in the groove. If the ring was free to rotate, an end might very well catch on the port opening and break.

slowly lower a cylinder down over the piston crown and rings, with the cylinder exhaust port on the opposite side to the rotary valve housing.

CRITICAL WORDS

If difficulty is experienced in fitting the piston into the cylinder, **DO NOT** force the piston. Such action might result in a broken piston ring. Check to be sure the ring end gap straddles the locating pin in the groove.

17- Install and tighten the four Allen head bolts securing each cylinder to the crankcase. These bolts are to be installed **DRY** -- with no sealing compound applied to the threads. Tighten the bolts alternately and evenly to a torque value of 16 ft lb (22Nm).

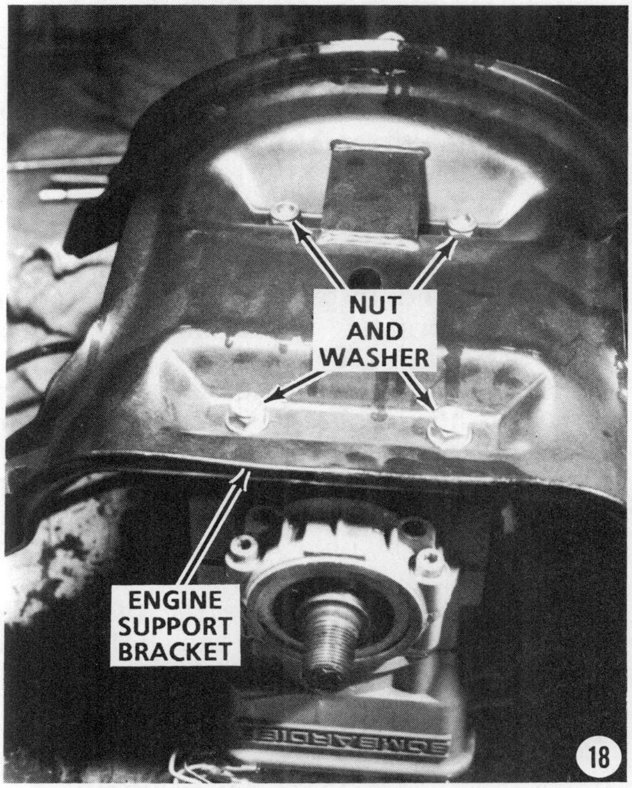

Engine Support Bracket

18- Apply a coating of Loctite 242 to the threads of the four studs on the bottom of the lower crankcase. Install the engine support bracket. Install the washers and locknuts. Tighten the nuts alternately and evenly to a torque value of 26 ft lb (35Nm).

Stator Plate Installation

19- Apply a coating of Loctite Primer N to the crankcase and stator plate mating surfaces. Allow about twenty minutes for the primer to dry. Apply a coating of Loctite 515 Gasket Eliminator over the primer. Apply Loctite 242 onto the Allen bolt threads. Install the stator plate and magneto assembly over the crankshaft. Align the bolt holes and install the bolts. Tighten the bolts alternately and evenly to a torque value of 16 ft lb (22Nm).

20- Coat the threads of the two attaching Allen head bolts with Loctite 242. Install the magneto assembly onto the stator plate. Align the marks made during disassembling. Install the bolts, washers, and lockwashers. Tighten the bolts alternately and evenly to a torque value of 16 ft lb (22Nm). Check to be sure the two marks remained aligned while the bolts were tightened.

Flywheel Installation

21- Place a dab of thick lubricant onto the curved surface of the Woodruff key to hold it in place while the flywheel is being installed. Press the Woodruff key into place in the crankshaft recess. Wipe away any excess lubricant to prevent the flywheel from "walking" during engine operation. Check the flywheel magnets to ensure they are free of any metal particles. Double check the taper in the flywheel hub and the taper on the crankshaft to verify they are clean and free of any oil.

Now, slide the flywheel onto the crankshaft, with the Woodruff key in place. Rotate the flywheel **COUNTERCLOCKWISE** to be sure it does not contact any part of the magneto assembly or wiring on the stator plate.

22- Attach Puller Plate P/N 290-876-065 to the flywheel with three M8x25mm bolts. Install the extension handle onto the puller arm. Apply Loctite 242 onto the crankshaft taper and threads. Install the crankshaft nut and washer. Tighten the flywheel nut to a torque value of 77 ft lb (105Nm). Leave the puller plate in place over the flywheel for the next three steps.

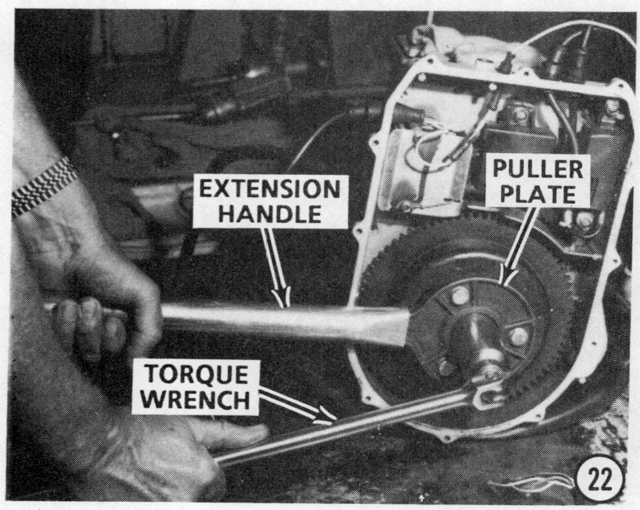

23- Apply anti-sieze compound to the threads of the aft end of the crankshaft. Thread the aft flywheel onto the crankshaft.

24- Insert Impeller Remover P/N 295-000-001 into the splines in the aft flywheel.

25- Tighten the aft flywheel to a torque value of 81 ft lb (110Nm).

Cranking Motor Installation

26- Position a new gasket over the mounting flange and insert the cranking motor into the stator plate.

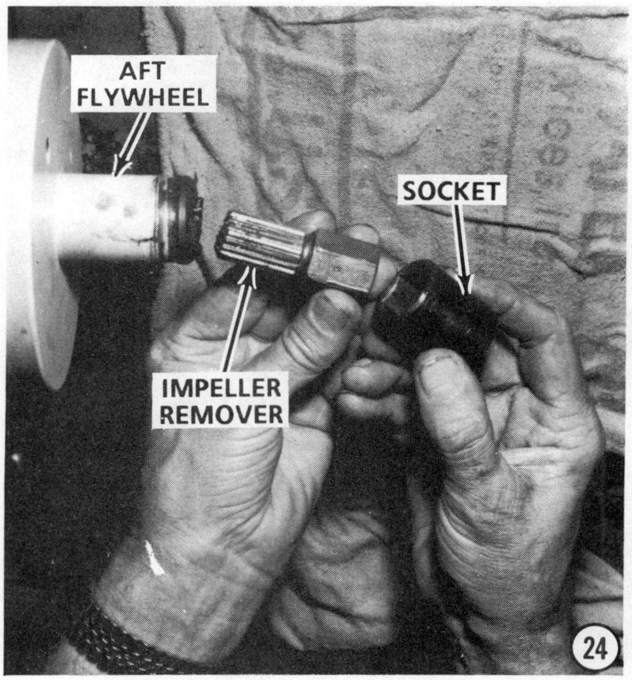

27- Apply Loctite 242 onto the threads of the three mounting bolts. Position the Black ground lead over the upper mounting bolt. Install the other two bolts, and then tighten all three to a torque value of 16 ft lb (22Nm).

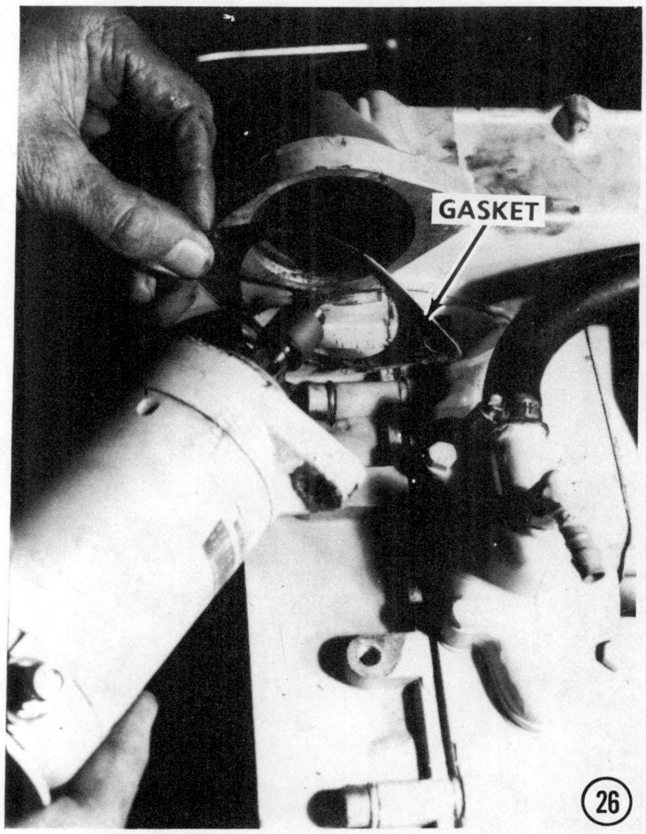

Rotary Valve Installation Original Crankcase

28- Make a setup with a dial indicator gauge over the forward piston. Rotate the crankshaft until the dial indicator shows the piston to be at TDC. If the original crankcase is being used, observe the notch cast into the top of the rotary valve housing, to the left of the water fitting, as shown.

Coat the rotary valve disc with Sea-Doo injection oil. Install the disc over the splines on the rotary valve shaft to align one edge of the disc with the notch in the housing. If the disc edge is not perfectly aligned, remove it, turn it over, and then install it a second time. The disc is asymmetrical to allow the best fit to be made for one face, but not the other.

The edge aligned with the notch is the opening edge and the edge below the forward port is the closing edge.

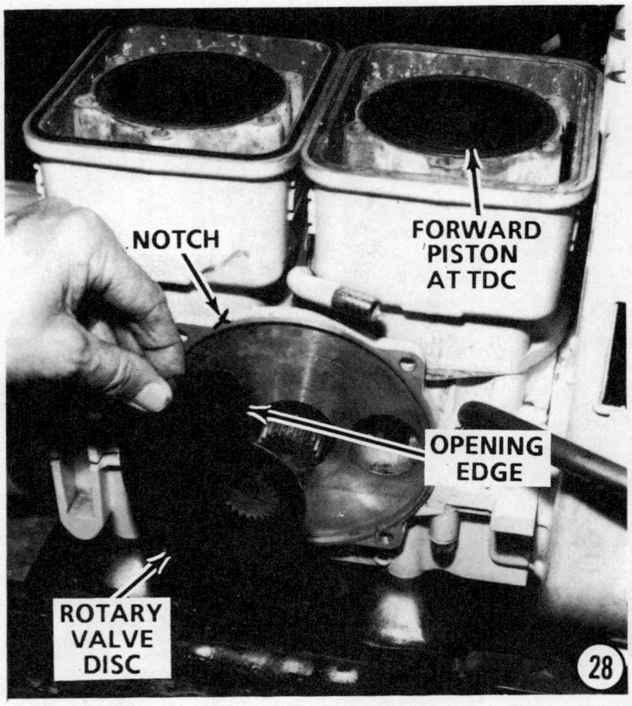

28

Rotary Valve Installation -- Replacement Crankcase and Verification of Timing for Original Crankcase

29- Obtain factory Degree Wheel P/N 295-000-007, or cut out one of the degree wheels included for this purpose in the last portion of the Appendix.

Make a setup with a dial indicator gauge over the forward piston. Rotate the crankshaft until the dial indicator indicates the piston to be at the TDC position. Place the degree wheel over the rotary valve shaft. If the paper cutout from the Appendix section of this manual is used, cut out the center, using the circles, to permit the "wheel" to slide over the shaft with barely a snug fit.

Align the 360° line with the lowest point of the forward inlet port. Identify the 132° line on the degree wheel, and scribe a matching mark on the rotary valve

27

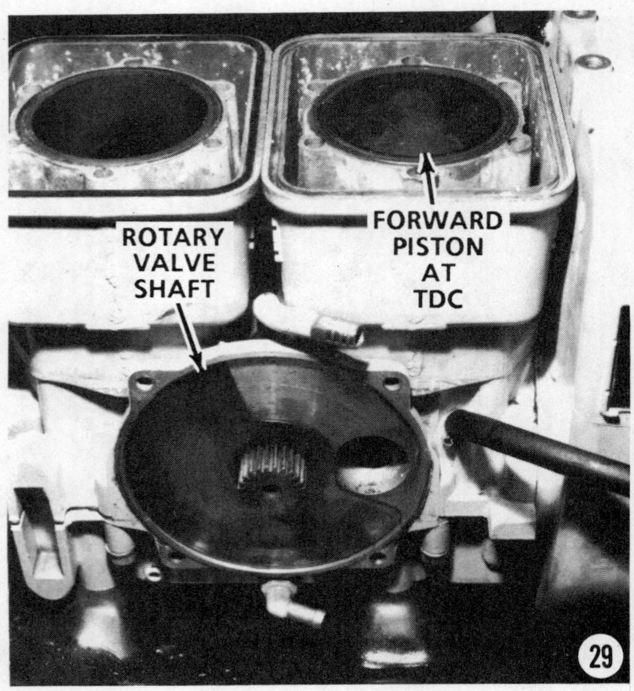

29

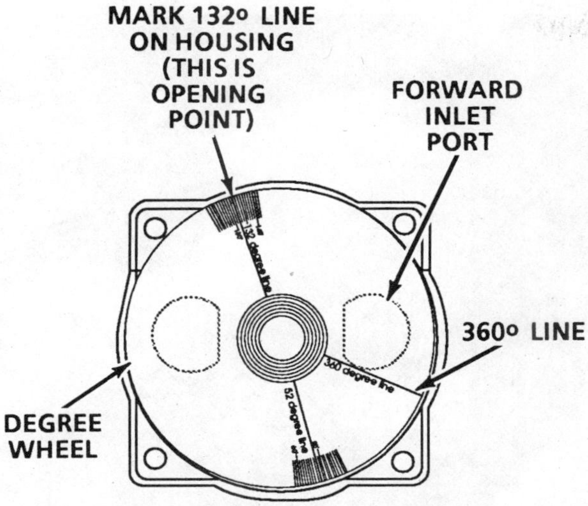

Determining the opening point of the rotary valve disc, using a degree wheel. See illustration below for closing point.

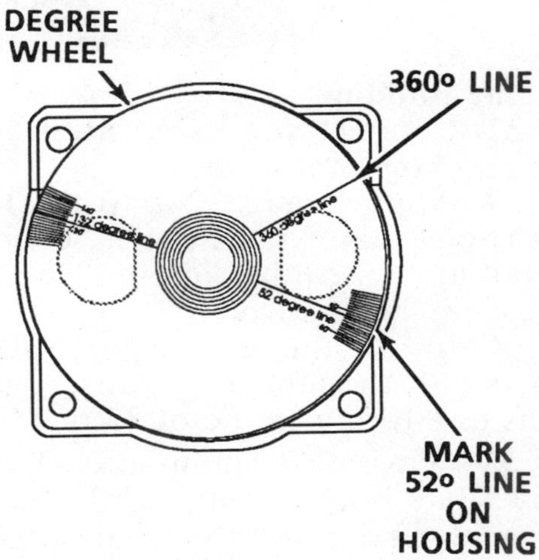

Determining the closing point of the rotary valve disc using a degree wheel, as explained in detail in the text. A couple degree wheels, suitable for cutting out and using, are included in the Appendix.

housing. This mark will represent the port opening point and will be at approximately the 11 o'clock position, as depicted in the accompanying line drawing -- with the degree arrangement just described identified.

Change the position of the degree wheel on the rotary valve shaft, to place the 360° line at the highest point of the forward inlet port. Identify the 52° line and scribe a matching mark on the rotary valve housing. This mark will represent the port closing point and will be at approximately the 4 o'clock position, just below the inlet port. This arrangement is depicted in the other accompanying line drawing.

Coat both sides of the rotary valve disc with Sea-Doo injection oil. Remove the the degree wheel and install the rotary valve disc onto the shaft. Align one edge of the disc with the opening mark and the other edge with the closing mark. If the disc is not perfectly aligned with the two marks, remove the disc, turn it over and install it a second time. The disc is asymmetrical to allow the best fit to be made for one face, but not the other.

For more information on this type of rotary valve operation, see Chapter 3 -- **TUNING.**

ENGINE INSTALLATION

ENGINE MOUNT GOOD WORDS

The manufacturer states engine alignment procedures are not necessary. Proper engine alignment is performed at the factory and if the engine mounts and pieces of shim material are left **UNDISTURBED**, the engine will always remain in perfect alignment.

30- Check to be sure the shim material is still in place on top of the engine mounts. Enlist the aid of an assistant, if necessary, and lower the engine down through the opening and onto the engine mounts. Align the holes in the engine support bracket with the holes in the mounts.

31- Apply a coating of Loctite 242 onto the threads of the four mounting bolts. Install the bolts and washers. Tighten the bolts alternately and evenly to a torque value of 18 ft lb (25Nm).

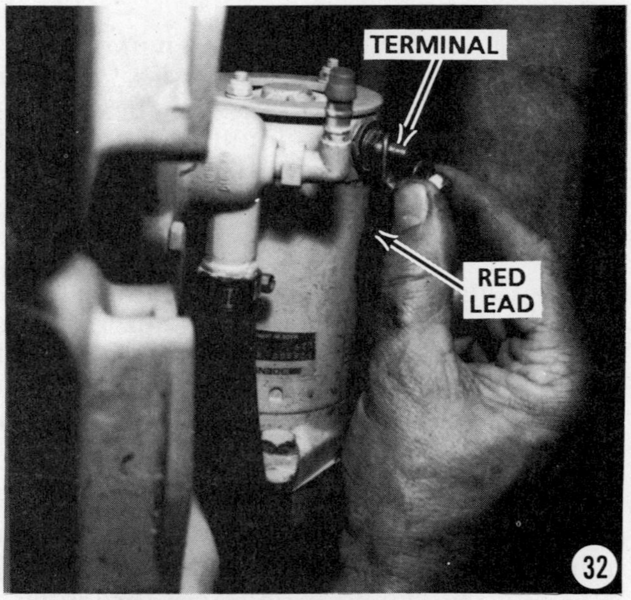

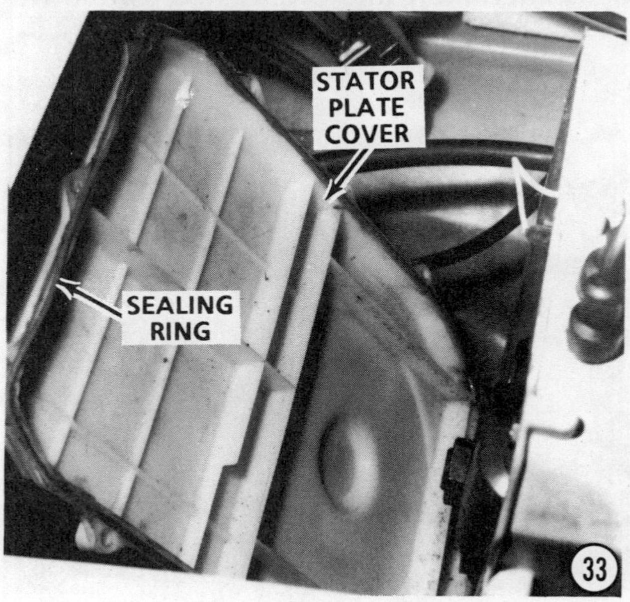

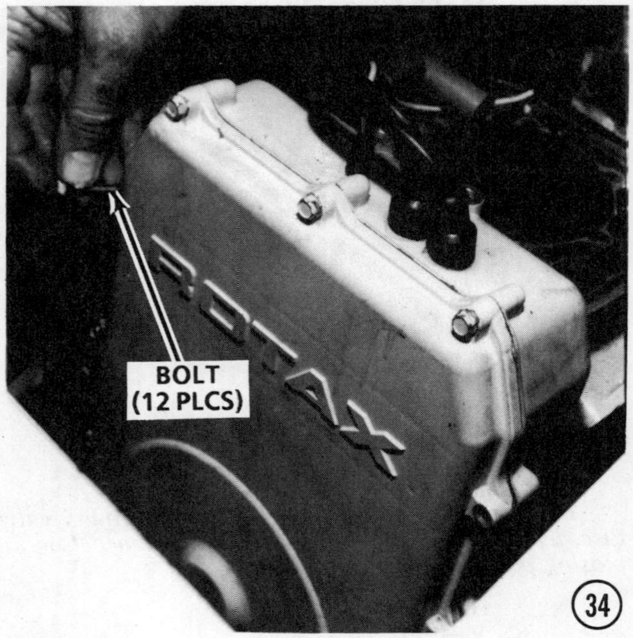

Engine Buildup

32- Connect the large Red lead onto the cranking motor terminal.

33- Apply Loctite 242 to the threads of the twelve stator plate cover securing bolts. Install a new sealing ring into the groove of the stator plate cover.

34- Install the cover and secure it in place with the attaching bolts. Tighten the bolts to a torque value of 35 in lb (4Nm).

35- Position a clamp at each end of the driveshaft boot. Install the boot between the seal carrier and the aft flywheel. Tighten both hose clamps just **SNUG**, at this

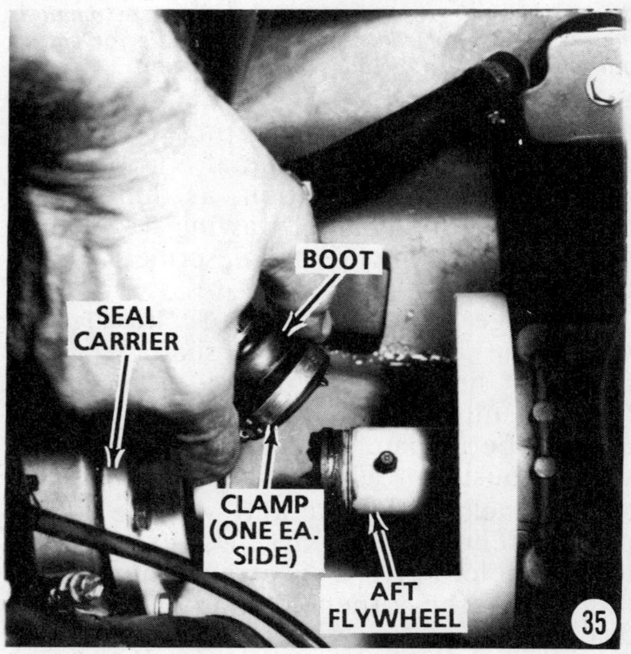

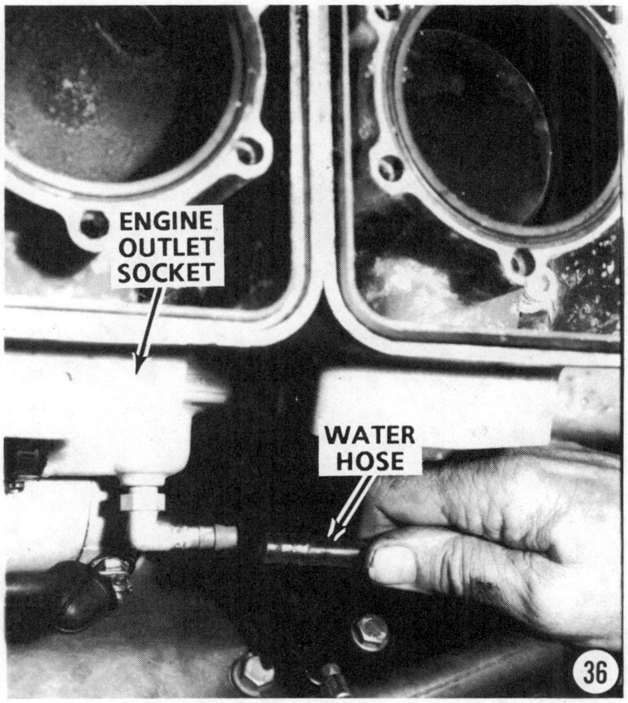

time. If the clamps are tightened securely now, it could affect engine alignment.

36- Slide a clamp over the end of the water return hose, and then slide the hose onto the fitting on the engine outlet socket. Tighten the clamp to secure the hose in place.

Exhaust Manifold Installation

37- Slide a clamp over the end of the inlet cooling water hose. Work the hose onto the fitting at the base of the exhaust manifold. Tighten the clamp to secure the hose. Position new gaskets over the ex-

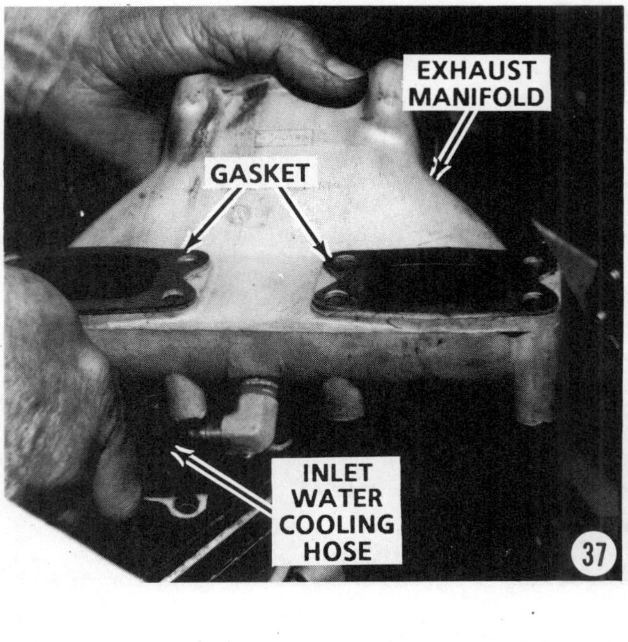

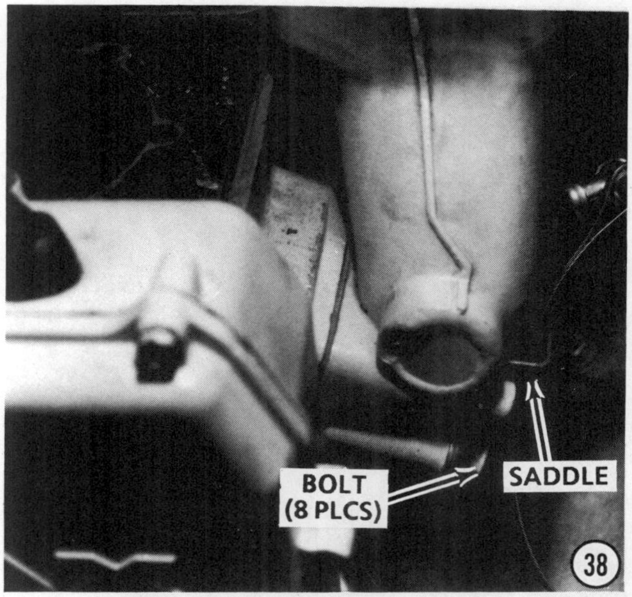

haust ports on the intake manifold. Align the gaskets and manifold with the engine exhaust ports.

38- Apply Loctite 242 to the eight exhaust manifold securing bolts. **FIRST**, install and **HAND** tighten the four bolts and washers securing the exhaust manifold to the **AFT** cylinder. Next, place the rubber pad, four washers, and saddle over the forward set of bolt holes. Install the four bolts and washers through all the parts held in place, and then **HAND** tighten the bolts. Now, tighten all eight bolts alternately and evenly to a torque value of 15 ft lb (20Nm).

39- Install the two small water passage O-rings and the single large exhaust pas-

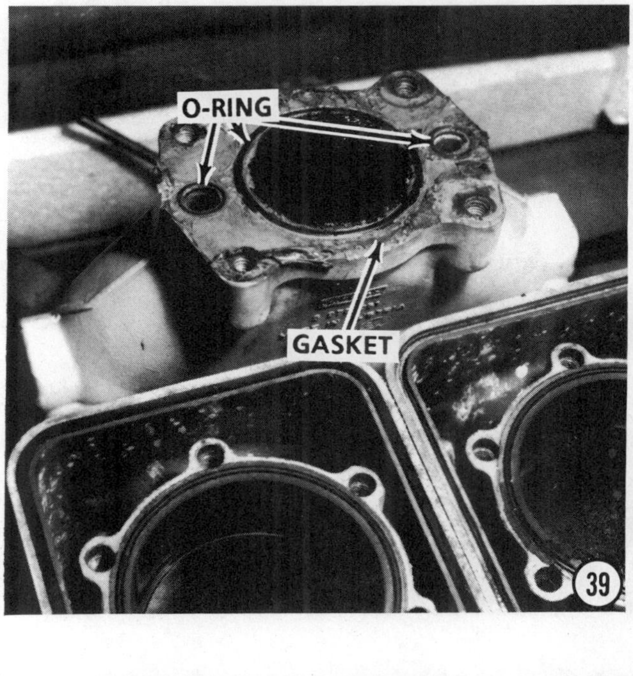

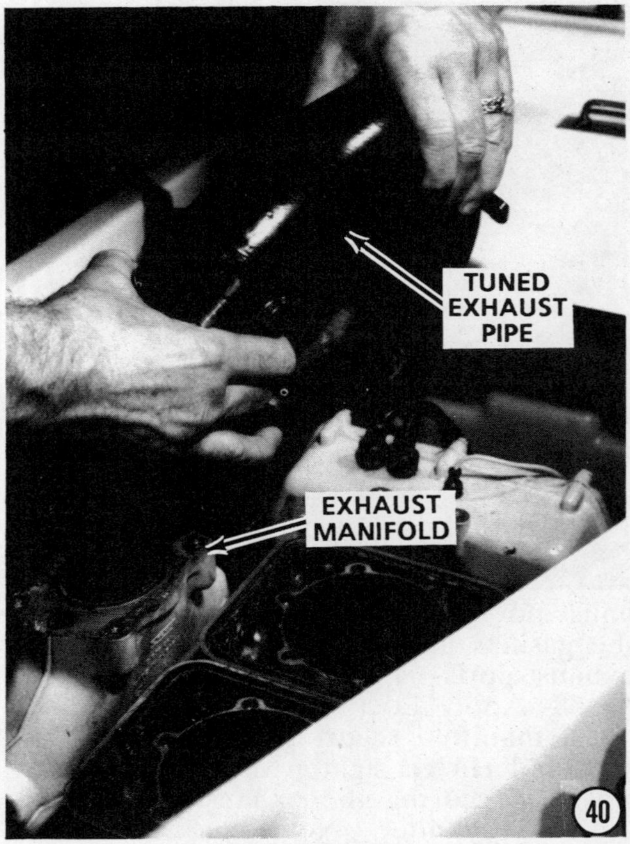

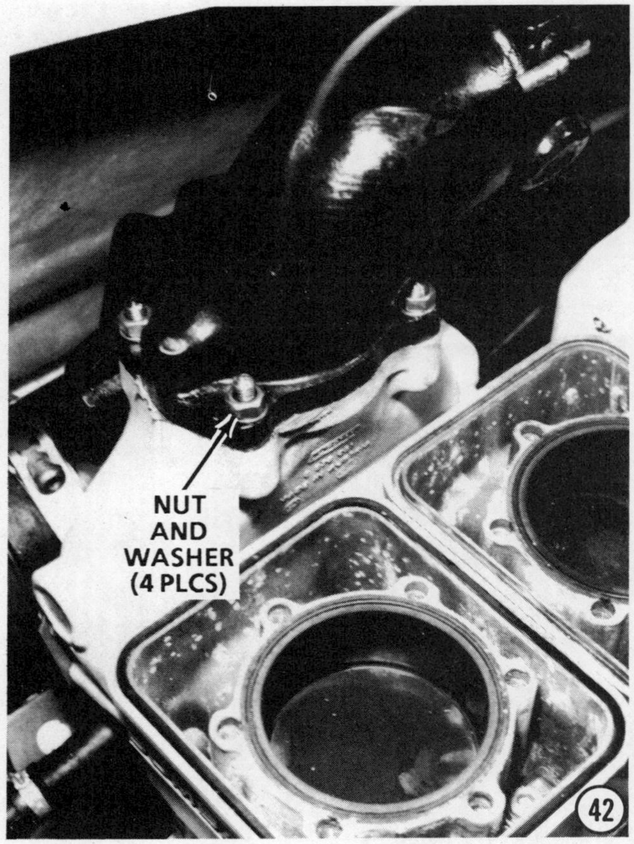

sage **O**-ring onto the top of the exhaust manifold. Install a new gasket over the **O**-rings.

40- Ease the tuned exhaust pipe into the engine compartment and slightly forward of the exhaust manifold mounting flange.

41- Apply Loctite 242 to the manifold stud threads. Install a stud into the exhaust manifold mounting flange. Thread two nuts onto the stud, and then tighten them against each other. Tighten the upper nut to install the stud into the exhaust manifold flange. Remove both nuts and use them to install the other three studs in the same manner.

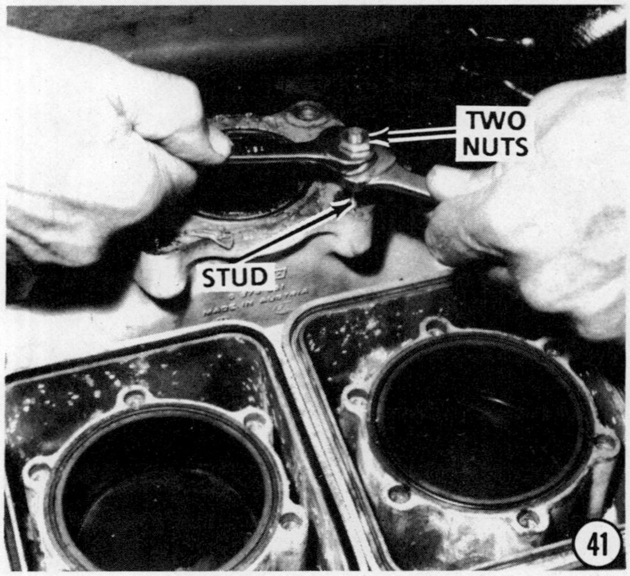

42- Move the tuned exhaust pipe over the studs just installed. Install and tighten the nuts and washers to a torque value of 18 ft lb (25Nm).

43- Tighten the two clamps on the hose connecting the tuned exhaust pipe to the muffler.

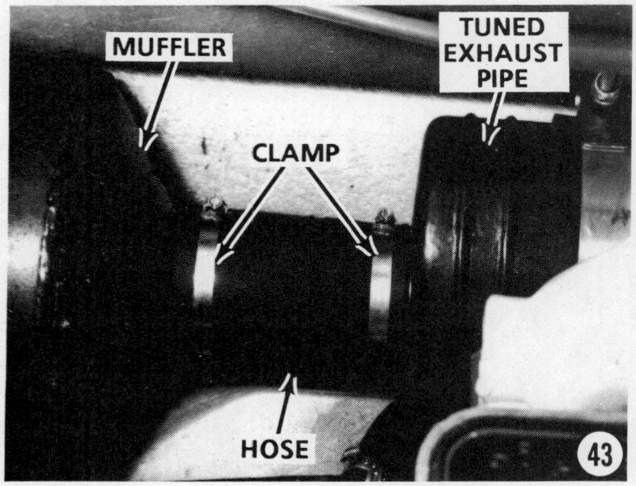

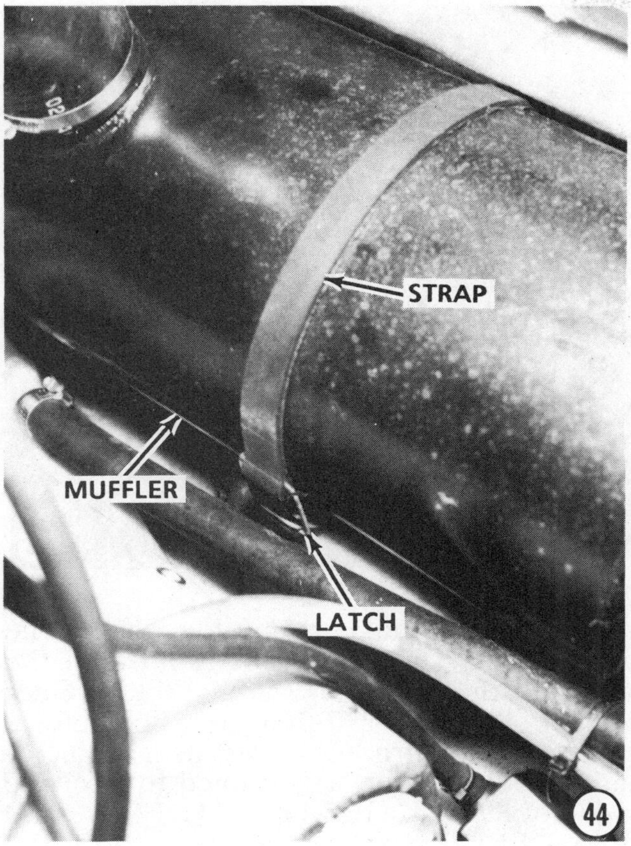

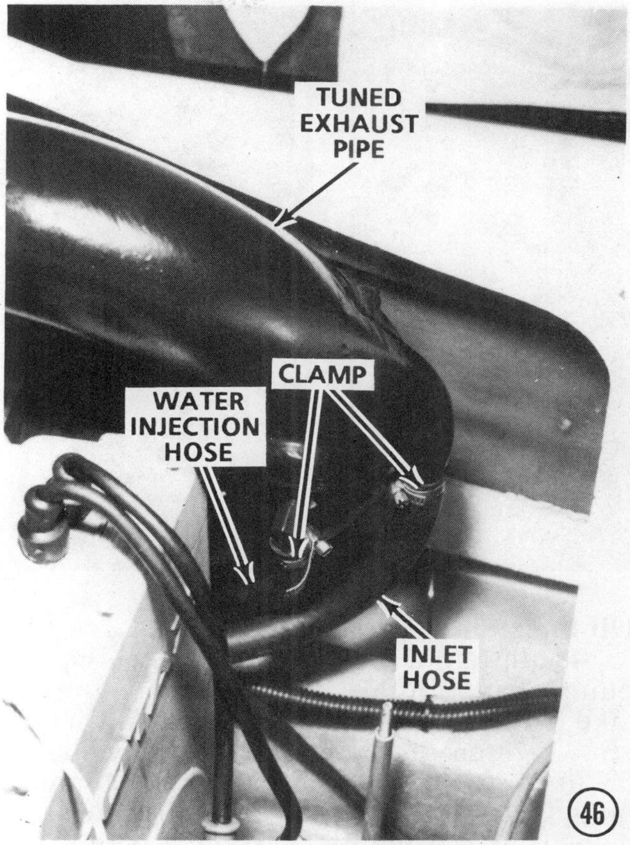

44- Hook the end of the muffler strap onto the latch on the hull.

45- Install the rubber cushioning strip between the tuned exhaust pipe and the large clamp. Tighten the clamp securely.

46- Slide clamps over the water injection hose and the water inlet hose. Work the hoses onto the fittings on the tuned exhaust pipe. Tighten the clamps securely.

47- Connect the drain tube to the tuned exhaust pipe fitting. Use a new tie wrap over the tube to secure it in place.

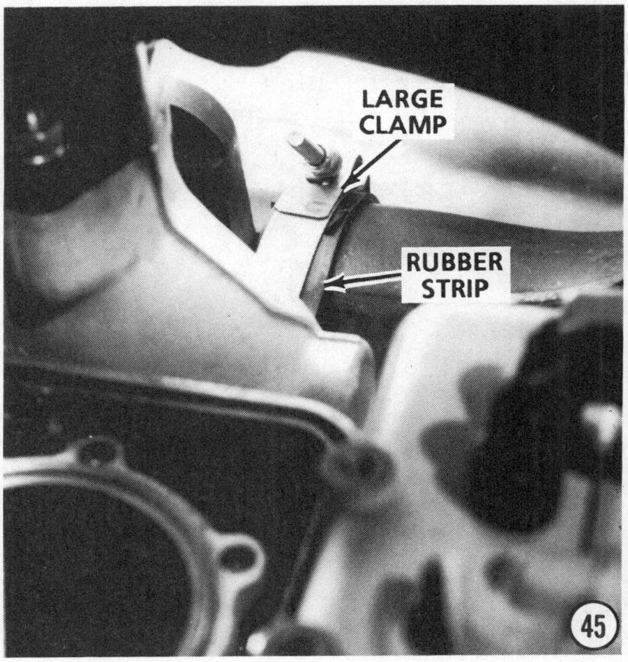

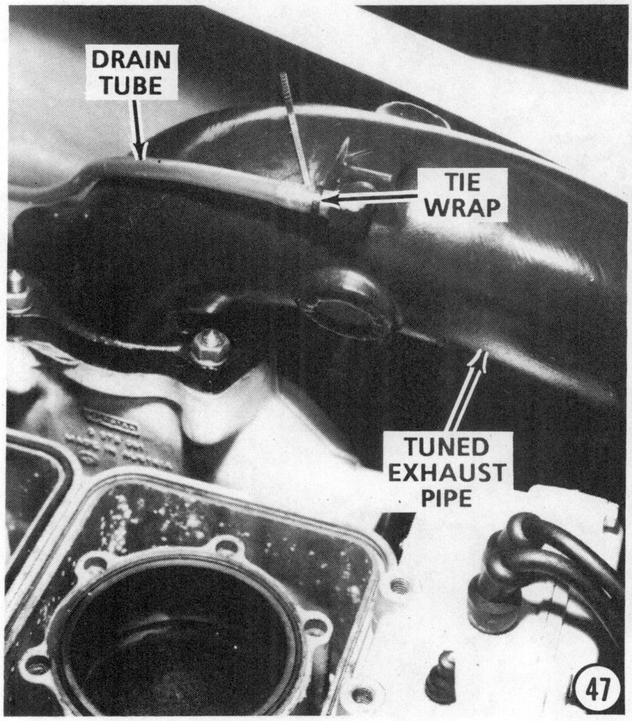

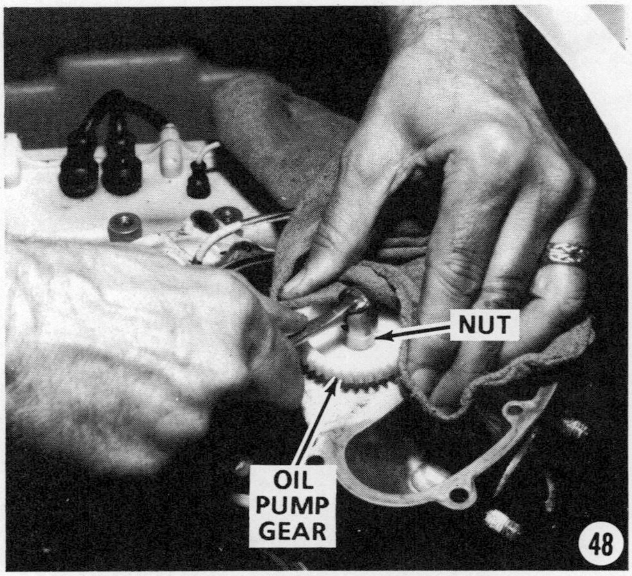

48

50

Oil Injection Pump Installation

48- Install a new **O**-ring over the oil pump shaft. Install the pump onto the intake manifold with the two bolts and washers. Tighten the bolts to a torque value of 53 in lb (6Nm). Install the two "banjo" bolts and fittings, with the oil lines attached, to the pump. Apply Loctite 242 to the pump shaft threads.

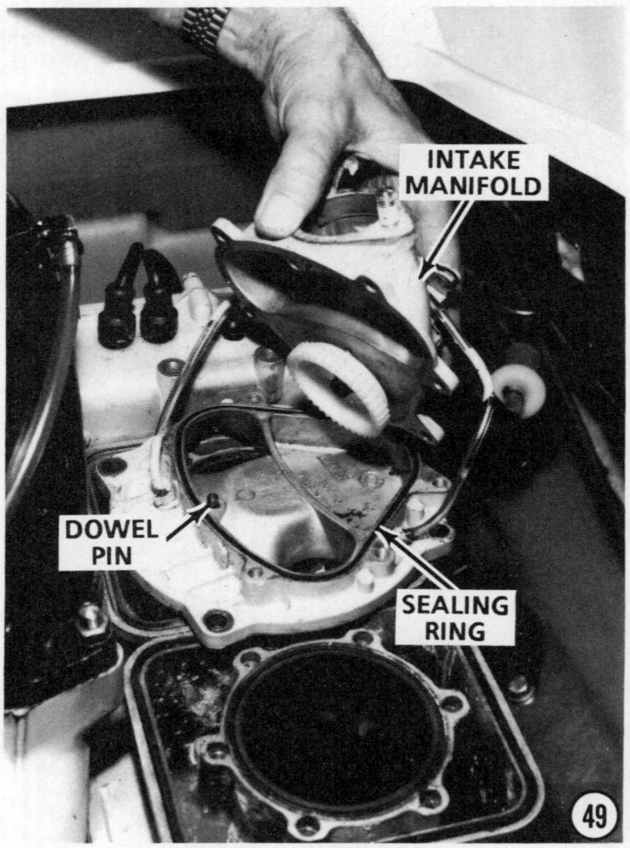

49

Inside the intake manifold: install the washer, gear, another washer, and then the nut onto the oil pump shaft. Obtain Gear Holder P/N 290-277-905 or shop cloth. Hold the gear firmly to prevent it from rotating while the nut is tightened to a torque value of 89 in lb (10 Nm).

49- Fit a new sealing ring into the groove of the rotary valve cover. Install the intake manifold onto the rotary valve cover, with the manifold indexed over the single dowel pin in the cover.

50- Install the six bolts and washers securing the intake manifold to the rotary valve cover. Tighten the bolts alternately and evenly to a torque value of 80 in lb (9Nm).

Throttle Cable Installation

51- Hook the end of the throttle cable into the pump lever. Push the cable into

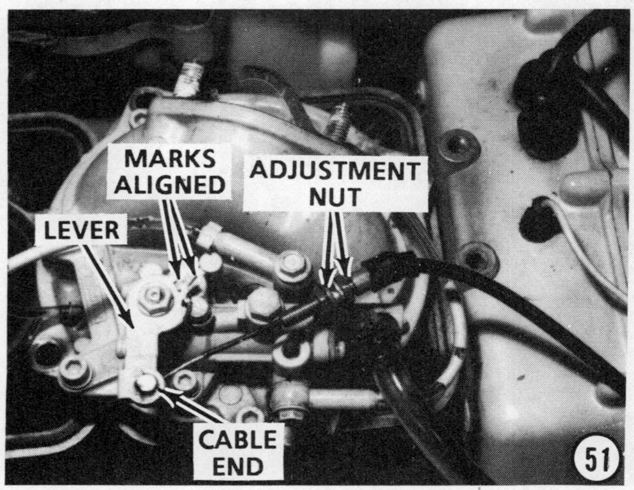

51

the slot with an adjustment nut on each side of the slot.

Oil Pump Synchronization

Observe the mark on the pump lever and the mark on the pump boss. These two marks **MUST** align for oil pump synchronization. Adjust the position of the two nuts to change the length of the cable, and therefore, the position of the lever until the two marks align.

Rotary Valve Cover Installation

52- Slide a clamp over the end of the oil supply hose. Remove the pincer tool (or plug in the end of the hose), and quickly work the hose onto the fitting at the base of the rotary valve. Tighten the clamp securely.

53- Fit a new sealing ring into the groove in the rotary valve housing. Install the housing against the upper and lower crankcase. Install the four securing bolts and washers. Tighten the bolts to a torque value of 15 ft lb (20Nm).

Slide a clamp over the end of the oil return hose. Remove the pincer tool (or plug in the end of the hose), and quickly work the hose onto the fitting at the top of the rotary valve. Tighten the clamp securely.

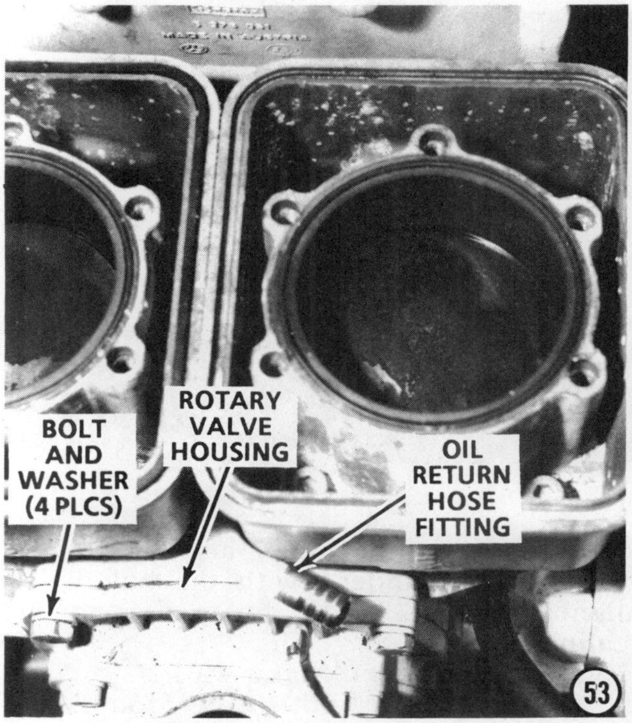

Cylinder Head Installation

54- Apply a coating of Loctite Primer N into the cylinder and water jacket grooves for each cylinder. Apply the primer into the corresponding grooves in the cylinder head. Coat both sides of the 12 cylinder head bolts and washers with the Loctite Primer. Allow about 20 minutes for the primer to dry. Apply Loctite 515 gasket eliminator over all the surfaces coated with the primer, including the grooves.

Install the sealing rings into the cylinder and water jacket groove for each cylinder.

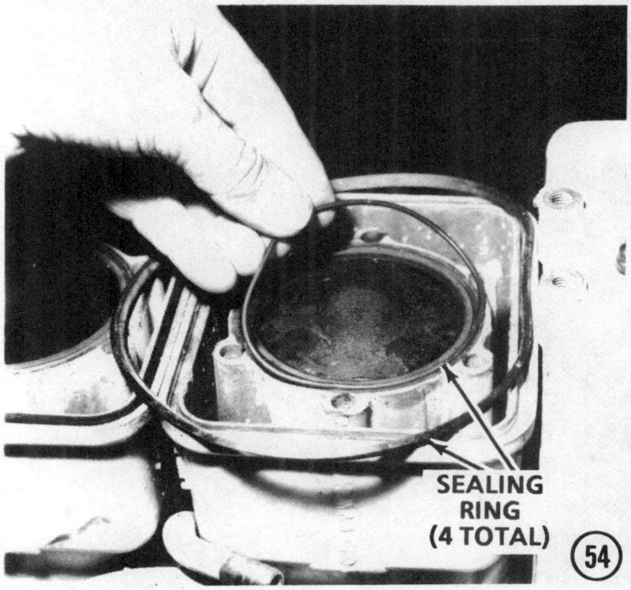

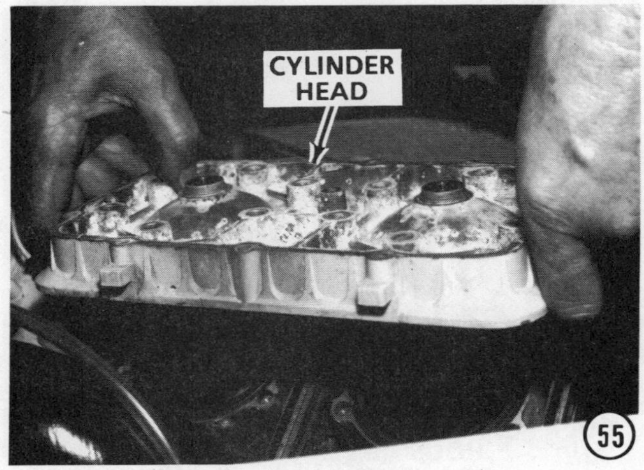

55- Lower the cylinder head into place on the two cylinders, with the bolt holes aligned. **TAKE CARE** not to deform the sealing rings.

56- Install the bolts and washers -- coated with the primer and Loctite 515. Tighten the bolts just **HAND** tight, at this time. Insert new **O**-rings into place in each spark plug opening. Place a new small **O**-ring into the temperature sensing hole, indicated in the accompanying illustration.

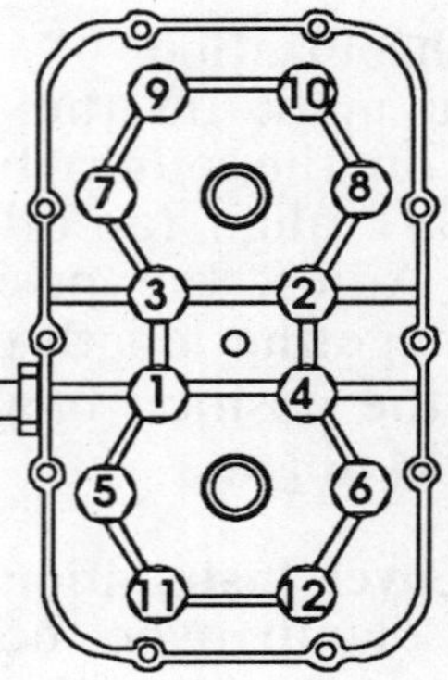

Tightening sequence for the cylinder head securing bolts. Tightening sequence for the head cover is given on Page 8-52.

57- Tighten the cylinder head bolts to a torque value of 15 ft lb (20Nm), in the sequence indicated in the accompanying captioned line drawing.

58- Install a new sealing ring into the groove in the cylinder head. Lower the cover into place on the head.

59- Position the stator plate support bracket over the two forward bolt holes

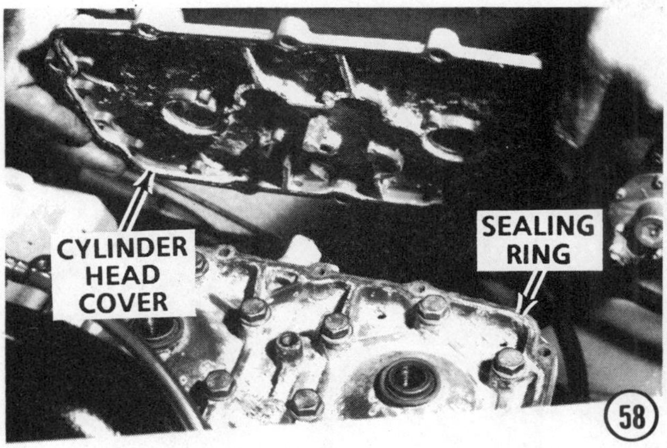

and the flame arrestor support bracket over the two outer starboard bolt holes. Install the nine Allen head bolts and washers securing the cover to the head. Tighten the bolts to a torque value of 80 in lb (9Nm), in the sequence shown in the accompanying captioned line drawing.

60- Install the two Allen head bolts securing the stator plate support bracket to the plate. Tighten the bolts alternately and evenly to a torque value of 80 in lb (9Nm).

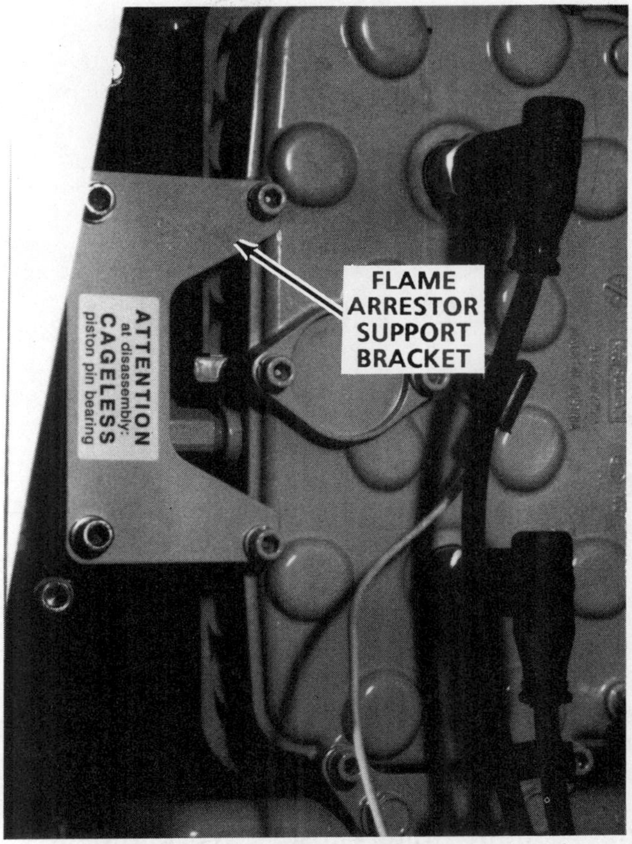

Location of the flame arrestor support bracket on the cylinder head cover.

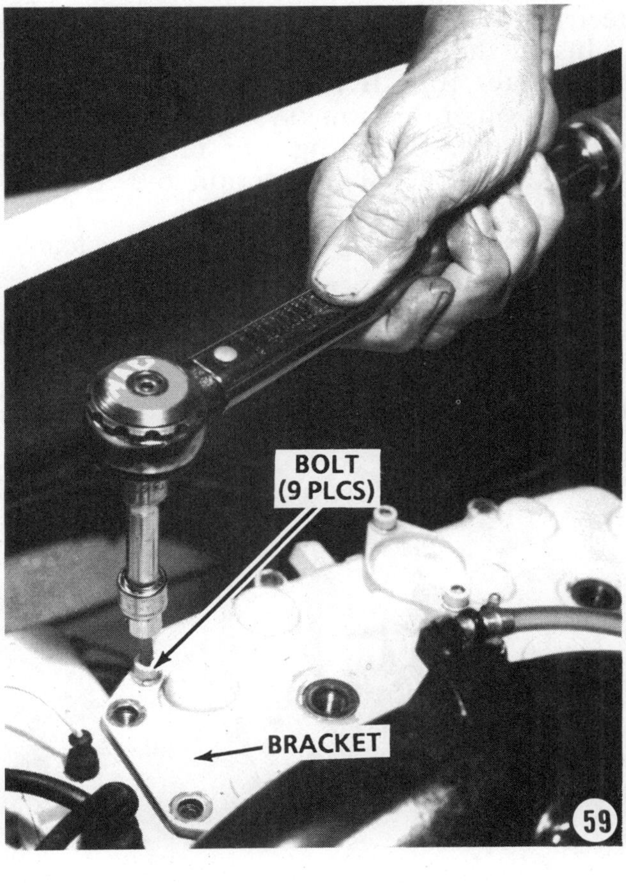

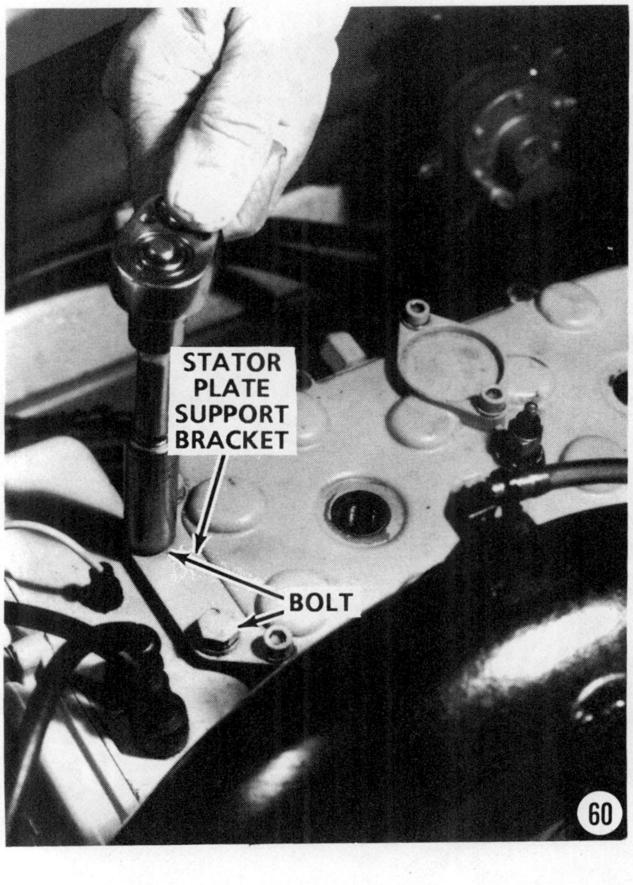

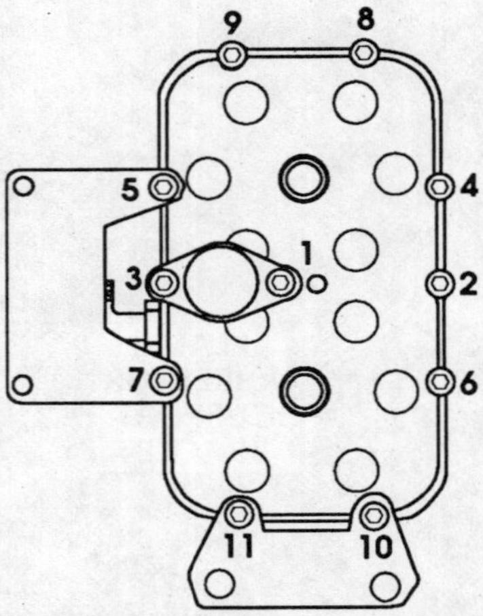

Tightening sequence for the cylinder head ***COVER*** *securing bolts.*

61- Install the spark plugs. Connect the lead to the temperature sensor.

62- Connect the spark plug leads to the spark plugs.

63- Tighten the two clamps on the boot between the aft flywheel and the seal carrier.

64- Slide a clamp over the end of engine cooling water hose. Work the hose onto the fitting on the jet pump. Tighten the clamp securely.

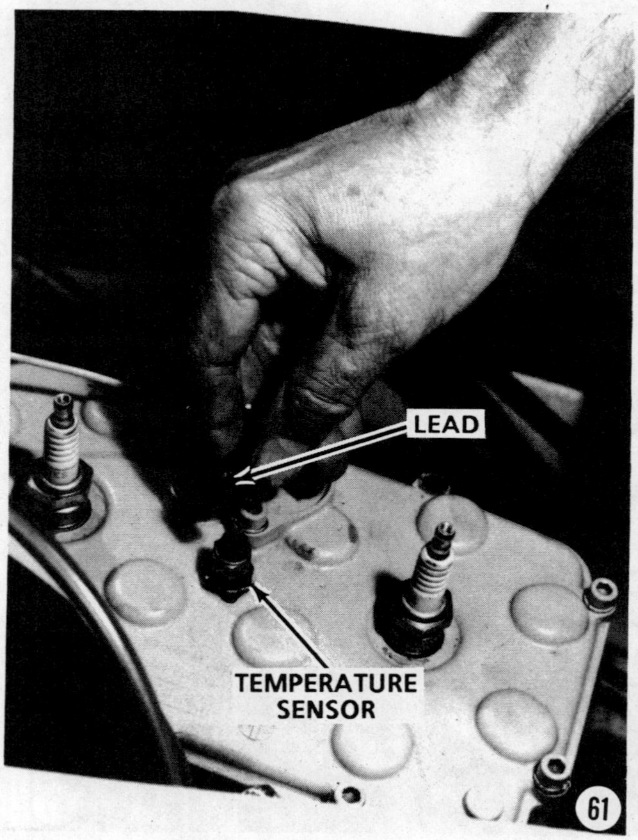

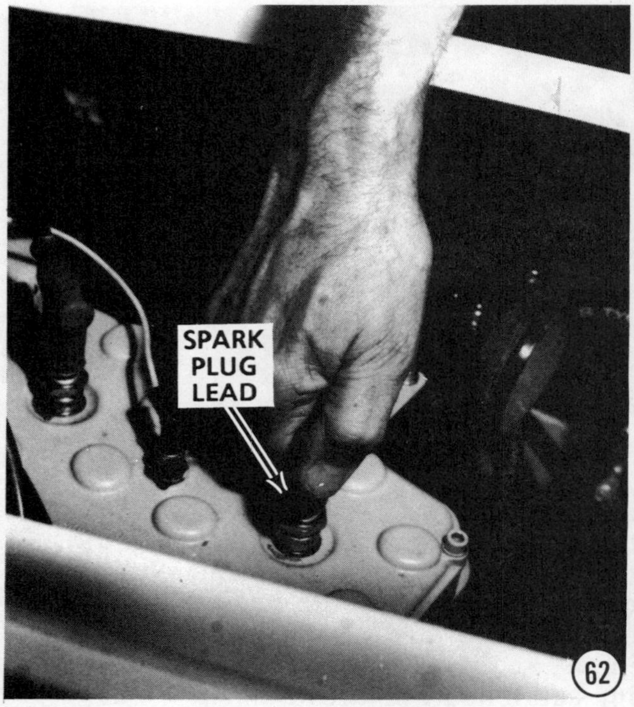

CLOSING TASKS

Install the carburetor, associated hoses, throttle and choke cables, see Chapter 6 -- **FUEL.**

Connect the stator harness lead at the electrical box.

Install the driveshaft into the jet pump, see Chapter 10 -- **JET PUMP**.

The oil injection system **MUST** be "bled" while operating the engine on a premix, as directed in Section 6-9, beginning on Page 6-37.

Connect the negative and positive leads to the battery.

Move the craft into a body of water or connect a garden hose to the engine flush system.

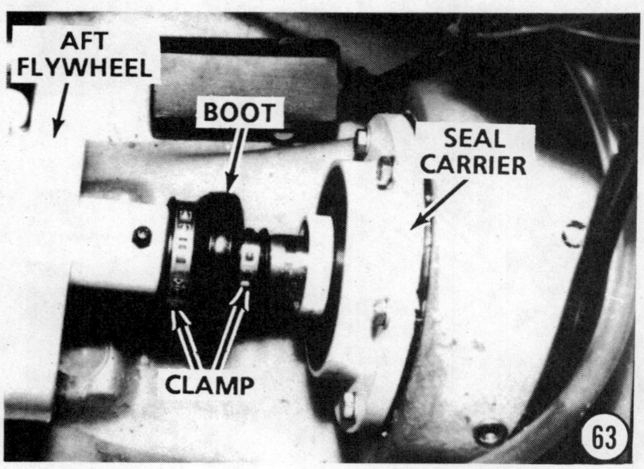

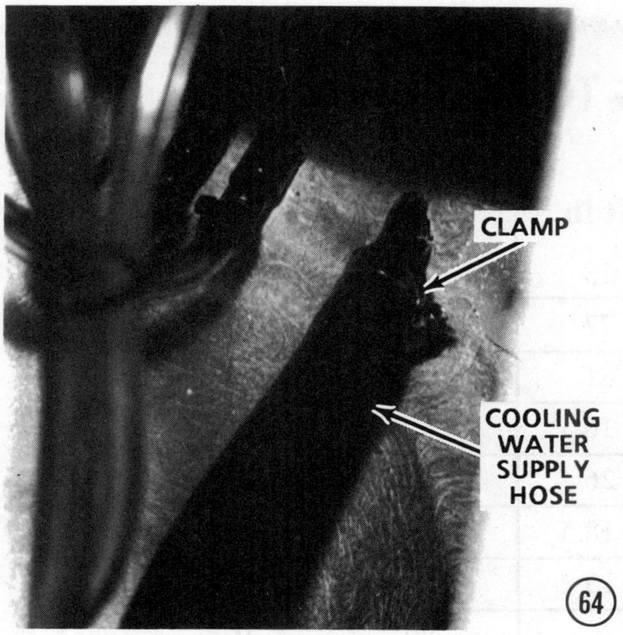

CAUTION

The following procedure MUST be followed whenever the engine is started with a hose attached for cooling the engine. Start the engine, and then IMMEDIATELY turn the water on. The engine MUST be started first to prevent water from backing up into the tuned exhaust pipe and into the engine. Water must circulate anytime the engine is operating to prevent overheating -- which could cause damage to moving parts and possibly cause engine seizure. NEVER allow the engine to operate without cooling water for more than 15 seconds. NEVER run the engine with a flushing attachment for more than 5 minutes. The driveline seal is not provided with cooling water if it is not immersed in water. To shut the engine down, first shut the water supply off, and then shut down the engine. If this sequence of tasks is not followed as stated, the engine could be severely damaged.

REMEMBER

Start the engine.
Turn the water on.
Turn off the water.
Shut down the engine.

Attempt to start and run the engine, without the engine compartment cover in place. This will provide the opportunity to check for water, exhaust, or fuel leaks.

After the engine is operating properly, install the engine cover. Follow the break-in procedures with a load on the jet pump.

BREAK-IN PROCEDURES

Do not operate the engine at full throttle, except for very short periods, until after ten hours of operation.

a- Operate at half throttle -- approximately 2500 to 3500 rpm for two hours.

b- Operate at any speed after two hours, but not at sustained full throttle until after another eight hours of operation.

c- Mix gasoline and oil during the break-in period -- total of ten hours -- at a ratio of 50:1.

d- While the engine is operating during the initial period, check the fuel, exhaust, and water system for leaks.

Make a final check, while the engine is operating, for water, fuel, and exhaust leaks, before stowing the tools, cleaning the engine compartment, and "squaring away" the work area.

ENGINE TIGHTENING TORQUE VALUES

	ft lb	in lb	Nm
Exhaust manifold nuts	15		20
Forward flywheel nut	77		105
Aft flywheel	81		110
Crankcase bolts	16		22
Engine bracket support nuts	26		35
Engine mount bolts	18.5		25
Cylinder head bolts	15		20
Cylinder head cover bolts		80	9
Tuned exhaust pipe bolts	18.5		25
Crankcase-to-cylinder bolts	15		20
Carburetor intake nuts	18.5		25
Cranking motor saddle nuts		9	1
Flame arrestor screws		89	10

9 ELECTRICAL

9-1 INTRODUCTION

The battery, cranking system, stop "kill" switch, and temperature warning system, are all considered subsystems of the electrical system. Each of these areas will be covered in this chapter beginning with the battery. The charging system is considered a subsystem of ignition and is covered in detail in Chapter 7 -- **IGNITION.**

The battery, charging system, and the cranking system are considered subsystems of the electrical system. Each of these units will be covered in detail in this chapter beginning with the battery.

9-2 BATTERIES

The battery is one of the most important parts of the electrical system. Because of its job and the consequences, (failure to perform when needed), the best advice is to purchase a well-known brand, with an extended warranty period, from a reputable dealer.

Personal Watercraft Batteries

The battery developed for use in a personal watercraft is the same kind used for many years on motorcycles. Personal watercraft batteries are required to perform under the most rigorous conditions.

Personal watercraft batteries have a much heavier exterior case than the usual automobile battery to withstand the violent pounding and shocks imposed on it as the craft moves through rough water. The battery case must also be vibration and seepage resistant. Therefore, a personal watercraft battery should always be the best the owner can afford.

The caps of marine batteries are "spill proof" to prevent acid from spilling into the bilges when the craft is subjected to violent maneuvers. These special caps are designed to vent gases regardless of the battery position, even when upside down.

Because of these features, the personal watercraft battery will recover from a low charge condition and give satisfactory service over a much longer period of time.

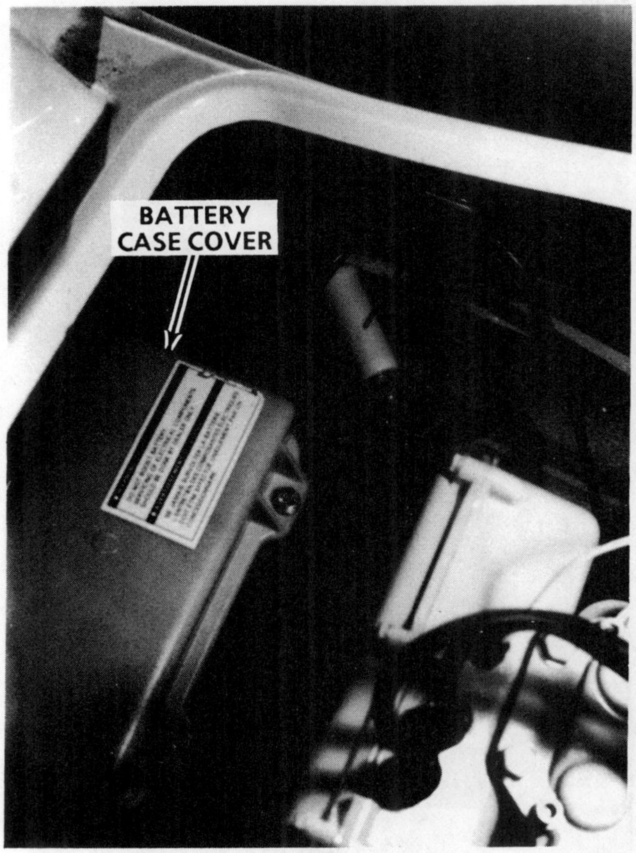

The battery is housed inside a tough, well-constructed case. Many of the ignition and charging components are secured inside the cover where they are protected from damage and corrosion.

Battery Construction

A battery consists of a number of positive and negative plates immersed in a solution of diluted sulfuric acid. The plates contain dissimilar active materials and are kept apart by separators. The plates are grouped into what are termed elements. Plate straps on top of each element connect all of the positive plates and all of the negative plates into groups.

The battery is divided into cells which hold a number of the elements apart from the others. The entire arrangement is contained within a hard plastic case. The top is a one-piece cover and contains the filler caps for each cell. The terminal posts protrude through the top where the battery connections for the craft are made. Each of the cells is connected to its neighbor in a positive-to-negative manner with a heavy strap called the cell connector.

Battery Ratings

Four different methods are used to measure and indicate battery electrical capacity:

After the protective boot is moved back on the battery cable, the cable can be removed using a slotted screwdriver or a socket.

1- Ampere-hour rating
2- Cold cranking performance
3- Reserve capacity
4- Watt-hour rating

The **ampere-hour** rating of a battery refers to the battery's ability to provide a set amount of amperes for a given amount of time under test conditions at a constant temperature of 80° (27°C). Amperes x hours equals ampere-hour rating. Therefore, if the battery is capable of supplying 3 amperes of current for 5 consecutive hours, the battery is rated as a 15 ampere-hour battery.

The ampere-hour rating is useful for some service operations, such as slow charging or battery testing.

The recommended ampere-hour rating for a personal watercraft battery is 15-19 ampere-hours.

Cold cranking performance is measured by cooling a fully charged battery to 0°F (-18°C) and then testing it for 5 seconds to determine the maximum current flow. In this manner the cold cranking amperes rating is the number of amperes available to be drawn from the battery before the voltage drops below 9.8 volts. A typical cold cranking ampere rating for a personal watercraft battery is 240 amps.

Reserve capacity of a battery is considered the length of time -- in minutes -- at 80°F (27°C) a 25 ampere current can be maintained before the voltage drops below 10.5 volts. This test is intended to provide an approximation of how long the engine, including electrical accessories such as bilge pump, could operate satisfactorily if the stator assembly or charging coil did not produce sufficient current. A typical rating is 25 minutes.

Watt-hour is a very useful rating of battery power. It is determined by multiplying the number of ampere hours times the voltage. Therefore, a 12-volt battery rated at 16 ampere-hours would be rated at 192 watt-hours (16 x 12 = 192).

If possible, the new battery should have a power rating equal to or higher than the unit it is replacing.

Battery Installation

Every battery installed in a personal watercraft must be secured in a well-protected ventilated area. If the battery area lacks adequate ventilation, hydrogen gas which is given off during charging could become very explosive. This is especially true if the gas is concentrated and confined.

Battery Service

The battery requires periodic servicing and a definite maintenance program will ensure extended life. If the battery should test satisfactorily, but still fails to perform properly, one of five problems could be the cause.

1- Slow speed engine operation for long periods of time resulting in an undercharged condition.

2- A defect in the charging system. A faulty stator assembly or charging coil, defective rectifier, or high resistance somewhere in the system could cause the battery to become undercharged.

3- Failure to maintain the battery in good order. This might include a low level of electrolyte in the cells; loose or dirty cable connections at the battery terminals; or possibly an excessively dirty battery top.

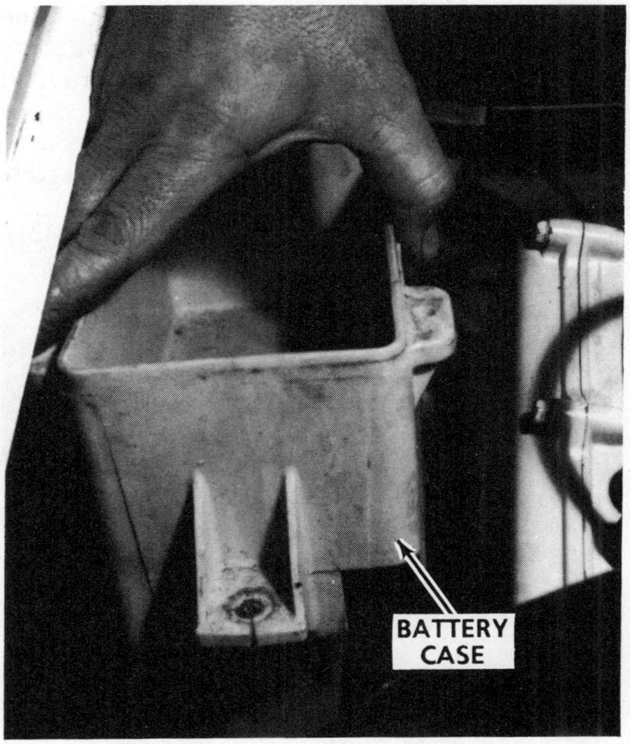

Removing the battery case from the hull for cleaning.

Location of the breather pipe -- on the starboard side of the battery.

Remove the battery from the engine compartment for service. Inspect and service the battery, cables and connections. Check for signs of corrosion. Inspect the battery case for cracks or bulges, dirt, acid, and electrolyte leakage.

Make sure the breather pipe is attached to the battery and not pinched by any part of the engine compartment.

Cleaning

Dirt and corrosion should be cleaned from the battery just as soon as it is discovered. Any accumulation of acid film or dirt will permit current to flow between the terminals. Such a current flow will drain the battery over a period of time.

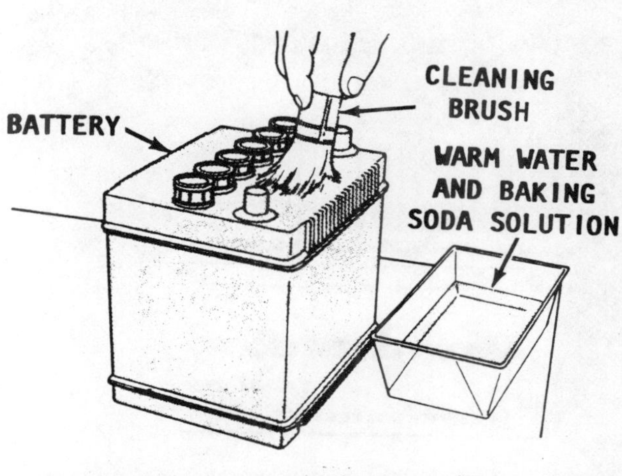

The battery top should be kept clean. A water and baking soda solution applied with a brush will neutralize battery electrolyte and clean the top nicely.

Clean the exterior of the battery with a solution of diluted ammonia or a soda solution to neutralize any acid which may be present. Flush the cleaning solution off with clean water. **TAKE CARE** to prevent any of the neutralizing solution from entering the cells, by keeping the caps tight.

A poor contact at the terminals will add resistance to the charging circuit. This resistance will cause the voltage regulator to register a fully charged battery, and thus cut down on the stator assembly or charging coil output adding to the low battery charge problem.

Scrape the battery posts clean with a suitable tool or with a stiff wire brush. Clean the inside of the cable clamps to be sure they do not cause any resistance in the circuit.

Battery Sulfation

Battery sulfation can be identified by the formation of white crystals forming on top of exposed battery plates. This usually occurs when a battery has been stored in a discharged condition over a period of time and the electrolyte has evaporated leaving the tops of the plates exposed to the air. This condition may also occur if the battery is not serviced at regular intervals during the season and the electrolyte not replenished when low.

Once white crystals form, the plates are permanently damaged and the battery will never hold a charge. It must be replaced.

Filling the Battery

Remove each filler cap using a pair of pliers. Fill each cell to the proper level with distilled water. The level of electrolyte should be between the upper and lower level marks during operation. When filling the battery, **WITHOUT CHARGING,** fill to the upper line and allow the battery to stand for twenty minutes. Check the level again and replenish, as necessary. If filling the battery **PRIOR** to charging, fill to the lower lever only, because the electrolyte level will rise during charging.

During hot weather and periods of heavy use, the electrolyte level should be checked more often than during normal operation. Add distilled water to bring the level of electrolyte in each cell to the proper level. **TAKE CARE** not to overfill, because adding an excessive amount of water will cause loss of electrolyte and any loss will result in poor performance, short battery life, and will contribute quickly to corrosion. **NEVER** add electrolyte from another battery. **NEVER** add sulphuric acid because such acid would make the electrolyte too stong and permanently damage the battery. **NEVER** add ordinary tap water. The minerals in tap water will shorten battery life.

Battery Testing

A hydrometer is a device to measure the percentage of sulphuric acid in the battery electrolyte in terms of specific gravity. When the condition of the battery drops from fully charged to discharged, the acid leaves the solution and enters the plates,

*The battery **MUST** be filled correctly. The proper procedure for this task is explained in the text.*

Pay attention to the "Poison/Danger" warning decal on the back of the battery.

causing the specific gravity of the electrolyte to drop.

Use a temperature corrected hydrometer to test the specific gravity of the electrolyte. At 20°C (68°F) the hydrometer reading should be 1.20-1.26 on the scale.

It may not be common knowledge, but hydrometer floats are calibrated for use at 80°F (27°C). If the hydrometer is used at any other temperature, hotter or colder, a correction factor must be applied. Remember, a liquid will expand if it is heated and will contract if cooled. Such expansion and contraction will cause a definite change in the specific gravity of the liquid, in this case the electrolyte.

A quality hydrometer will have a thermometer/temperature correction table in the lower portion, as shown in the accompanying illustration. By knowing the air temperature around the battery and from the table, a correction factor may be applied to the specific gravity reading of the hydrometer float. In this manner, an accurate determination may be made as to the condition of the battery.

The following six points should be observed when using a hydrometer.

1- **NEVER** attempt to take a reading immediately after adding water to the battery. Allow at least 1/4 hour of charging at the recommended rate to thoroughly mix the electrolyte with the new distilled water. This time will also allow for the necessary gases to be created.

2- **ALWAYS** be sure the hydrometer is clean inside and out as a precaution against contaminating the electrolyte.

3- If a thermometer is an integral part of the hydrometer, draw liquid into it several times to ensure the correct temperature before taking a reading.

4- **BE SURE** to hold the hydrometer vertically and suck up liquid only until the float is free and floating.

5- **ALWAYS** hold the hydrometer at eye level and take the reading at the surface of the liquid with the float free and floating.

Disregard the light curvature appearing where the liquid rises against the float stem. This phenomenon is due to surface tension.

6- **DO NOT** drop any of the battery fluid on the boat or on your clothing, because it is extremely caustic. Use water and baking soda to neutralize any battery liquid that does accidentally drop.

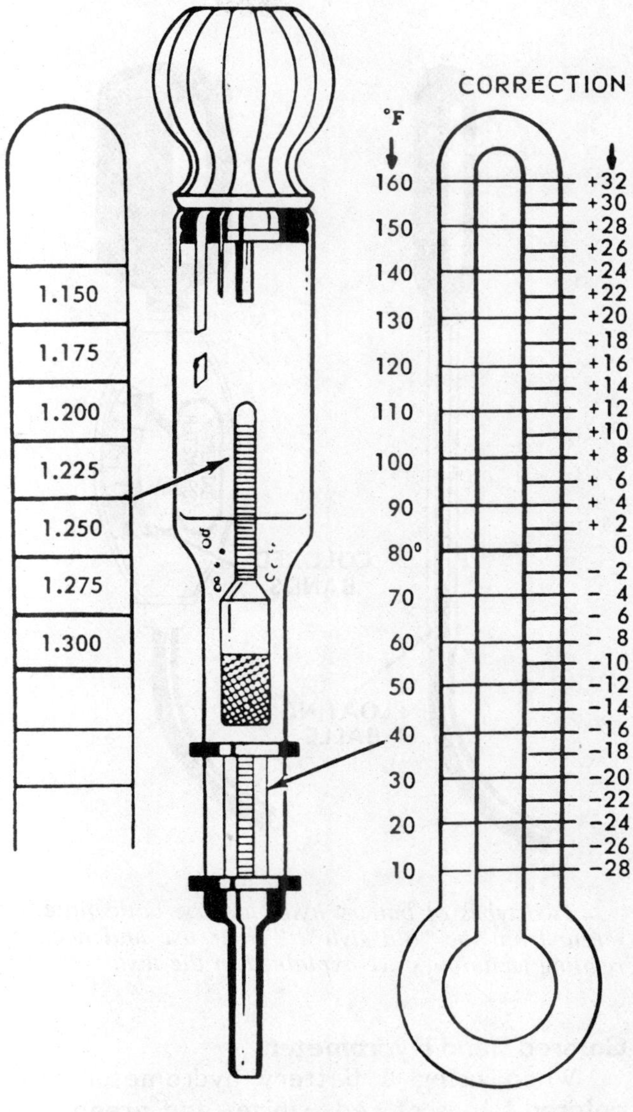

A check of the electrolyte in the battery should be on the maintenance schedule for any craft. A hydrometer reading of 1.300, or in the Green band, indicates the battery is in satisfactory condition. If the reading is 1.150, or in the Red band, the battery MUST be charged. Observe the six safety points listed in the text, when using a hydrometer.

After withdrawing electrolyte from the battery cell until the float is barely free, note the level of the liquid inside the hydrometer.

Reading the Hydrometer

Different manufacturers of battery hydrometers use different methods of displaying specific gravity readings. Three methods are used:

Red, white, and green bands
Small floating balls
Graduated scale

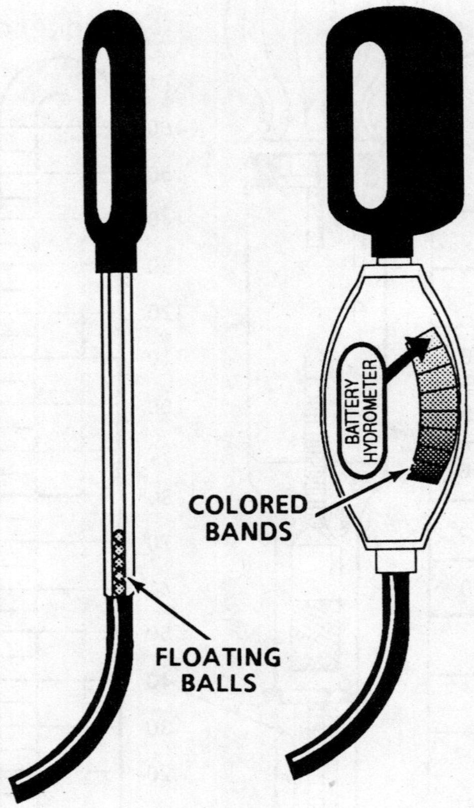

Two types of battery hydrometers, both quite different from the "old style". Their use and accurate reading techniques are explained in the text.

Colored Band Hydrometer

When using a battery hydrometer with colored bands of red, white, and green, the electrolyte level is interpreted as follows:

Draw electrolyte from the battery cell until the float is barely free, note the level of the liquid inside the hydrometer.

If the level is within the green band range for all cells, the condition of the battery is satisfactory. If the level is within the white band for all cells, the battery is in fair condition.

If the level is within the green or white band for all cells except one, which registers in the red, the cell is shorted internally. No amount of charging will bring the battery back to satisfactory condition.

If the level in all cells is about the same, even if it falls in the red band, the battery may be recharged and returned to service. If the level fails to rise above the red band after charging, the only solution is to replace the battery.

Small Float Ball Hydrometer

If using a battery hydrometer with small floating balls, the electrolyte level is interpreted as follows:

If four or five balls are floating on top of the electrolyte drawn into the hydrometer, the battery is 100% charged. If three balls are floating, the battery is 75% charged. If two balls are floating, the battery is 50% charged. If only one ball is floating the battery is 25% charged. If no balls float, the battery is 0% charged.

If the same number of balls float for each cell, the condition of the battery is satisfactory.

If all the cells but one float balls, the cell is shorted internally. No amount of charging will bring the battery back to satisfactory condition.

If the number of floated balls for all cells is about the same, even if it is only one or two, the battery may be recharged and returned to service. If the number of balls fails to increase after charging, the only solution is to replace the battery.

Graduated Scale Hydrometer

If using a battery hydrometer with a graduated scale, note the level of the liquid inside the hydrometer. If the level is between the 1.280 and 1.265 mark, the battery is 100% charged. If the level is closer to the 1.210 mark, the battery is 75% charged. If the level is closer to the 1.160 mark, the battery is 50% charged. If it is closer to the 1.120 mark the battery is only 25% charged. If the level is below 1.100 the battery is at 0% charge.

If the level is the same and above 1.160 for all cells, the condition of the battery is satisfactory. If the level is the same and above 1.120 for all cells, the battery is in fair condition.

If the level is the same for all cells except one, which registers low on the scale, the cell is shorted internally. No amount of charging will bring the battery back to satisfactory condition.

If the level in all cells is about the same, even if it registers low on the scale, the battery may be recharged and returned to service. If the level fails to rise above 1.210 after charging, the only solution is to replace the battery.

Charging the Battery

If it is necessary to charge the battery, leave the filler caps lightly resting on the cell openings to allow the gases to escape.

The charging current should not exceed 1.9 amps for 10 hours.

NEVER use a fast charger, such as those found at automotive service stations, to charge a personal watercraft battery. These chargers operate at too high an output, ideal for automotive batteries, but very damaging to personal watercraft batteries. If such a battery is fast charged, excessive heat will be generated which will warp the plates and cause internal shorting. The plates will also shed active material which will accumulate at the bottom of the battery and cause internal shorting.

If, under the recommended charging procedures, the temperature of the electrolyte rises above 115°F (45°C) during charging, reduce the charging rate and increase the charging time proportionately. If possible, stop, allow the battery to cool and then resume charging for the recommended length of time.

After charging, push each filler cap into place and wipe the top of the battery clean before installation.

Installing the Battery

Check to be sure the battery is fastened securely in position. The battery case and cover **MUST** be fastened securely to the hull to prevent any movement of the battery in the holder.

Clean the battery posts and battery cable ends with a wire brush to ensure good, clean connections. If the battery posts or cable terminals are corroded, the cables should be cleaned separately with a baking soda solution and a wire brush. Apply a thin coating of Multi-purpose Lubricant to the posts and cable clamps before making the connections. The lubricant will help to prevent corrosion.

Identify the "POS" or "+" and "NEG" or "-" embossed symbols. Correctly connect the battery cables to the battery terminals. If the cables are connected incorrectly, the ignition system **WILL** be destroyed the first time the engine is started.

Check to be sure the breather pipe is attached to the battery and not pinched by any part of the engine compartment.

Jumper Cables

If a booster battery is used for starting an engine, the jumper cables must be connected correctly and in the proper sequence to prevent damage to either battery, or to the diodes in the regulator/rectifier.

ALWAYS connect a cable from the positive terminal of the dead battery to the positive terminal of the good battery **FIRST. NEXT**, connect one end of the other cable to the negative terminal of the good battery and the other end to a good ground on the engine. **DO NOT** connect the negative jumper from the good battery to the negative terminal of the low battery. Such action will almost always cause a spark which could ignite gases escaping through the vent holes in the battery filler caps. Igniting the gases may result in an explosion destroying the battery and causing severe personal **INJURY.**

By making the negative (ground) connection on the engine, if an arc is created, it will not be near the battery.

DISCONNECT the battery ground cable before replacing a stator assembly or charging coil, or before connecting any type of meter to the ignition system.

If it is necessary to use a trickle charger on a dead battery, **ALWAYS** disconnect one

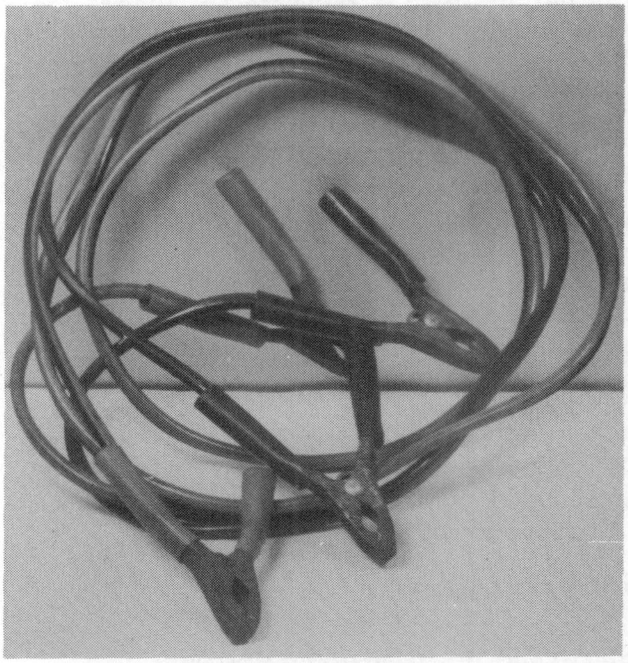

A common set of heavy-duty jumper cables. The booster battery must be connected correctly and in the proper sequence, as outlined in the text. Proper procedure will prevent damage to the battery or the diodes in the rectifier/regulator.

of the cables from the battery **FIRST**, to prevent burning out the diodes in the regulator/rectifier.

NEVER use a trickle charger as a booster to start the engine because the diodes will be **DAMAGED.**

Storage

If the craft is to be laid up for the winter or for more than a few weeks, special attention must be given to the battery to prevent complete discharge or possible damage to the terminals and wiring. Before putting the craft in storage, disconnect and remove the battery. Clean the battery thoroughly of any dirt or corrosion, and then charge it to full specific gravity reading. After the battery is fully charged, store the battery in a clean cool dry place where it will not be damaged or knocked over, preferably on a couple blocks of wood. Storing the battery up off the deck, will permit air to circulate freely around and under the battery and will help to prevent condensation.

NEVER store the battery with anything on top of it or cover the battery in such a manner as to prevent air from circulating around the filler caps. All batteries, both new and old, will discharge during periods of storage, more so if they are hot than if they remain cool. Therefore, the electrolyte level and the specific gravity should be checked at regular intervals. A drop in the specific gravity reading is cause to charge the battery back to a full reading.

One of the largest personal watercraft battery manufacturers recommends that a battery in storage should be checked every two weeks and charged if necessary. The electrolyte level can also be replenished at this time to prevent sulfation occurring.

In cold climates, care should be exercised in selecting the battery storage area. A fully charged battery will freeze at about 60° below zero. A discharged battery, almost dead, will have ice forming at about 19° above zero.

9-3 TACHOMETER

An accurate tachometer can be connected to any engine. Such an instrument provides an indication of engine speed in revolutions per minute (rpm). This is accomplished by measuring the number of electrical pulses per minute generated in the primary circuit of the ignition system.

The meter readings range from 0 to 6,000 rpm, in increments of 100. Tachometers have solid-state electronic circuits which eliminates the need for relays or batteries and contributes to their accuracy. The electronic parts of the tachometer susceptible to moisture are coated to prolong their life.

The manufacturer recommends the use of a Bombardier Electronic Tachometer. Instructions on how to connect the leads are provided with the meter. However, as a general rule, on a two stroke engine with CDI type ignition, the tachometer leads are usually connected across the charging coil leads, refer to Chapter 7 for charging coil lead color identification.

9-4 ELECTRICAL SYSTEM GENERAL INFORMATION

The electrical system consists of three circuits:

Charging circuit
Cranking motor circuit
Ignition circuit

Most ignition components are secured behind or above the flywheel on the stator plate.

Charging Circuit

Components of the charging circuit include:

- Permanent magnets attached to the inside perimeter of the flywheel.
- Battery charging coil installed on the magneto assembly -- a cluster of three coils.
- Regulator/rectifier located elsewhere on the engine.
- External battery.
- Necessary wiring to connect it altogether.

The negative side of the rectifier/regulator is grounded. The positive side of the rectifier/regulator is connected to the battery. The negative side of the battery is connected to a good ground on the engine.

The alternating current generated in the stator windings passes to the rectifier/regulator. The rectifier/regulator changes the alternating current (AC) to direct current (DC) to charge the 12-volt battery.

The magneto assembly is located behind and protected by, the flywheel. Therefore, the assembly, including the battery charging coil, seldom causes problems in the charging

Once the flywheel has been "pulled", the magneto assembly, with battery charging coils, and the generating coils are exposed.

circuit. Most problems in the charging circuit can usually be traced to the rectifier/regulator or to the battery. If either the magneto assembly or the rectifier/regulator fails the troubleshooting tests, the defective unit cannot be repaired, it **MUST** be replaced.

See Chapter 7 -- **IGNITION** to service and test components in the charging system.

Cranking Motor Circuit

The cranking motor circuit consists of a cranking motor and a starter-engaging mechanism. A starter solenoid is used as a heavy-duty switch to carry the heavy current from the battery to the cranking motor. The starter solenoid is actuated by either depressing the **START** button.

Ignition Circuit

The ignition circuit is covered extensively in Chapter 7.

9-5 CRANKING MOTOR CIRCUIT SERVICE

DESCRIPTION

As the name implies, the sole purpose of the cranking motor circuit is to control operation of the cranking motor to crank the engine until the air/fuel mixture ignites and the flywheel develops enough torque to continue under its own steam. The circuit includes a solenoid or magnetic switch to connect or disconnect the motor from the battery. The operator controls the switch with a push button.

The cranking motor is a series wound electric motor which draws a heavy current from the battery. It is designed to be used only for short periods of time to crank the engine for starting. To prevent overheating the motor, cranking should not be continued for more than 30-seconds without allowing the motor to cool for at least three minutes. Actually, this time can be spent in making preliminary checks to determine why the engine fails to start.

Theory of Operation

Power is transmitted from the cranking motor to the engine flywheel through meshed gears or clutch mechanisms. All Bombardier models are equipped with a Bendix drive on the end of the armature shaft. When the motor is operated, the pinion gear moves forward and meshes with the teeth on the flywheel ring gear.

On a cranking motor equipped with a Bendix drive: When the engine starts, the pinion gear is driven faster than the armature shaft, and as a result, it screws out of mesh with the flywheel. A rubber cushion is built into the Bendix drive to absorb the shock when the pinion meshes with the flywheel ring gear. The parts of the drive **MUST** be properly assembled for efficient operation. If the drive is removed for cleaning, **TAKE CARE** to assemble the parts in the correct order. If the screw shaft assembly is reversed, it will strike the splines and the rubber cushion will not absorb the shock.

Cranking Motor Noises

The sound of the motor during cranking is a good indication of whether the cranking motor is operating properly or not. Naturally, temperature conditions will affect the speed at which the cranking motor is able to crank the engine. The speed of cranking a cold engine will be much slower than when cranking a warm engine. An experienced operator will learn to recognize the favorable sounds of the engine cranking under various conditions.

Faulty Symptoms

If the cranking motor spins, but fails to crank the engine, the cause is usually a corroded or gummy Bendix drive. The drive should be removed, cleaned, and given an inspection.

If the cranking motor cranks the engine too slowly, the following are possible causes and the corrective actions that may be taken:

a- Battery charge is low. Charge the battery to full capacity.

b- High resistance connections at the battery, solenoid, or motor. Clean and tighten all connections.

c- Undersize battery cables. Replace cables with sufficient size.

Maintenance

The cranking motor does not require periodic maintenance or lubrication, except for lubrication of the pinion gear. If the motor fails to perform properly, the checks outlined in the previous paragraph should be performed.

Naturally, the motor will have to be removed if the corrective actions outlined under **Faulty Symptoms** above, does not restore the motor to satisfactory operation.

Otherwise, the cranking motor is not normally removed from the engine until it fails.

CRANKING MOTOR TROUBLESHOOTING

Before wasting too much time troubleshooting the cranking motor circuit, the following checks should be made. Many times, the problem will be corrected.

a- Battery fully charged.
b- Main fuse is "good" (not blown).
c- All electrical connections clean and tight.
d- Wiring in good condition, insulation not worn or frayed.

The following troubleshooting procedures are presented in a logical sequence.

Do not operate the cranking motor for more than 15 seconds. Prolonged cranking motor operation will cause overheating and damage the motor.

After each test allow the cranking motor to cool for a minute or so.

Never press the start button to activate the cranking motor while the engine is operating. Such action will damage the pinion and/or flywheel gears.

Cranking Circuit Tests

1- Remove both spark plug leads from the plugs and ground the ends to the engine to prevent accidental engine start. Keep the leads grounded for the following tests.

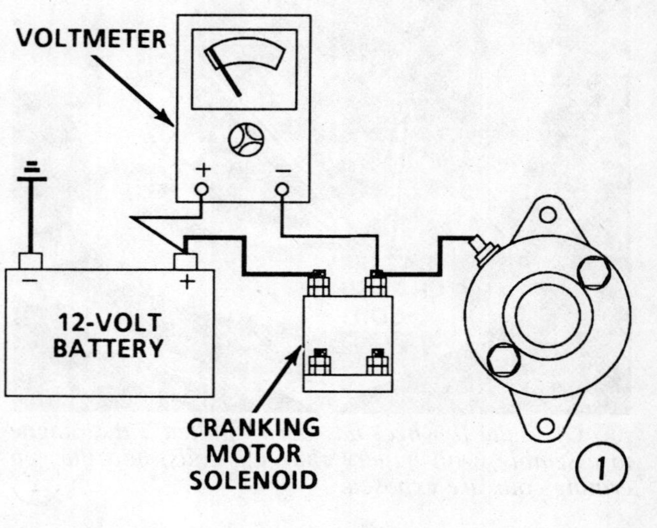

1

Select the Vx1 DC scale on the meter. Observe the starter solenoid. Both large terminals have Red leads connected to them. One terminal is connected to the battery and the other terminal is conected to the cranking motor. Make contact with the negative Black meter lead to the terminal on the starter solenoid which has a Red lead from the battery connected to it.

Make contact with the positive Red meter lead to the positive battery terminal. The voltmeter should register less then 0.25V DC. If the reading exceeds 0.25V DC, clean the connections at both ends of the positive battery cable and repeat the test. If the reading does not change, replace the cable.

2- Keep the meter on the same scale. Make contact with the negative Black voltmeter lead to the negative terminal on the battery. Make contact with the positive Red voltmeter lead to a good engine ground. The voltmeter should register less then 0.25V DC. If the reading exceeds 0.25V DC, clean the connections at both ends of the negative battery cable and repeat the test. If the reading does not change, replace the cable.

CRITICAL WORDS

During the following tests -- Step 3 and Step 4, both meter leads must be placed on the solenoid terminals **WHILE** the engine is being cranked. Both meter leads must be removed from the solenoid terminals **WHILE** the engine is being cranked and **BEFORE** the engine has stopped cranking. If these precautions are not followed, the voltmeter may be damaged.

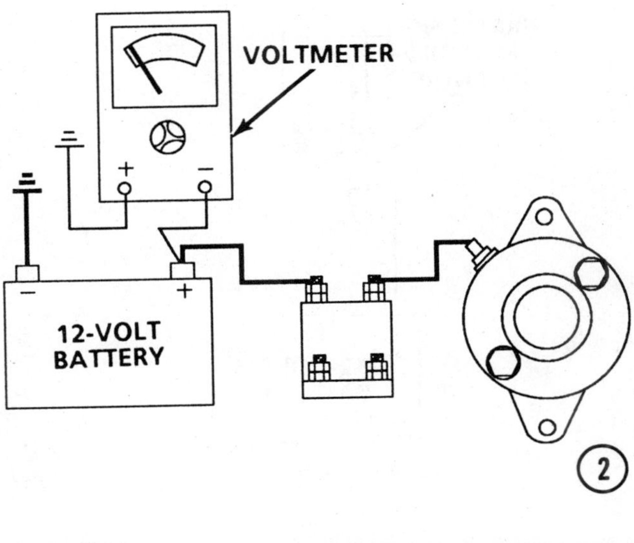

2

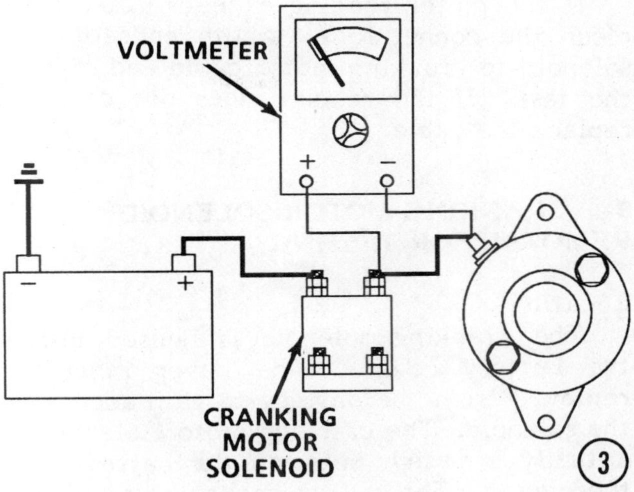

3

3- Keep the meter on the same scale. Obtain the services of an assistant. Have the assistant crank the engine, while the negative Black voltmeter lead makes contact with the solenoid terminal which is also connected to the cranking motor. Then, make contact with the positive Red voltmeter lead to the solenoid terminal which is also connected to the battery.

If the meter reading exceeds 0.25V DC, replace the cranking motor solenoid.

4- Keep the meter on the same scale and the assistant at the **START** button. Observe the same precautions as before -- in Step 3. With the engine cranking, make contact with the negative Black voltmeter lead to terminal on the cranking motor. Make contact with the positive Red voltmeter lead to the solenoid terminal which also has the Red lead from the cranking motor attached to it.

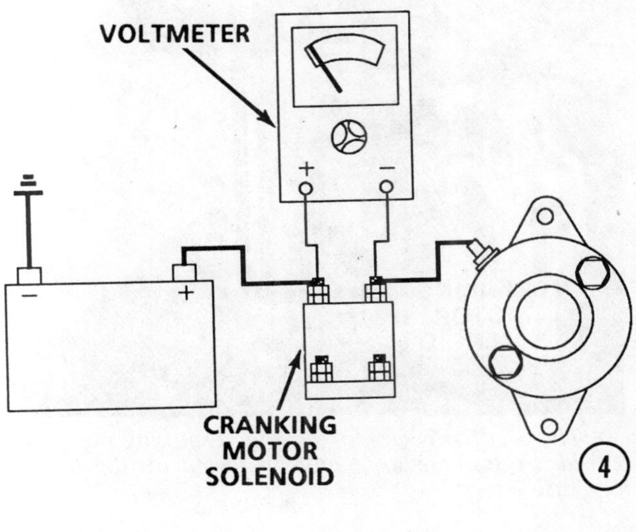

4

If the meter reading exceeds 0.25V DC, clean the connections at both ends of the solenoid-to-cranking motor cable and repeat the test. If the reading does not change, replace the cable.

9-6 CRANKING MOTOR SOLENOID REMOVAL FOR TESTING

Location

The cranking solenoid is housed inside the battery cover. The cover must be removed from the battery to gain access to the solenoid. The cranking motor solenoid is actually a switch between the battery and the engine. The switch cannot be serviced. Therefore, if testing indicates the switch to be faulty, it **MUST** be replaced. The solenoid must be removed from the battery cover for testing purposes.

Preliminary Tasks

Remove the battery cover.

Disconnect the three Red leads from the upper left terminal.

Disconnect the single Red large lead from the upper right terminal.

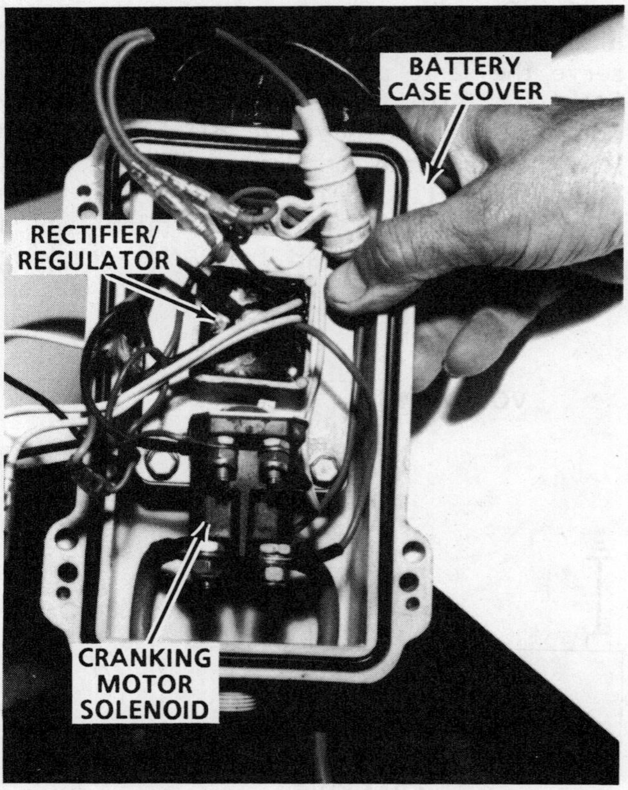

The rectifier/regulator and the cranking motor solenoid are both secured and protected inside the battery case cover.

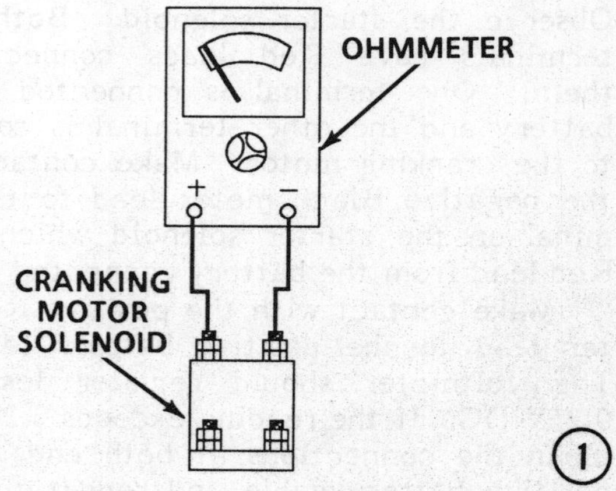

1

Disconnect the Yellow/Red lead from the lower left terminal.

Disconnect the two Brown leads from the lower right terminal.

Cranking Solenoid Testing

1- Select the Rx1 ohm scale on the ohmmeter. Make contact with the two meter leads across the two upper solenoid terminals. The meter should read an infinite resistance. If the meter indicates a resistance of less than infinity, the solenoid is defective and must be replaced.

2- Keep the ohmmeter on the same scale. Obtain a 12V battery and a pair of jumper leads. Using the set of jumper leads, connect the battery across the two lower terminals. Make contact with the two meter leads across the two upper solenoid

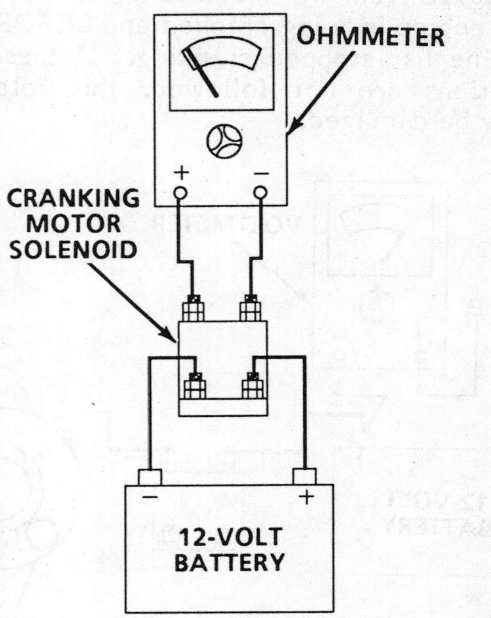

2

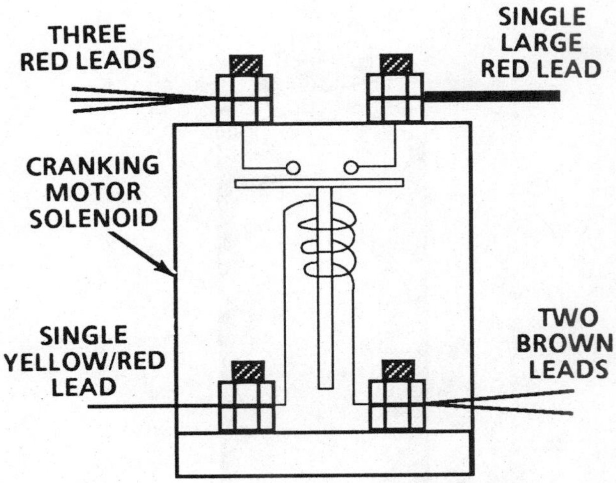

Simple diagram to indicate the wiring connections on the cranking motor solenoid housed inside the battery case cover.

terminals. If the solenoid emits an audible "click" and the ohmmeter registers zero ohms, the solenoid is good. If no "click" is heard, or if the meter registers a high or infinite resistance, the solenoid is defective and must be replaced.

If the solenoid failed the test, it must be replaced. Two bolts secure the solenoid to the battery cover. When installing the replacement solenoid, apply Loctite 242 to the threads of the bolts and install the new solenoid in place. Tighten both bolts securely.

Connect the two Brown leads to the lower right terminal.

Connect the Yellow/Red lead to the lower left terminal.

Connect the single large Red lead to the upper right terminal.

Connect the three smaller Red leads to the upper left terminal.

Place the cover over the battery. Install and tighten the four securing screws.

9-7 CRANKING MOTOR SERVICE

Description

Marine cranking motors are very similar in construction and operation to the units used in the automotive industry.

NEVER operate a cranking motor for more than 30-seconds without allowing it to cool for at least three minutes. Continuous operation without the cooling period can cause serious damage to the cranking motor.

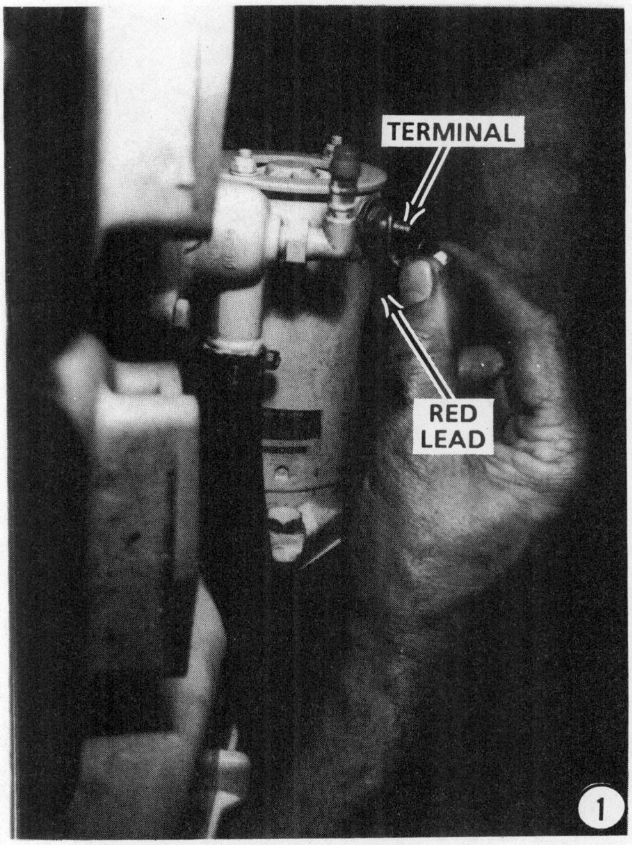

Cranking Motor Removal

1- Disconnect the large Red lead from the terminal on the cranking motor.

2- Remove the upper mounting bolt and the large Black ground lead.

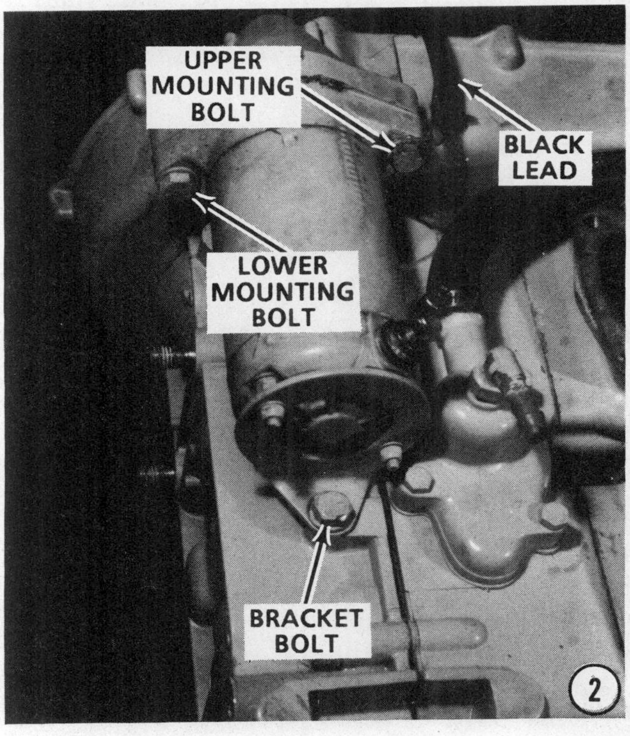

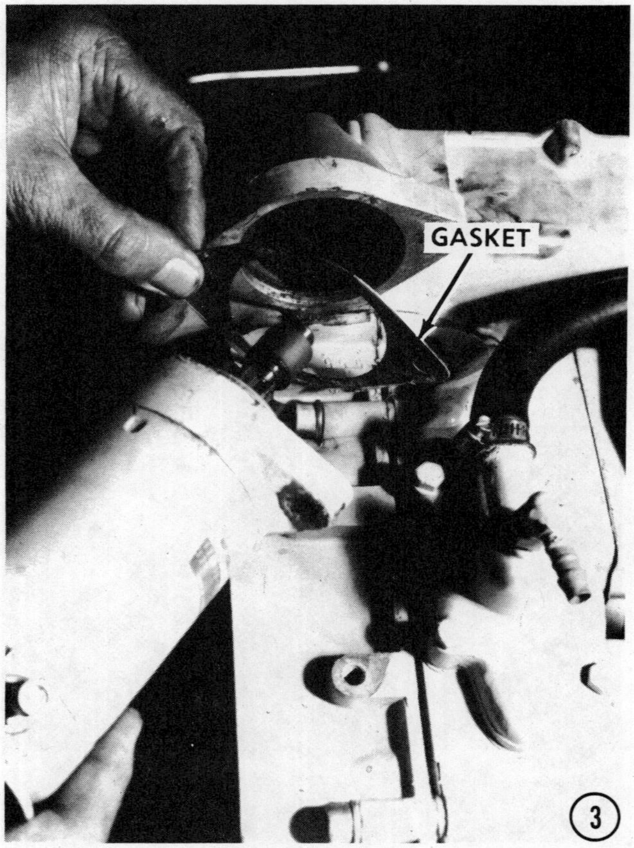

3

Remove the lower mounting bolt and a third bolt securing the aft end of the motor to the lower crankcase.

3- Pull the cranking motor away from the stator plate. If the motor is stuck, tap the mounting flange with a soft head mallet to jar it loose. Remove and discard the gasket.

SPECIAL WORDS

If the cranking motor cannot be removed from the engine compartment, because it will not clear the tuned exhaust pipe -- see Chapter 8 -- **ENGINE**, Steps 17 thru 25 to remove the pipe.

Cranking Motor Disassembling

1- Remove the two nuts and washers from the thru-bolts, and then lift off the support bracket. Remove another set of washers from the thru-bolts.

2- Make a mark across the mating surface of the clutch housing and the motor frame. Make another mark across the mating surface of the motor frame and the end cap. These marks will be most valuable later during assembling.

Without some type of alignment marks, the task of inserting the thru-bolts and

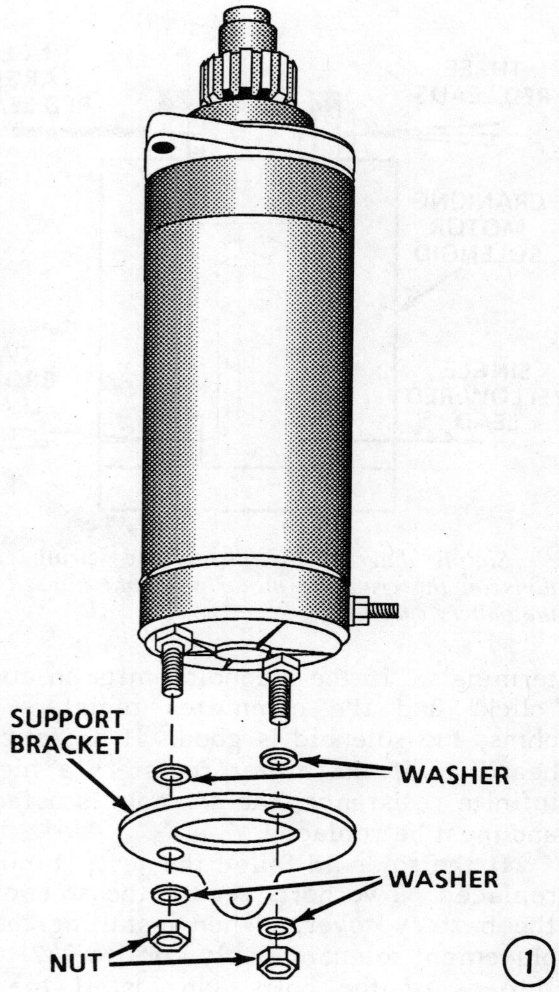

1

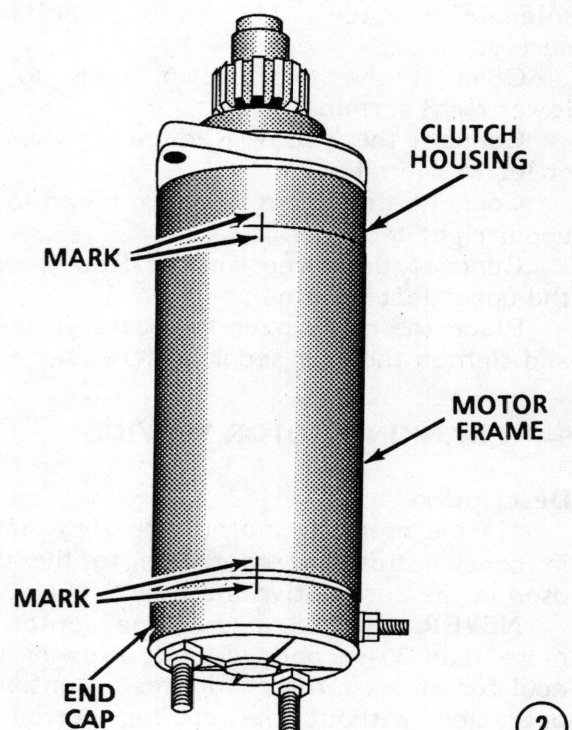

2

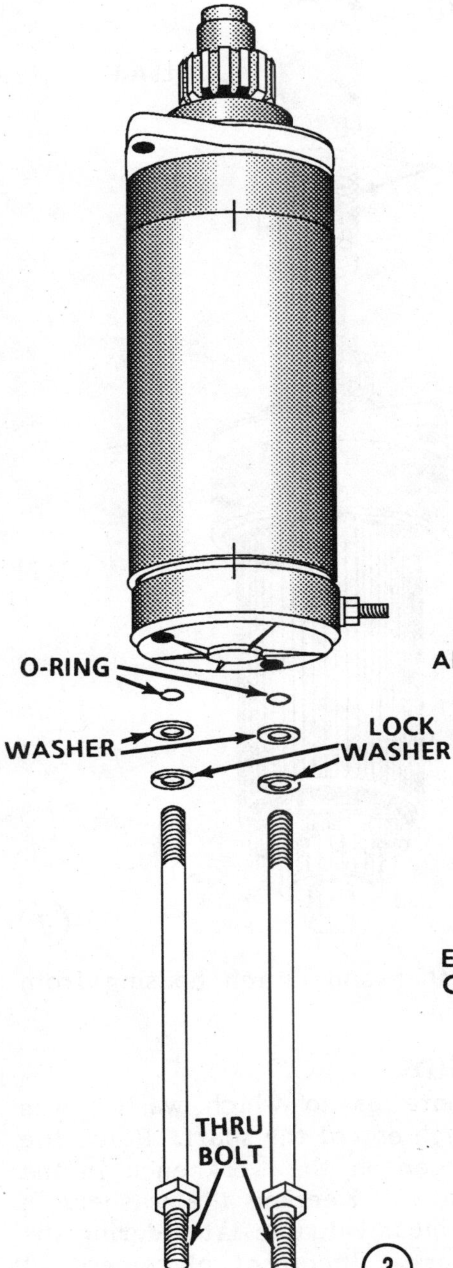

3

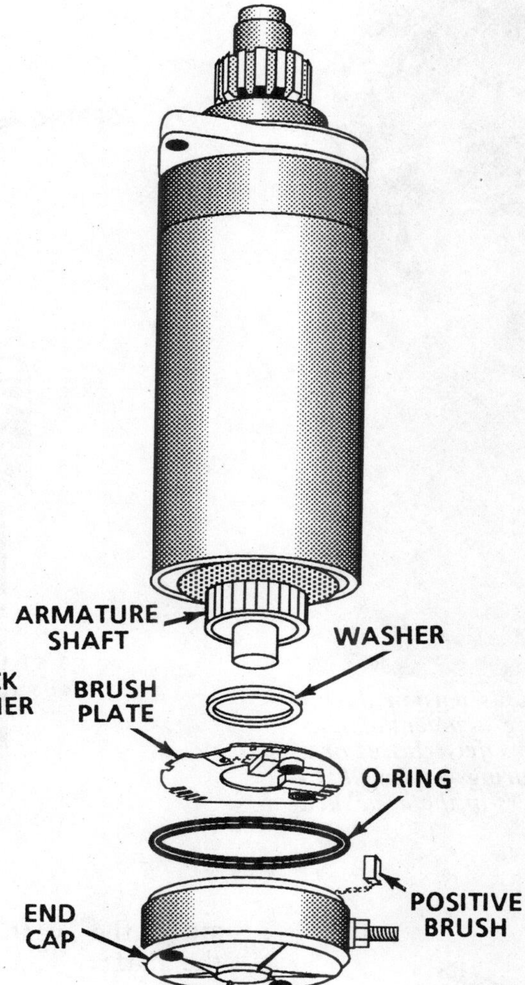

4

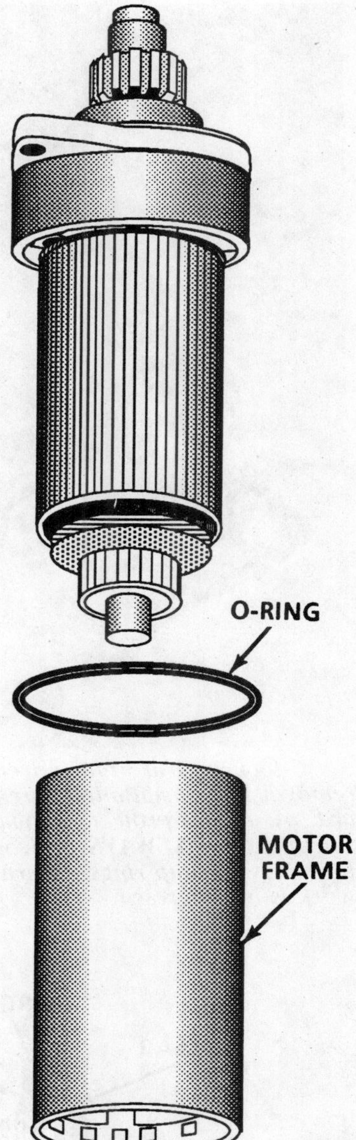

5

starting the threads into the end cap is almost impossible.

3- Remove the two long thru-bolts, lockwashers, plain washers, and the O-rings.

4- Slide the end cap free of the armature shaft, followed by an O-ring, brush plate, and a washer. Remove the positive brush from the brush holder on the plate to separate the brush holder from the end cap. The negative brush will remain in the holder on the plate.

5- Pull the motor frame from the armature shaft. Remove and discard the O-ring.

SAFETY WORDS

Snap rings and C-lockrings are made of spring steel and may pop out of the groove with considerable force. Therefore, warn others in the area and **WEAR** eye protection glasses or a shield while removing this type ring.

6- Push the collar down the shaft until a screwdriver can be inserted under the snap ring. Pry the snap ring out of the groove.

7- Slide the collar and spring off the shaft.

8- Rotate the pinion gear assembly **COUNTERCLOCKWISE** to remove the assembly from the splines on the shaft. Re-

*C-lockrings are under tremendous tension during removal and installation, presenting a potential hazard. As an eye protection measure, safety glasses or a shield should **ALWAYS** be worn during work with C-lockrings or snap rings. Warn others in the area such work is in progress.*

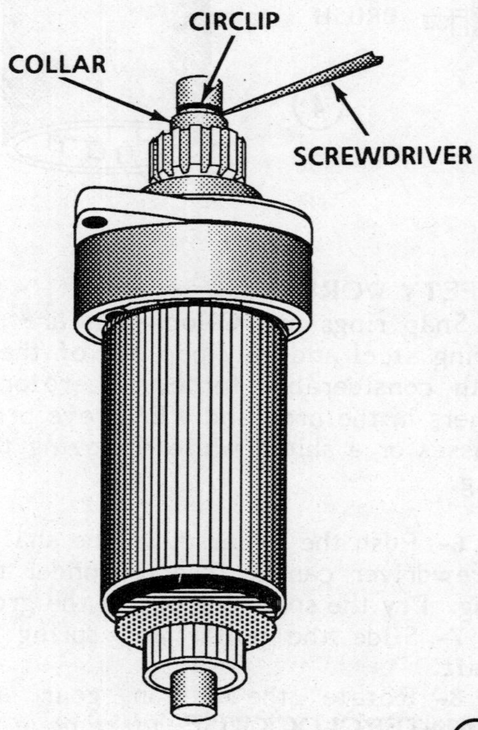

6

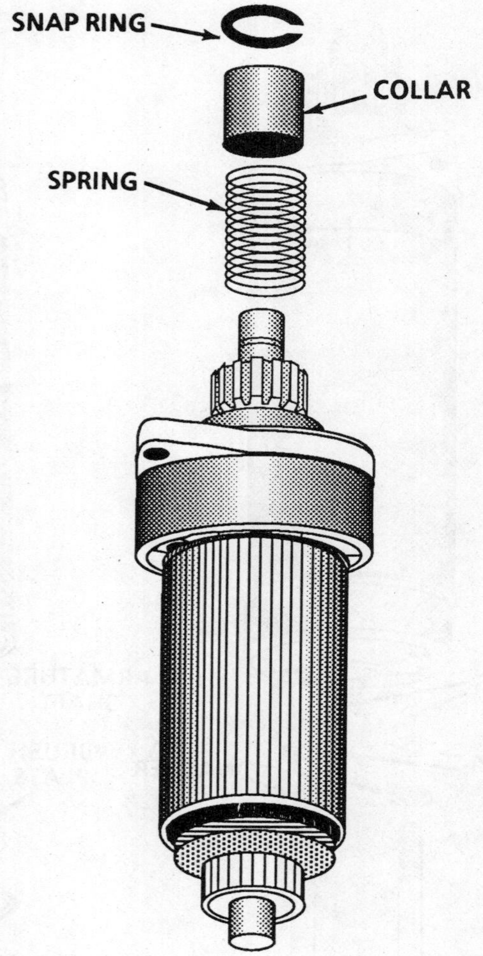

7

move the washer and clutch housing from the shaft.

SPECIAL WORDS

Make a note as to which washer was located at which end of the shaft. Place the washers in order on the workbench in the order of removal. Keeping the washers in order will be most valuable later during the assembling work. Incorrect placement of these washers will cause the armature to be misaligned with the brushes and the field magnets during assembling.

CLEANING AND INSPECTING

Inspect the pinion gear teeth for chips, cracks, or a broken tooth. Check the splines inside the pinion gear for burrs and to be sure the gear moves freely on the armature shaft.

BAD NEWS

This type of cranking motor pinion gear assembly cannot be repaired if the unit is

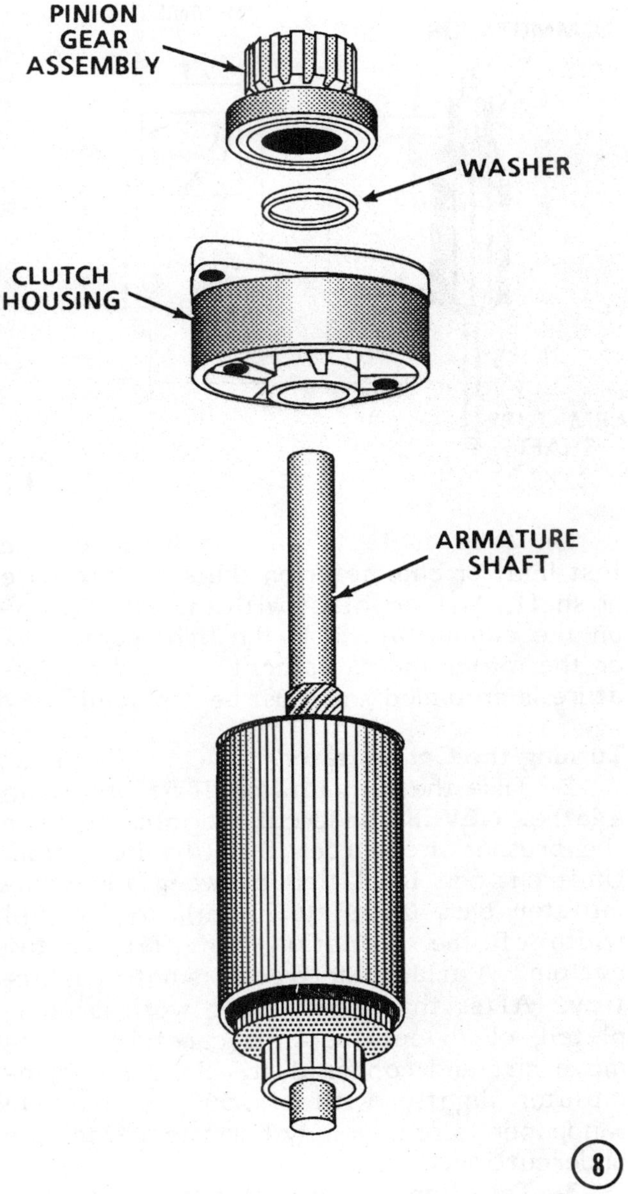

defective. Replacement of the pinion gear assembly as a unit is the only answer.

Check to be sure the return spring is flexible and has not become distorted. Check both ends of the spring for signs of damage.

Clean the armature shaft and check to be sure the shaft is free of any burrs. If burrs are discovered, they may be removed with crocus cloth.

Clean the commutator on the armature shaft and check to be sure the shaft is free of any burrs. If burrs are discovered, they may be removed with crocus cloth.

Clean the field coils, armature, commutator, armature shaft, brush assembly, and end caps with a brush or compressed air.

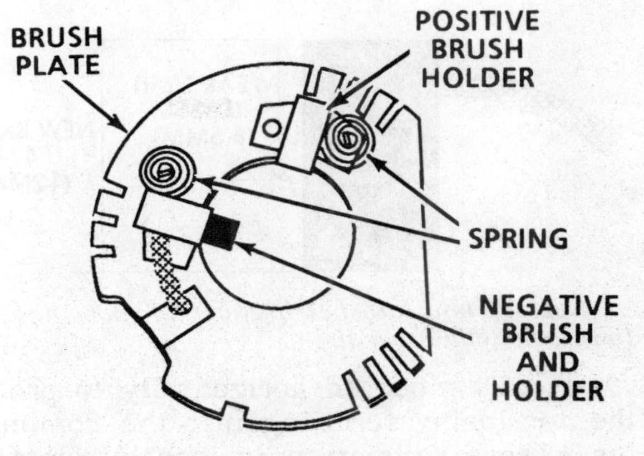

Replace both brushes if either brush has worn down below the wear limit given in the text.

Wash all other parts in solvent and blow them dry with compressed air.

Inspect the insulation and the unsoldered connections of the armature windings for breaks or burns.

Check the commutator for runout. Inspect the armature shaft and both bearings for scoring.

Turn the commutator in a lathe if it is out-of-round by more than 0.005" (0.13mm).

Measure the diameter of the commutator. If the measurement is less than 1.209" (30.7mm), the armature shaft **MUST** be replaced.

Check the springs in the brush holder to be sure none are broken. Check the insulated brush holder for a short to ground.

The armature, fields, and brush holders must be checked before assembling the cranking motor. See Section 9-8, beginning on next page, for detailed procedures to test cranking motor parts.

The negative brush is mounted on the brush plate. The positive brush is attached to the positive terminal. When both the brushes are installed onto the plate, each brush is held in place with wound spring

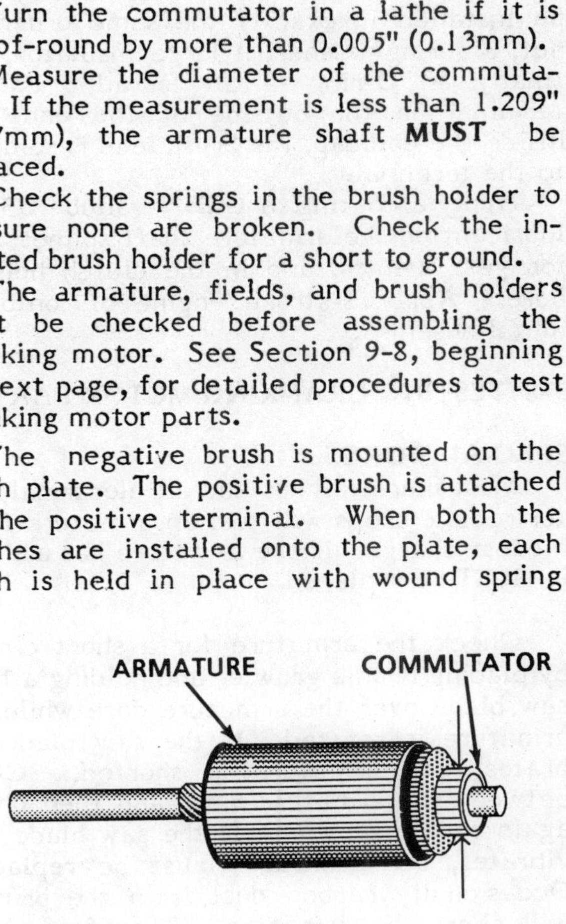

*If the diameter of the commutator is less than 1.209" (30.7mm), the armature **MUST** be replaced.*

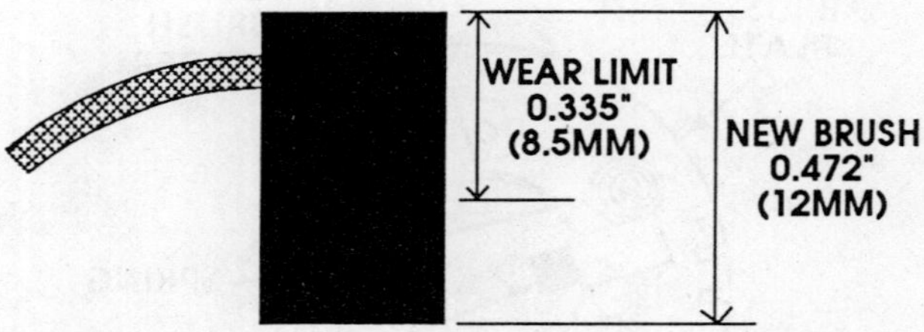

Replace both brushes if either of them has worn down below the wear limit.

steel coils mounted horizontally to provide the tensioning force against the commutator. These coils are permanently riveted to the brush plate and are not removable. If they are broken or missing, a new plate must be purchased.

Measure the length of each brush. A new brush measures 0.472" (12mm). If a brush is less than 0.335" (8.5mm) in length, it must be replaced.

To replace the positive brush, the positive terminal must be removed from the end cap. Remove the following parts and place them in order on the workbench so they may be installed in exactly the same order: a nut, a spring washer, a large insulator, and finally an **O**-ring -- all installed on the terminal on the outside of the end cap. Inside the end cap, the brush lead is soldered to the terminal.

Apply a coating of G.E. Versilube G341M lubricant on the armature shaft splines, pinion gear splines, and in the clutch housing bore. Apply 2-stroke engine oil onto all metal bushings.

9-8 TESTING CRANKING MOTOR PARTS

SPECIAL WORDS

Most marine shops and all electrical motor rebuild shops will test an armature for a modest charge. If the armature has a short, it **MUST** be replaced.

Check the armature for a short circuit by placing it on a growler and holding a hack saw blade over the armature core while the armature is rotated. If the saw blade vibrates, the armature is shorted. Clean between the armature bars, and then check again on the growler. If the saw blade still vibrates, the armature must be replaced. Occasionally carbon dust from the brushes will short the armature. Therefore, blow the slots in the armature clean with compressed air.

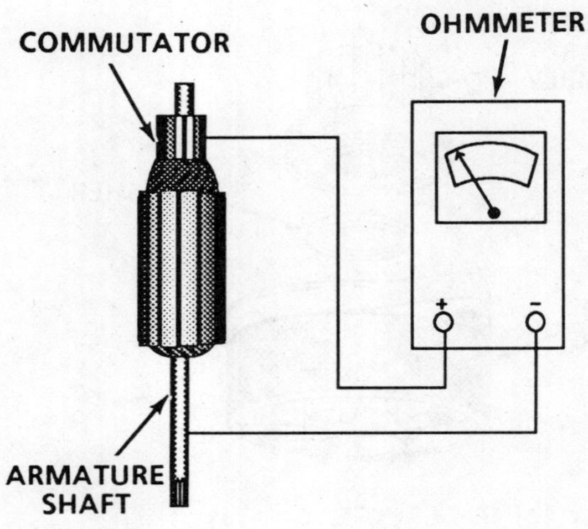

1

1- Make contact with one probe of the test light or ohmmeter on the armature core or shaft. Make contact with the other probe on the commutator. If the light comes on, or the meter indicates continuity, the armature is grounded and must be replaced.

Turning the Commutator

2- True the commutator, if necessary, in a lathe. **NEVER** undercut the mica because the brushes are harder than the insulation. Undercut the insulation between the commutator bars 0.008" (0.20mm), to the full width of the insulation and flat at the bottom. A triangular groove is not satisfactory. After the undercutting work is completed, clean out the slots carefully to remove dirt and copper dust. Sand the commutator lightly with No. 500 or No. 600 sandpaper to remove any burrs left from the undercutting.

3- Test light probes, placed on any two commutator bars, should light and indicate continuity.

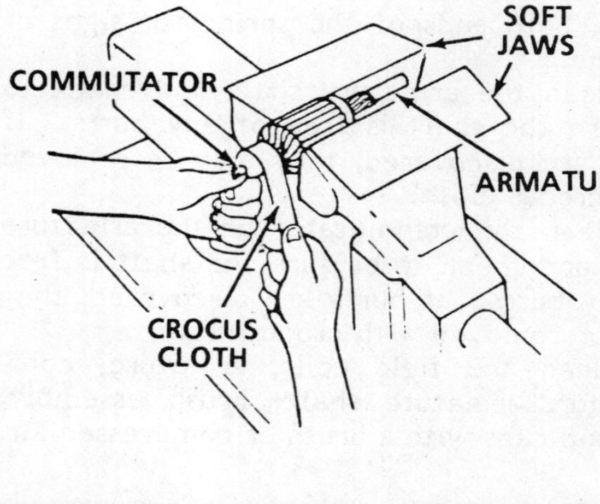

2

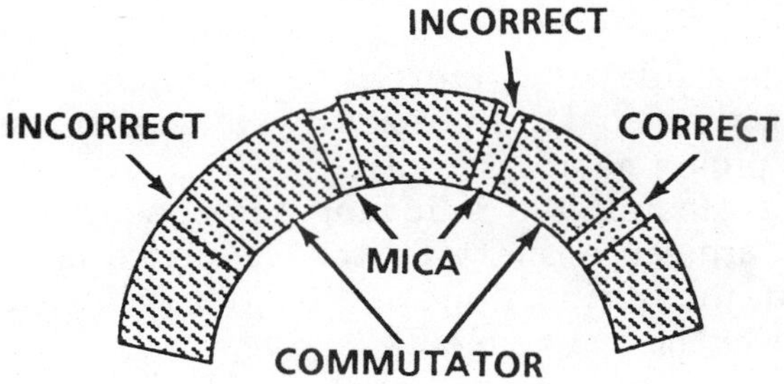

Simple line drawing to illustrate correct and incorrect mica cutting on a commutator.

4- Check the armature a second time on the growler for possible short circuits.

Positive Brush

5- Make contact with one ohmmeter lead to the positive brush. Make contact with the other ohmmeter lead to the positive terminal on the end cap. The meter should register continuity. If the meter registers no continuity, the positive brush has an open in the circuit and must be replaced.

Negative Brush

6- Make contact with one lead of the ohmmeter to the negative brush. Make contact with the other test lead to the brush plate. The ohmmeter should indicate continuity. If the meter indicates no continuity, there is an open in the negative lead or the lead is not grounded properly to the brush plate.

7- Make contact with the ohmmeter leads across the positive and negative brush holders on the brush plate. The meter should register no continuity. If the meter registers continuity, the positive brush holder is shorted to the plate and the plate must be replaced.

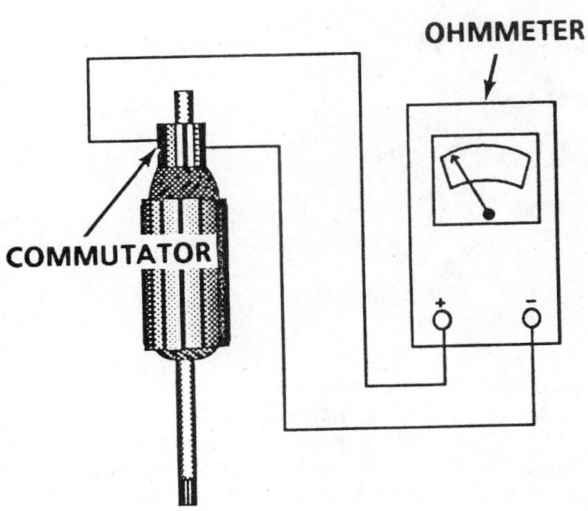

3

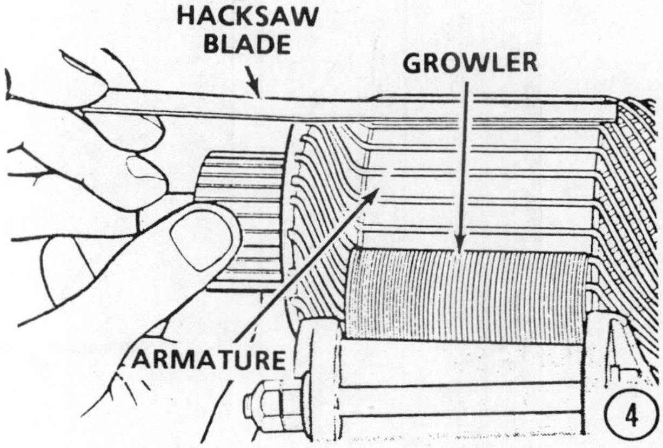

4

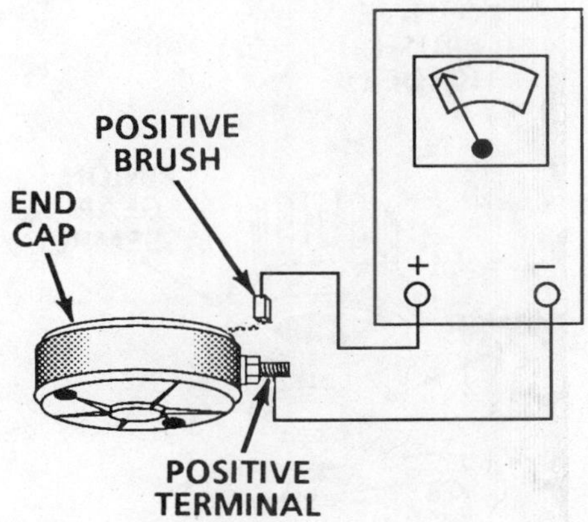

5

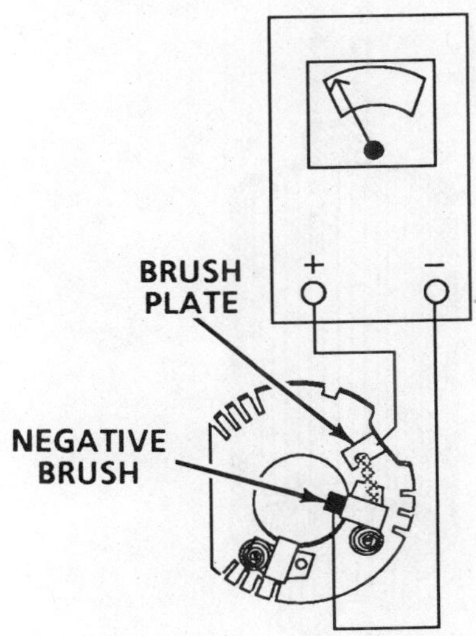

6

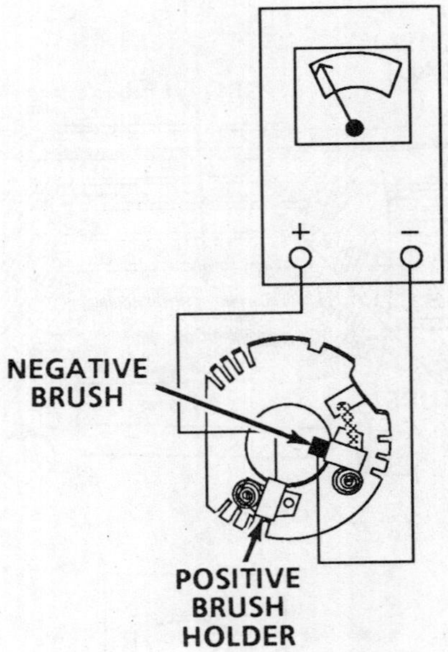

7

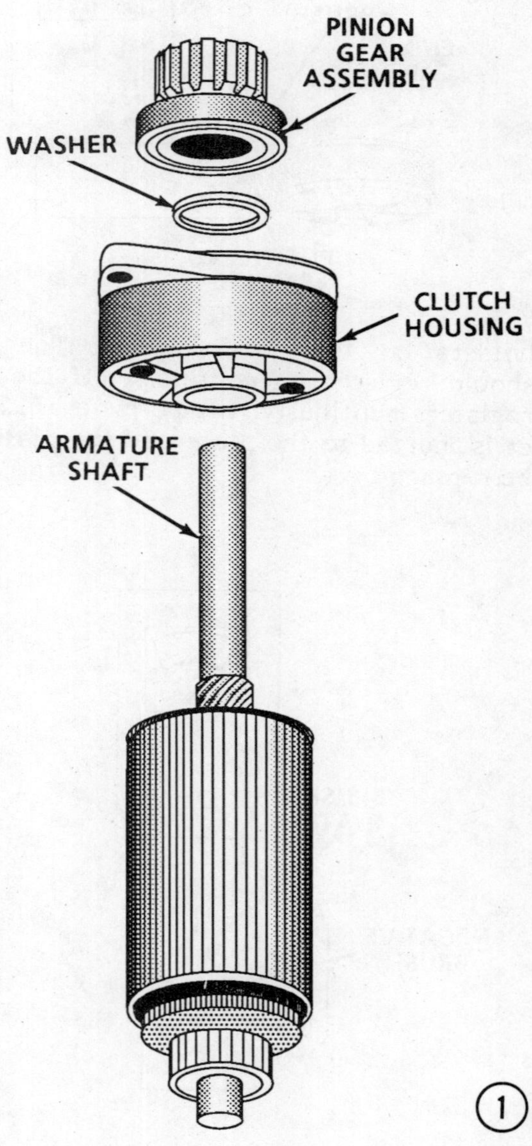

1

CRANKING MOTOR ASSEMBLING

1- Slide the clutch housing onto the armature shaft, followed by a washer, and the pinion gear assembly.

2- Install the following components onto the armature shaft in the sequence given: first the spring, followed by the collar, and finally the snap ring.

SAFETY WORDS

Snap rings and C-lockrings are made of spring steel and may pop out of the groove with considerable force. Therefore, warn others in the area and **WEAR** eye protection glasses or a shield while removing this type ring.

3- Hold the collar down against the spring while the snap ring is installed. The ring must be crimped around the groove in the shaft with a pair of pliers. When properly installed the collar will completely hide the spring. Place the end of the armature shaft on the workbench and use a small punch to tap down the collar, as shown.

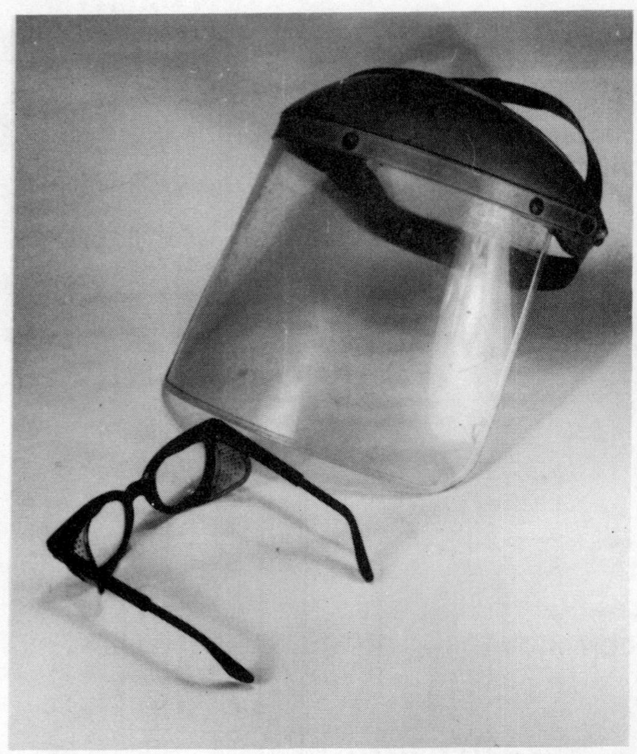

C-lockrings are under tremendous tension during removal and installation, presenting a potential hazard. As an eye protection measure, safety glasses or a shield should ***ALWAYS*** *be worn during work with C-lockrings or snap rings. Warn others in the area such work is in progress.*

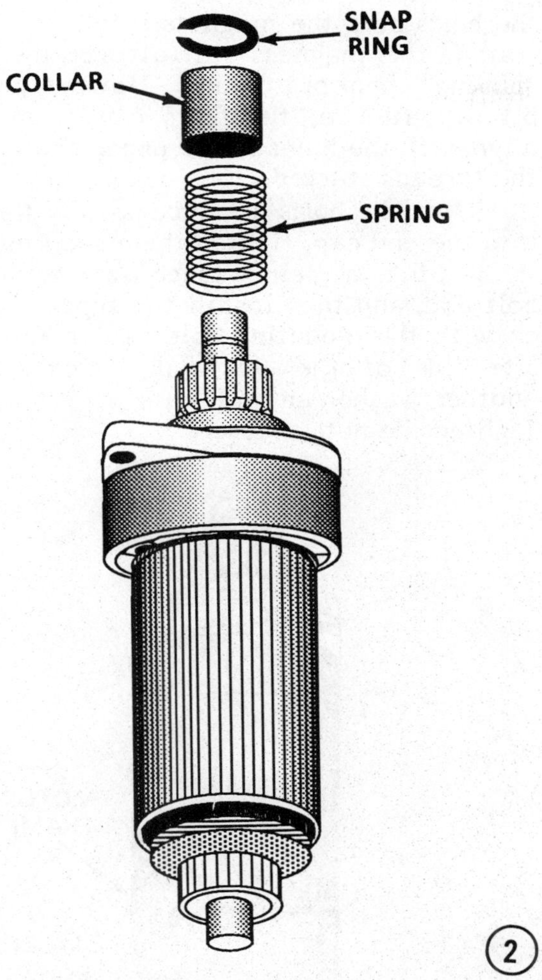

4- Slide a new O-ring over the armature. Install the motor frame over the armature with the protrusions inside the frame facing the commutator on the shaft when installed. Align the mark made on the motor frame with the mark on the clutch housing.

5- If the positive brush was removed or replaced, apply Loctite Superflex to the bolt hole in the end cap. Insert the terminal, with the positive brush attached, through the bolt hole from the inside. Install a washer and an insulator over the terminal and then secure it in place with a locknut.

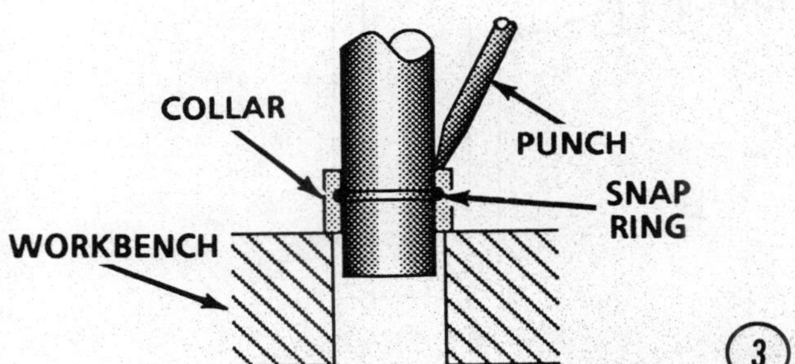

Place the washer over the armature shaft. Position a new O-ring around the end cap. Hold back the spring on the positive brush holder and insert the positive brush.

Push both brushes into their holders and slide the brush plate, with end cap attached, over the commutator.

Check the two brush springs for tension. Rotate the armature shaft. The brushes must stay in contact with the commutator surface without binding.

6- Align the highest protrusion inside the motor frame with the locating notch in the brush holder.

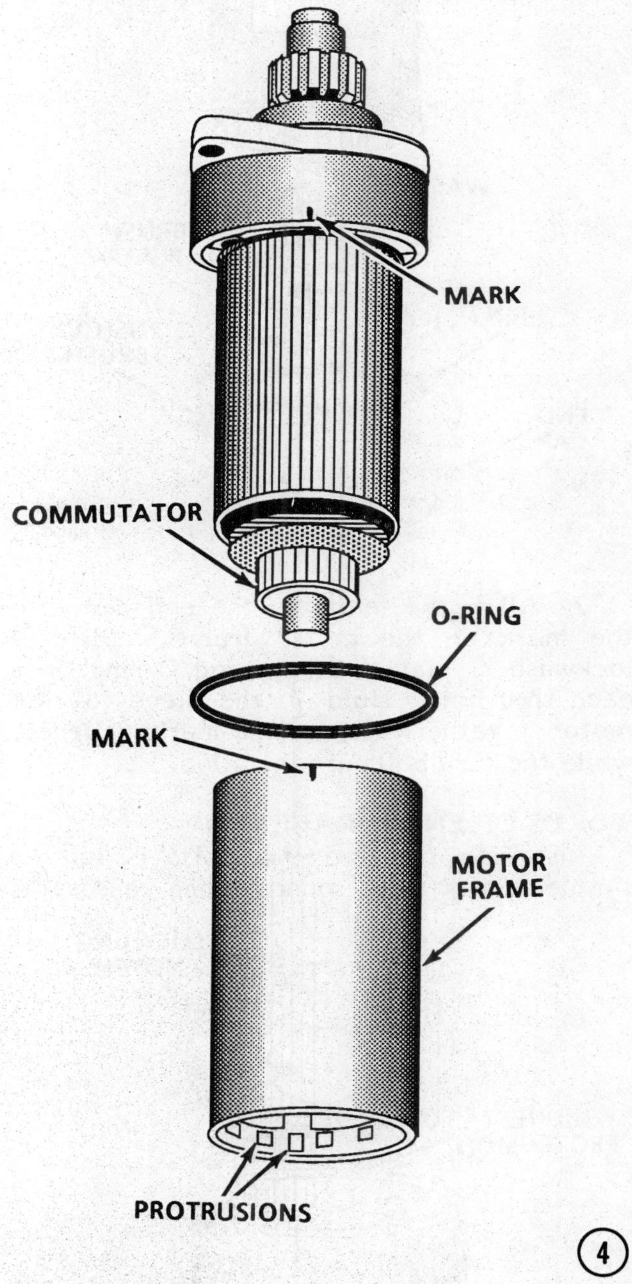

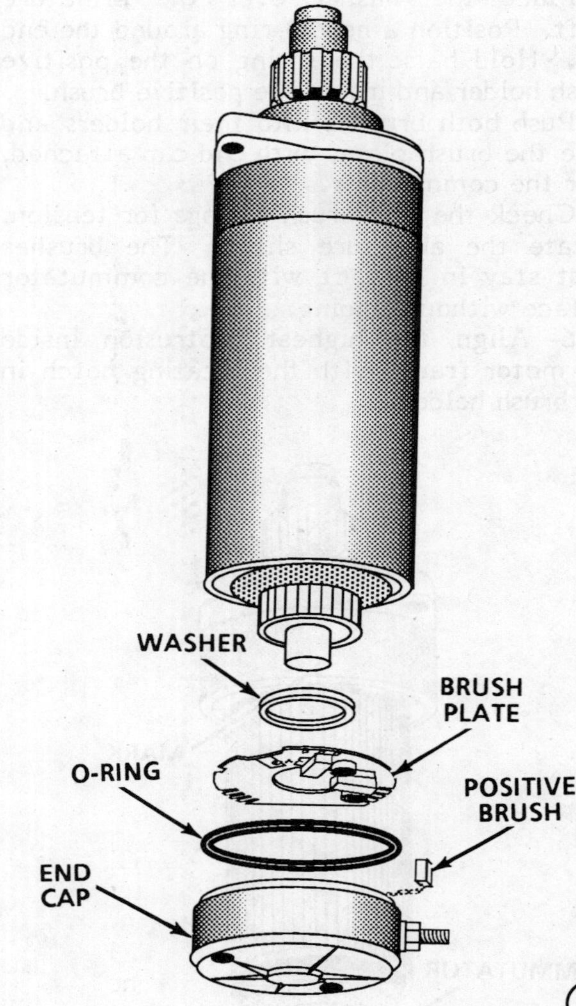

5

7- Align the mark on the end cap with the mark on the motor frame. Slide a lockwasher, plain washer, and O-ring onto each thru bolt. Hold all the pieces of the motor together with index marks aligned, while the thru bolts are installed.

WORDS OF ENCOURAGEMENT

Installing the two thru-bolts is not as simple a task as it sounds. The reason: as

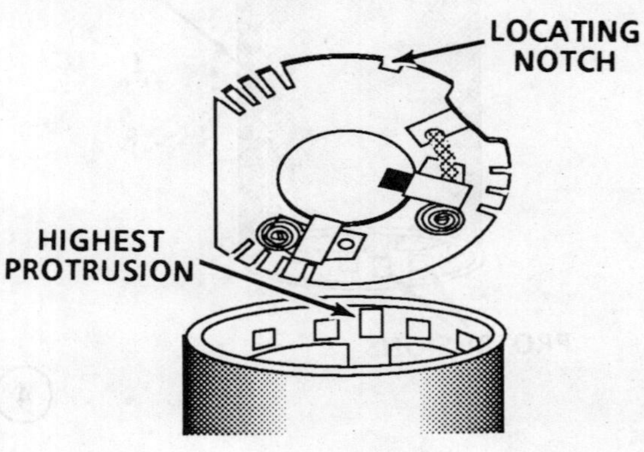

6

the bolts pass the magnets inside the motor frame, the magnets will attract the bolts, "pulling" them out of line. However, with a bit of patience, the bolts can be made to align with the holes in the upper end cap and the threads started.

After the bolts are successfully threaded into the end cap, tighten them securely.

8- Slide a plain washer over each thru-bolt end, and then install the support bracket with the mounting hole tab on the opposite side of the terminal. Next, install another washer and nut onto each thru-bolt. Tighten the nuts securely.

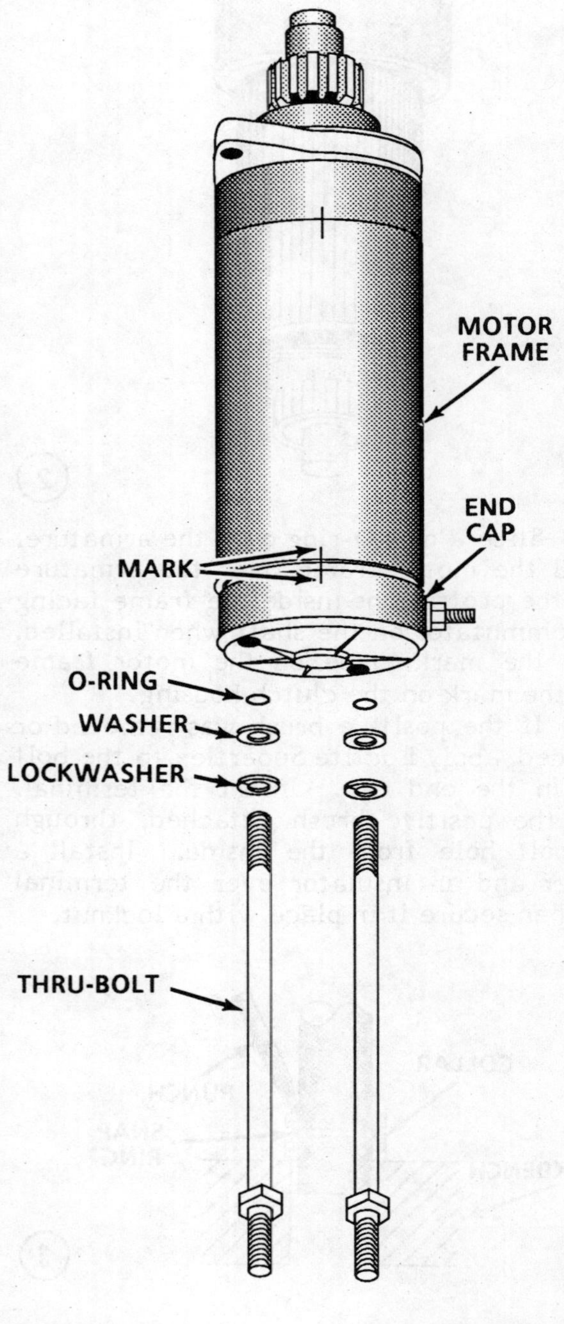

7

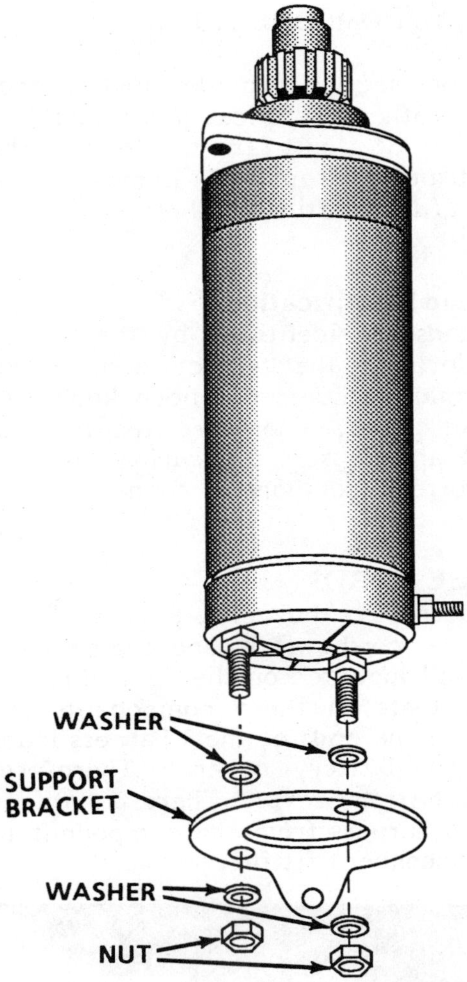

8

Cranking Motor Installation

1- Position a new gasket over the mounting flange and insert the cranking motor into the stator plate.

2- Apply Loctite 242 onto the threads of the three mounting bolts. Position the Black ground lead over the upper mounting bolt. Install the other two bolts, and then tighten all three to a torque value of 16 ft lb (22Nm).

Apply Dielectric lubricant onto the cranking motor terminal threads.

3- Connect the large Red lead onto the cranking motor terminal and tighten the securing nut to a torque value of 62 in lb (7Nm).

Check Completed Work

Secure the craft in a test tank or body of water. If a body of water or tank is not possible, connect a flush attachment and hose to the engine.

1

CAUTION

The following procedure MUST be followed whenever the engine is started with a hose attached for cooling the engine. Start the engine, and then IMMEDIATELY turn the water on. The engine MUST be started first to prevent water from backing up into

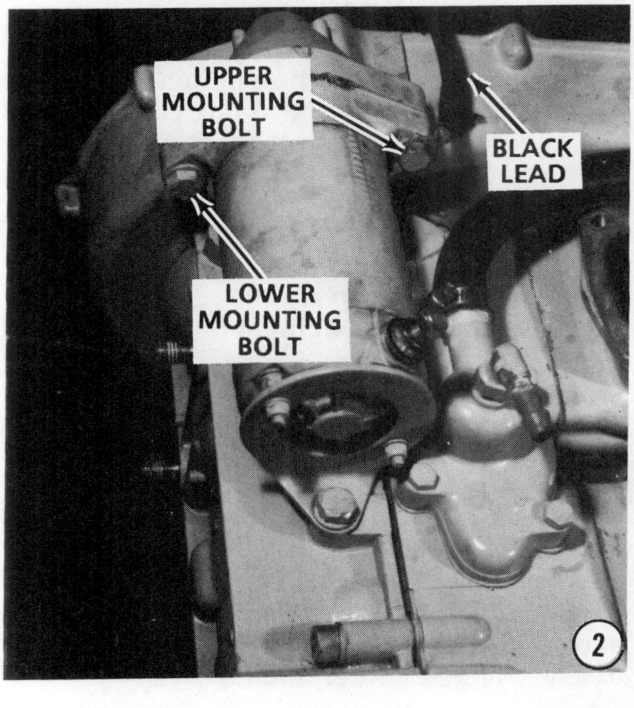

2

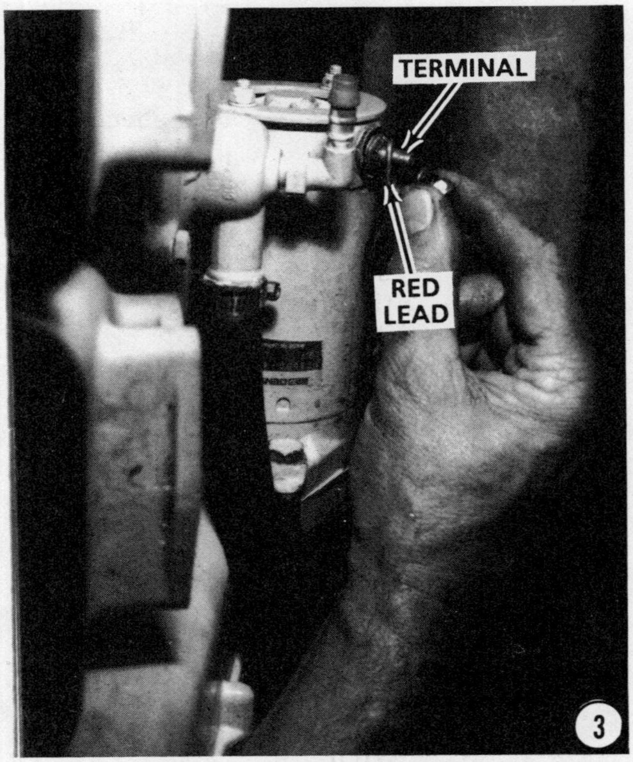

the tuned exhaust pipe and into the engine. Water must circulate anytime the engine is operating to prevent overheating -- which could cause damage to moving parts and possibly cause engine seizure. NEVER allow the engine to operate without cooling water for more than 15 seconds. NEVER run the engine with a flushing attachment for more than 5 minutes. The driveline seal is not provided with cooling water if it is not immersed in water. To shut the engine down, first shut off the water supply, and then shut down the engine. If this sequence of tasks is not followed as stated, the engine could be severely damaged.

REMEMBER
Start the engine.
Turn the water on.
Turn off the water.
Shut down the engine.

NEVER, again **NEVER** operate the engine at high speed with a flush device attached. The engine, operating at high speed with such a device attached, would **RUNAWAY** from lack of a load on the impeller, causing extensive damage.

Crank the engine with the cranking motor and start the unit. Shut the engine down and start it several times to check operation of the cranking motor and circuit.

9-9 TESTING OTHER ELECTRICAL COMPONENTS

This short section provides testing procedures for other electrical parts installed on the engine. If a unit fails the testing, the faulty part must be replaced. In most cases, removal and installation is through attaching hardware.

Colored Lead Identification

Test leads are identified by their color to aid in locating the correct lead. However, the manufacturer has been known to change lead colors. Therefore, identify the component and consult the wiring diagram for the correct component connections if colors vary.

NOW, THESE WORDS

The overheat buzzer, engine "kill" switch, tether cord switch, and engine start switch are all mounted on the control panel. Resistance tests on these components are performed at the ends of their harness leads -- inside the battery cover. Therefore, remove the battery cover. There is no need to trace the harness from the component to the quick disconnect fitting.

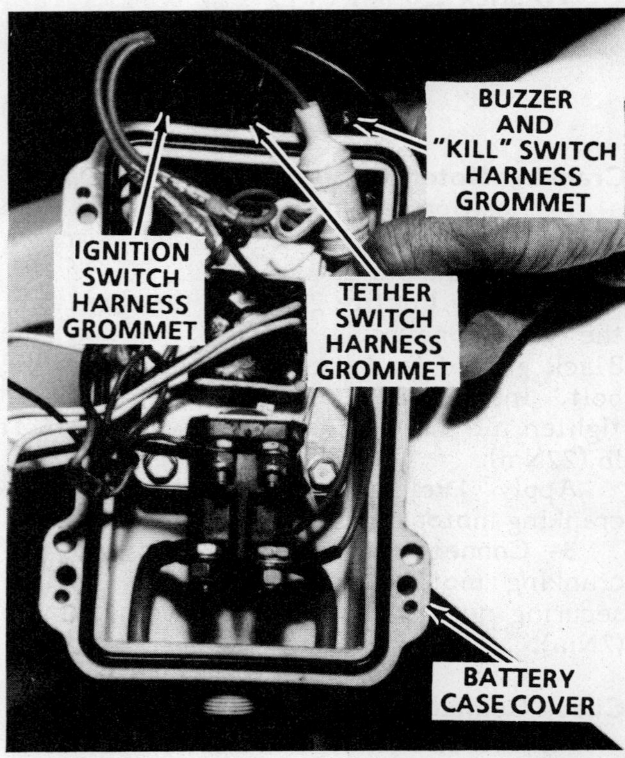

Identification of the ignition switch, tether switch, buzzer, and "kill" switch grommets in the side of the battery case cover.

The accompanying illustration identifies the three grommets on the battery cover and leads can be traced from there.

Stop "Kill" Switch Test

Identify the Gray and Black leads from the "kill" switch grommet in the side of the battery cover. Disconnect both leads at their quick disconnect fittings. Obtain an ohmmeter and set the meter to the Rx1000 ohm scale.

With the switch in the normal engine operation mode -- not depressed -- the resistance across the two leads should be approximately zero ohms -- continuity. Depress the switch and the meter should register infinity. If the switch fails both tests, the switch is defective and **MUST** be replaced.

Tether Switch Test

Identify the Purple and Red leads from the tether switch grommet in the side of the battery cover. Disconnect both leads at their quick disconnect fittings. With the tether in place behind the switch, the resistance across the two leads should be approximately 200 ohms -- continuity. Remove the tether cord from the switch -- the meter should register infinity. If the switch fails both tests, the switch is defective and **MUST** be replaced.

Overheat Buzzer Test

Disconnect the White lead from the thermosensor switch on the cylinder head cover. Ground the lead -- a small Allen wrench may be used, as shown in the accompanying illustration. The buzzer should sound when the lead is grounded and be silent when the lead is not grounded. If the buzzer fails this test, check the continuity of the lead from the terminal on the end of the disconnected White lead to the quick disconnect fitting at the rectifier/regulator. The lead changes color -- White to Brown. Check the continuity of the Brown lead from the rectifier/regulator to the small terminal on the cranking motor solenoid. If all the leads pass the test, obtain a 12-volt battery. Connect the negative battery terminal to the Black or White lead on the buzzer. Connect the positive battery terminal to the Red (changing color to Gray) lead between the buzzer and the "kill" switch. The buzzer should sound. If it fails to sound, the buzzer is defective and **MUST** be replaced.

Thermosensor Switch Test

The thermosensor switch is located in the cylinder head cover. The sensor measures the temperature of the cooling water circulating through the engine. If the cooling water temperature rises above 203°F (95°C), the sensor sends a signal to the warning buzzer -- the buzzer sounds and the craft operator is warned of the heating condition. If the warning buzzer sounds, the operator should immediately shut the engine down to prevent extensive and expensive damage to the engine from overheating.

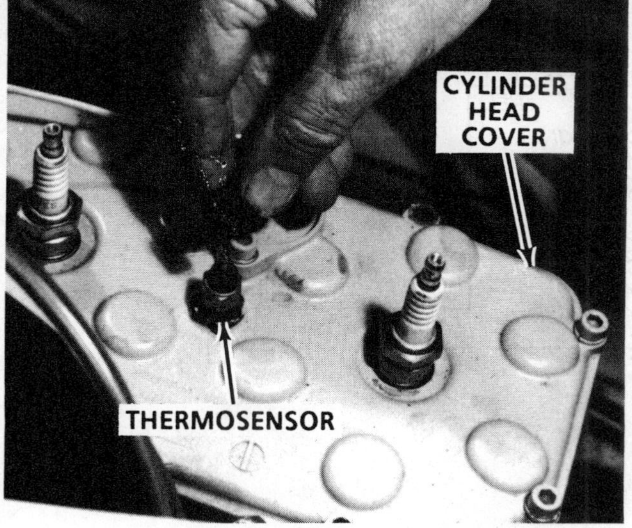

Disconnecting the White lead from the thermosensor, in preparation to testing the lead.

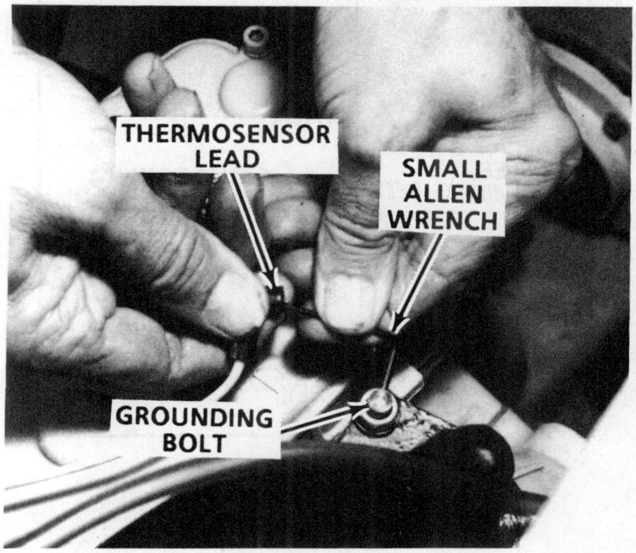

Testing the thermosensor lead, as explained in the text.

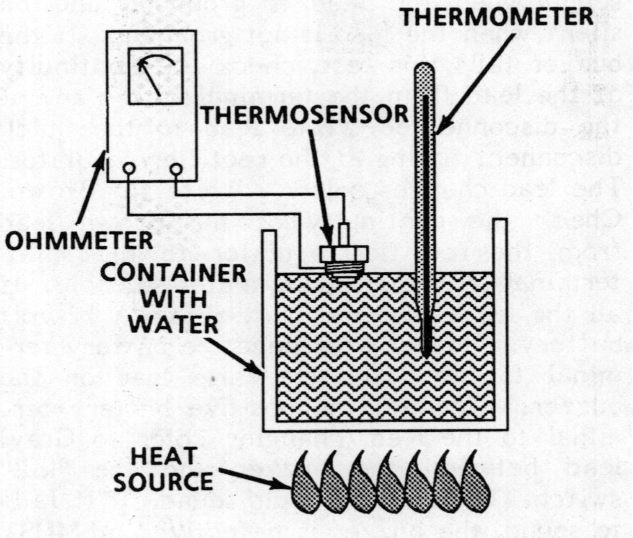

Simple drawing to depict testing operation of the temperature sensor.

To test operation of the temperature sensor, disconnect the White lead at the quick disconnect fitting. Remove the sensor from the cylinder head cover. Obtain an ohmmeter and select the Rx1000 (Rx1K) ohm scale. Connect the meter leads across the sensor terminal and the sensor threads. Heat a suitable container of water on the stove and have a thermometer handy to measure the water temperature.

Submerge only the sensor end in the water, as shown, without the sensor touching the sides or the bottom of the container. Observe the meter reading as the water temperature rises. Stir the water with the thermometer to evenly distribute the heat.

At temperatures above approximately 190°F (88°C), the meter should register more than 1,000,000 ohms (1,000K ohms). At temperatures less than 190°F (88°C), the meter should register almost zero ohms.

If the sensor does not perform or pass the tests, the sensor **MUST** be replaced.

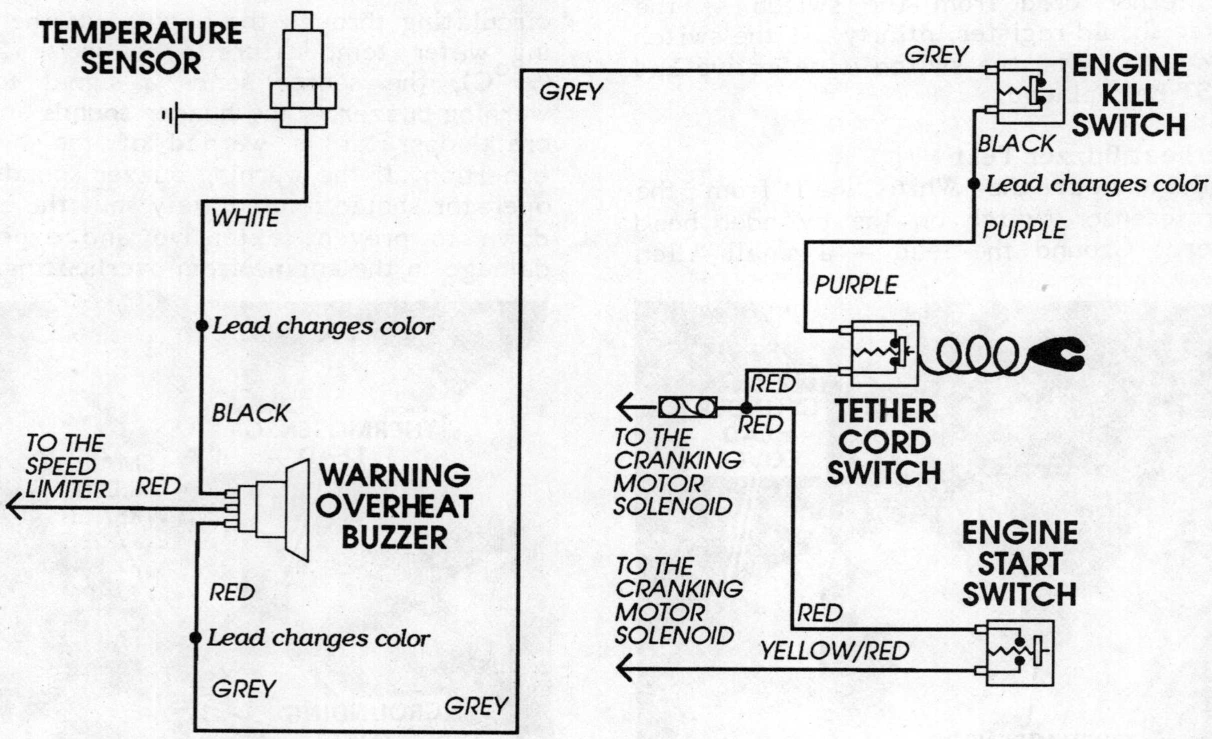

Wire identification and routing -- overheat buzzer, engine stop switch, tether switch, engine "kill" switch, and the start switch circuits.

10
JET PUMP

10-1 INTRODUCTION

Bombardier Sea-Doo watercraft use only axial type jet pumps. The axial pump used has a two-piece pump shaft -- a long driveshaft and a short impeller shaft. The driveshaft has splines at the forward end and is indexed with the aft flywheel. A rubber coupling is not used.

A single needle bearing in the seal carrier supports the drive shaft. The aft end of the driveshaft is indexed into the impeller. The impeller shaft is support by a needle bearing at each end. The forward end of the shaft indexes into the impeller.

10-2 PRINCIPLES OF OPERATION

One of the first lessons to be learned in any elementary physics class is Newton's basic law: "for every force, there is an opposite and equal force". This statement is the basic principle of the jet pump. Water is "sucked" and "scooped" in from under the craft by a powerful pump rotating at incredible speed, and then discharged, "blown" out, sternward in the opposite direction. In this manner the watercraft is propelled forward.

The personal watercraft covered in this manual are all equipped with a twin cylinder water-cooled two-stroke engine, matched with a single stage (one impeller) jet pump. The Sea-Doo GT is equipped with a "gate" which can be lowered down over the pump outlet nozzle forcing the exhausted water back in a forward direction, thus moving the craft sternward. Personal watercraft jet pumps may be classified as "axial flow" or "mixed flow".

All Sea-Doo Personal Watercraft are equipped with an axial flow jet pump.

Axial Flow

Water in an axial pump moves on a single axis, as depicted in the illustration on the following page, thus the word "axial". In simple terms, water is ingested and discharged parallel to the axis of impeller rotation. The inside of an axial pump case is parallel to the water flow -- unlike the inside of a mixed flow pump

EDUCATIONAL INFORMATION

Even though Bombardier Sea-Doo personal watercraft utilize only an axial flow pump, the following paragraphs are included in this chapter to acquaint the reader with a mixed flow pump -- how it differs from the axial flow pump and why.

Mixed Flow

Many experts consider the mixed flow system to the most efficient for personal watercraft application. The water flow is discharged at an inclined angle to the axis of impeller rotation, as indicated by the arrows in the accompanying illustration. The impeller blades are inclined in such a man-

A Bombardier Sea-Doo watercraft, driven by an axial flow pump and powered by a Rotax 587 water-cooled engine sits in the California sunshine waiting for a new owner.

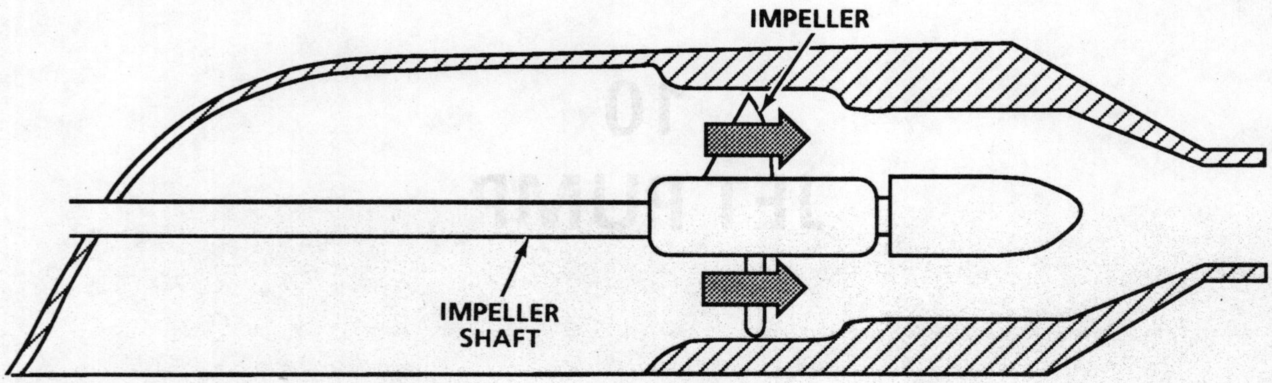

Simple cross-section line drawing to depict water flow through an axial flow jet pump. The water passes through parallel to the axis of impeller rotation.

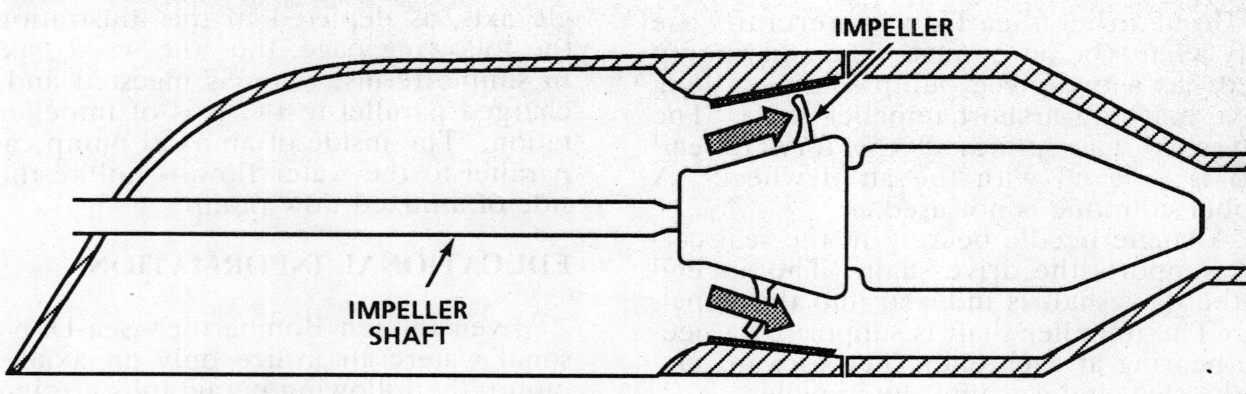

Line drawing to illustrate water flow through a mixed flow jet pump. Notice how the water is directed upward at an angle to the axis of impeller rotation.

ner to cause the water to be thrown off at a 45 degree angle to the impeller shaft within the pump case. From the pump case, the water is then forced out through the outlet nozzle and steering nozzle. The vanes in the pump case help straighten the water flow through the nozzle.

The inside of the pump case is inclined at an angle to the water flow. Therefore, this design pump incorporates shim material to space the impeller blades at a specified distance from the pump case. Mixed flow pump efficiency decreases by 30% as blade-to-pump case clearance increases by 0.04" (1mm).

Axial or Mixed?

Axial and mixed flow pumps are intended for different applications. One is not necessarily better than the other. The axial pump can be operated at higher engine rpm and is more suitable for racing. A mixed flow pump adds thrust at low engine rpm due to the pump design. Therefore, the mixed flow pump performs more efficiently at low engine rpm. This type pump is more suitable for a craft equipped to carry extra weight.

A jet pump seldom requires service except for the following:

Impeller-to-wear ring clearance becomes excessive.
Impeller is damaged.
The oil seal around the driveshaft fails -- allowing water to enter the hull.

10-3 ENGINE AND JET DRIVE

The basic principle and arrangement of engine and pump are almost identical for all manufacturers.

The engine crankshaft is coupled either directly to the pump shaft or through a short driveshaft without any gear reduction. An impeller mounted on the pump shaft draws water into an opening in the hull through a suction and intake casting. Once the craft has attained forward motion the intake also serves as a "scoop" adding to the volume of water moved through the pump.

Volume and Velocity

The outlet nozzle is slightly funnel shaped. This design coupled with the capacity of the pump (impeller), causes the volume of water entering the pump to exit the nozzle with increased velocity. This principle is similar to air passing through the venturi of a carburetor.

The amount of water ejected sternward from the nozzle at high velocity is the force propelling the craft through the water. This force can actually be measured and calculated in foot pounds or Newton meters. The greater the velocity of water mass moving through the nozzle, the greater the thrust to move the craft.

Cooling System

A fitting on the jet pump, aft of the impeller, syphons off cooling water. This water is delivered first to the exhaust manifold -- the hottest part of the engine. The water is then channeled around the cylinder walls, cylinder head, and exhaust pipe. Some water is mixed with exhaust gases in the tuned exhaust pipe and finally discharged from the exhaust outlet. This mixture of water and exhaust gases has the effect of cooling and somewhat quieting the exhaust emission.

10-4 IMPELLERS

Impellers provide thrust using a combination of water flow and water pressure in a closed environment.

The blades of an impeller overlap, thus water is trapped and forced through the impeller into the pump case and nozzle. The water moving past the impeller resists cavitation because it is under pressure. Blades in the pump case are angled exactly opposite to those of the impeller. These blades redirect the water to establish a concentrated flow through the nozzle.

The manufacturer recommends a radius of about 1/64" (0.3-0.5mm) for the leading edge of the impeller blade. A sharper radius will result in cavitation and a greater radius will reduce pump efficiency.

Cavitation Burns

Cavitation burns are the worst enemy of an impeller. These burns literally "eat away" material from the impeller, leaving holes and weakening the impeller structure. In extreme cases entire blades have been known to "depart" from the impeller hub because the cavitation burns at the base of the blade were so severe.

Cavitation burns are the result of imperfections (damage) on the impeller blades or air mixed with the water flow.

Cavitation burns may be caused by:

Wave jumping -- air is sucked into the pump as the craft leaves the surface of the water.

Bad seal between the pump case and outlet nozzle -- allowing air to enter.

Leading edge of the impeller becoming damaged.

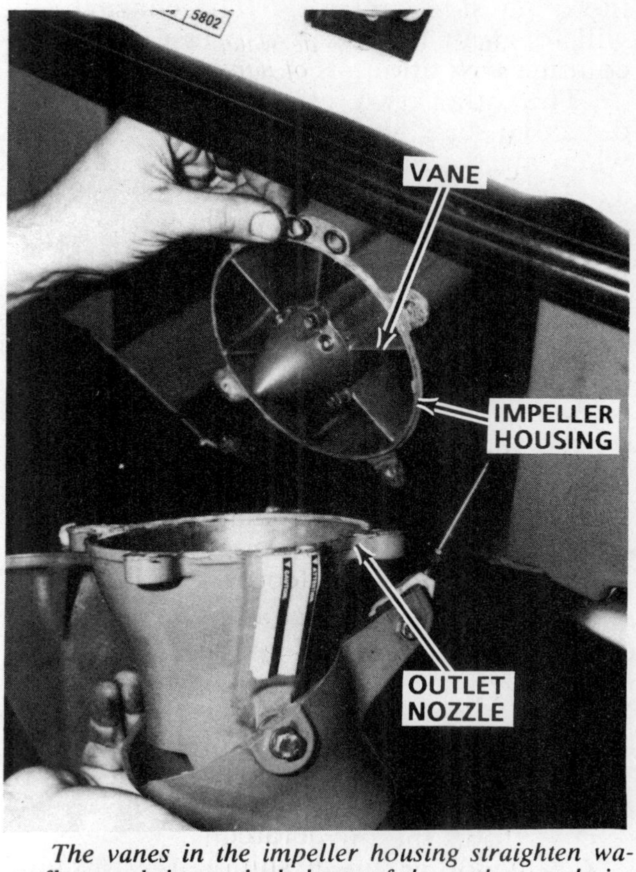

The vanes in the impeller housing straighten water flow and the conical shape of the outlet nozzle increases flow velocity.

To explain exactly what causes cavitation burns, first, let us examine the last cause. If the impeller leading edge is damaged and "mushrooms" over the blade, a low pressure area will form under the mushroomed lip.

At atmospheric pressure, water boils at 212 degrees F. Anytime there is an imperfection on the surface of an impeller blade and water passes over that imperfection, the water pressure is lowered. If the pressure is lowered, water will boil at a much lower temperature. Therefore, air bubbles will form in the water boiling under the mushroomed lip.

The bubbles will creep down the surface of the blade and accumulate at the impeller hub. A high pressure area is formed at the base of the blade. Here, the air bubbles will collapse and reform back into water with a release of energy. This energy is absorbed by the impeller and results in material being eaten away.

The impeller should be dressed to a slightly rounded edge. It is impossible to achieve a perfectly straight sharp edge, because one side will always be convex and the other side concave. The concave side will form an area of low pressure and encourage cavitation, as described above.

The other two causes of cavitation -- defective seals and wave jumping -- occurs when air is sucked in with the water flow. The same principles apply, as just described, in the previous paragraphs. Air bubbles will creep down the surface of the blade and accumulate at the impeller hub. A high pressure area is formed at the base of the blade. Here, the air bubbles will collapse and reform back into water with a release of energy. This energy is absorbed by the impeller and results in material being "eaten away".

Cooling Water and Bilge Hoses

On most pumps, three hoses are attached to the pump case and outlet nozzle. One hose channels some of the water flowing through the pump to the exhaust manifold -- the hottest part of the engine -- to cool the block during engine operation. The other two hoses siphon water out of the bottom of the hull. These hoses are referred to as the bilge system.

The cooling hose has water flowing **FROM** the pump. The bilge hoses carry

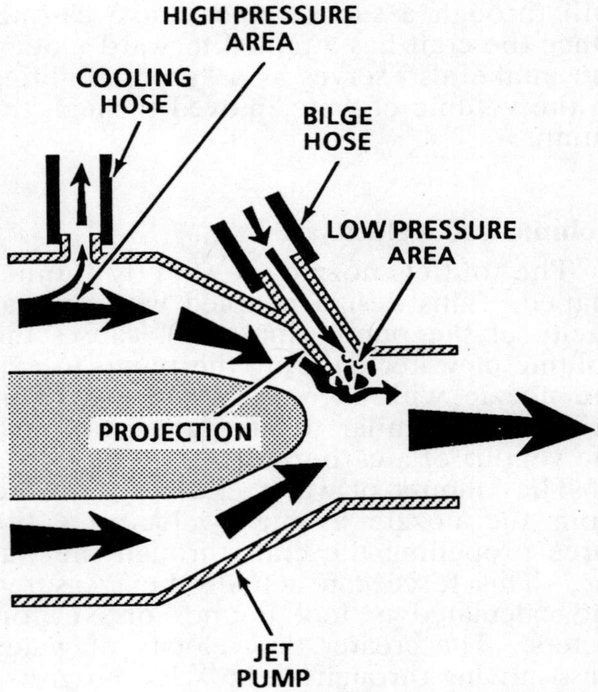

The direction of water flow through the cooling hose and the bilge hose is opposite. The cooling hose routes water from the pump to the engine. Water is "vacuumed" from the bilge through the bilge hose to the pump where it is forced out through the nozzle.

water **TO** the pump. All three hoses attach to the pump in a similar manner.

What determines the direction of water flow?

The answer to this question is in a simple explanation of high and low pressure areas along the inside surface of the pump.

Close inspection of the area inside the pump at the cooling water hose fitting reveals a smooth rounded shoulder with no obstruction or obstacle to impede the flow of water. A high pressure area develops here and draws the water down the hose attached to the fitting.

Further inspection of the area inside the housing at the bilge hose fitting reveals a small protrusion around the opening. This protrusion causes disturbance to the water flow and creates an area of low pressure. This low pressure area will have the effect of emptying air and water from the hose aft with the impeller water flow -- similar to the action of a vacuum cleaner, as depicted in the accompanying illustration. If the other end of the hose is submerged in water inside the hull, the water will be "vacuumed" out. In this manner, the bilge system keeps the bilge relatively "dry".

Many an owner -- with good intentions -- thinking his/her action would smooth water flow and increase pump performance -- has filed the protrusion described in the previous paragraph smooth with the surrounding area of the pump. During operation of the watercraft, the bilge system worked in **REVERSE.** Water was pumped from the impeller housing into the bilge area, quickly filling the bilge and possibly sinking the craft before the rider had time to realize what was happening.

10-5 CHECKING IMPELLER TO WEAR RING CLEARANCE

The following procedure may be performed with the watercraft elevated on a set of saw horses enabling a person to work underneath, or with the watercraft "laid over" on its **PORT** side, allowing access to the rock grate and the impeller.

SPECIAL WORDS

A decal on the transom includes a warning regarding which direction the craft must be "laid over" -- **ALWAYS** to port.

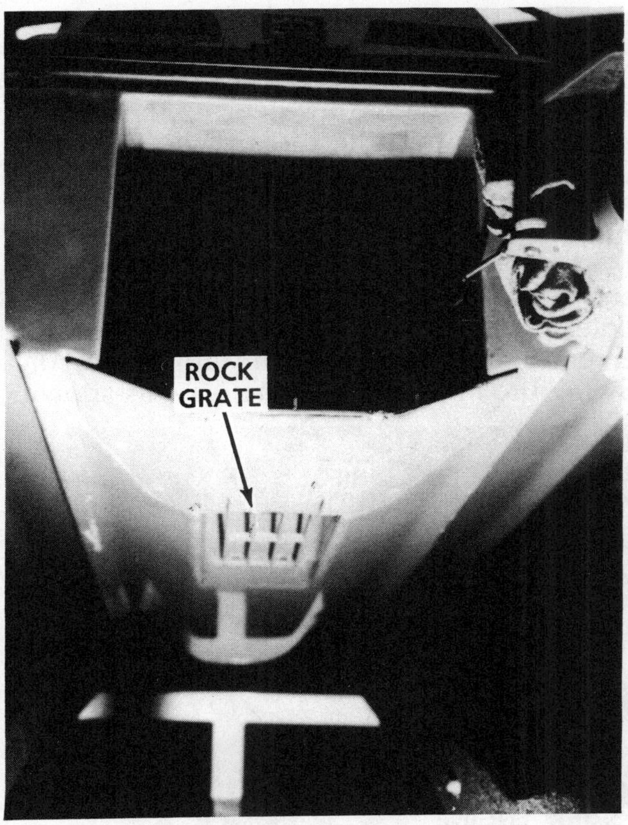

With the watercraft elevated on a set of saw horses and the bow down, the rock grate can be easily removed to measure clearance between the impeller and the wear ring, as explained in the text.

The accompanying illustration shows the craft elevated with the bow down, giving clearance to the rock grate.

If the impeller-to-wear ring clearance is to be determined with the watercraft raised on saw horses, the battery may be left secured in the craft. **HOWEVER**, if the watercraft is to be positioned on its side, first disconnect the electrical leads at both battery terminals, and then remove the battery.

Once the craft is secured in a favorable position, remove the four bolts securing the rock grate to the jet pump. If the seal between the hull and the pump housing is broken, or the foam packing is damaged, reseal the housing and intake area with silicone sealant to prevent impeller cavitation.

Insert a feeler gauge in the center between each impeller blade and the wear ring. Determine the clearance.

If the clearance is less than 0.040" (1.00mm) for each blade, the condition is considered satisfactory. No action is necessary.

If the clearance is more than the service limit, inspect the condition of the wear

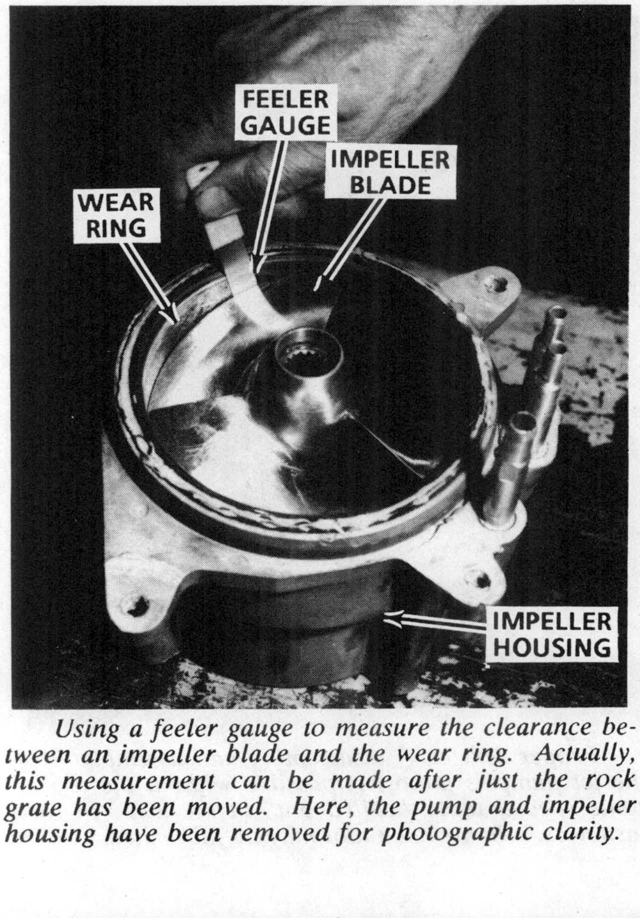

Using a feeler gauge to measure the clearance between an impeller blade and the wear ring. Actually, this measurement can be made after just the rock grate has been moved. Here, the pump and impeller housing have been removed for photographic clarity.

ring. If the wear ring has deep scratches, replace the wear ring. If the wear ring appears in good condition, the problem must be with the impeller. Visually inspect the impeller for nicks, scratches, pitting or a "mushroomed" edge.

If the clearance is not kept to a minimum, the jet pump will lose efficiency by allowing excess water to pass between the blades and the housing. This water has no thrust and will actually slow the stream of water under pressure from the impeller.

The jet pump must be removed from the craft to replace the wear ring or to service the impeller.

Install the rock grate and tighten the attaching bolts securely.

10-6 JET PUMP SERVICE

The procedures in this section include all the necessary steps to disassemble, inspect, and assemble virtually all parts of the jet pump. **THEREFORE**, if a part is found to be in satisfactory condition for further service, simply skip the steps involved for such part and continue with the necessary work to restore the pump to first class condition.

Proper service -- disassembly and assembly -- of the jet pump is almost impossible without the use of some special tools, "called out" in the text. In rare instances, a homemade tool may be fabricated.

Preliminary Tasks

Disconnect both battery cables from the battery. Remove the battery from the craft. If the craft is to be raised on a set of saw horses or similar blocking or fixture, with the stern higher than the bow, the battery may remain in place.

CAUTION

The battery **MUST** be removed to avoid spilling electrolyte into the bilge through the battery vent holes when the craft is moved over on its port side.

Rotate the fuel control valve to the **OFF** position. If the craft is to be rolled over on its port side, drain and remove the fuel tank and the oil tank.

JET PUMP REMOVAL

1- Loosen the clamps and remove the two bilge hoses and the engine cooling water hose from inside the engine compartment. Work the hoses free of the fittings on the jet pump.

2- Remove the four bolts and washers securing the nozzle to the impeller housing.

3- Separate the nozzle from the impeller housing, with the steering nozzle attached. Remove and discard the two small O-rings from the water passages at the top of the impeller housing. Leave the steering cable attached to the nozzle.

4- Obtain a "bungee" cord. Hook one end of the cord into one of the mounting holes in the nozzle, and then hook the other end of the cord into the upper tie ring. Swing the nozzle out of the way, as shown.

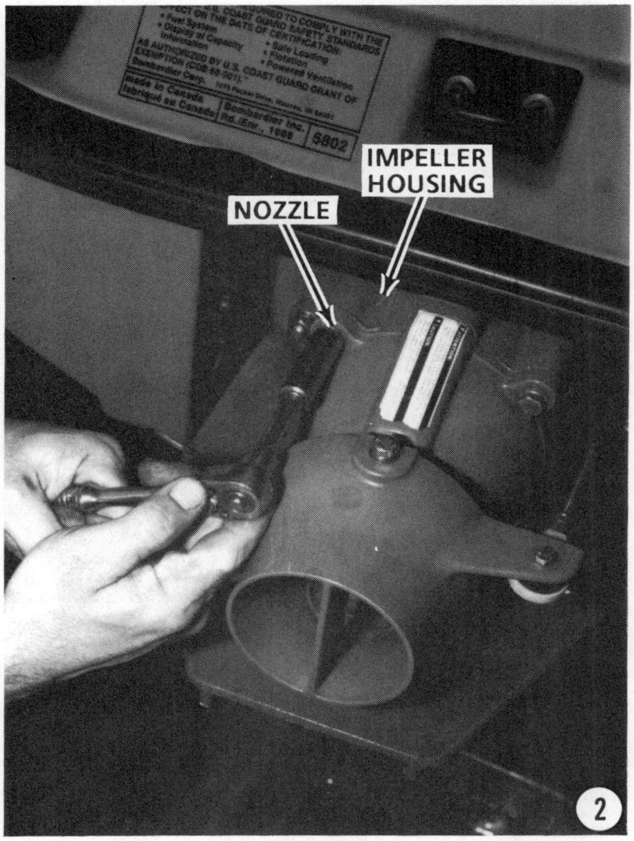

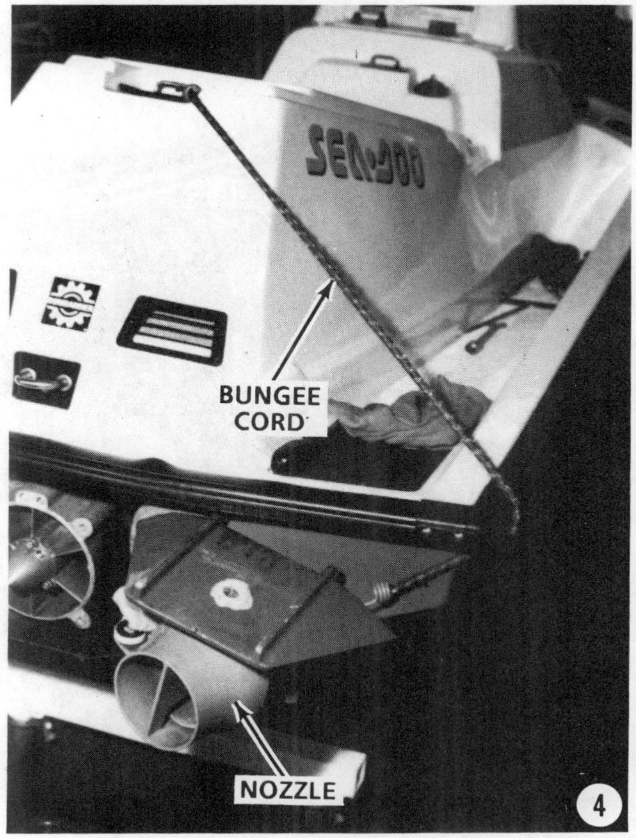

5- Position a suitable container under the jet pump to catch the oil from the reservoir in the end cap. Remove the three bolts securing the end cap to the impeller housing.

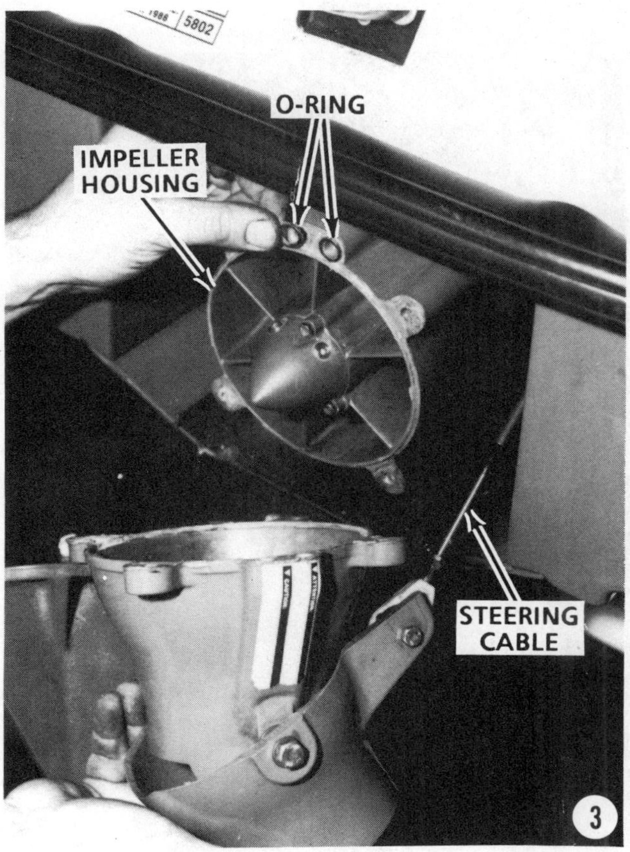

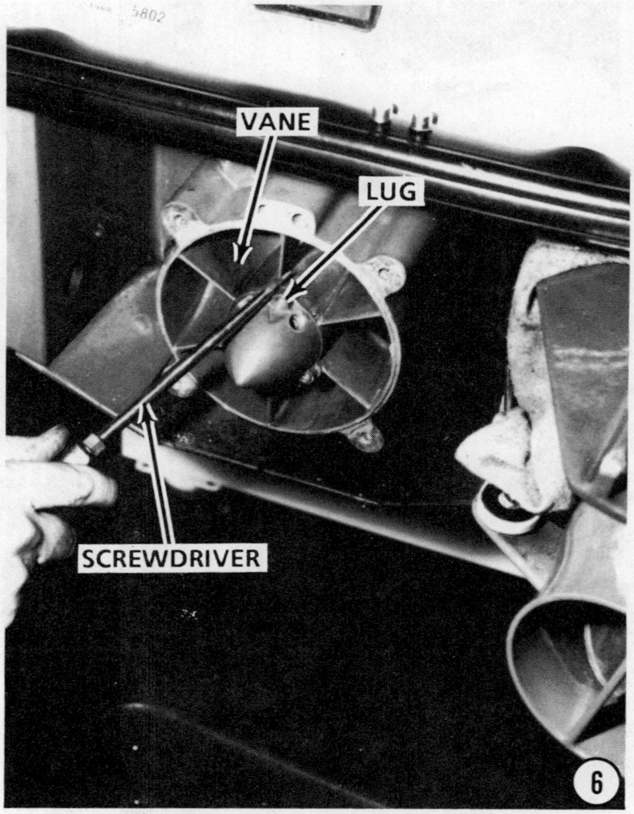

6

6- Insert a slotted screwdriver between one of the lugs on the end cap and adjacent vane on the impeller housing. Use the the screwdriver to rotate the end cap and break the seal between the cap and the housing. Continue to rotate the end cap until the hub is clear of the vane.

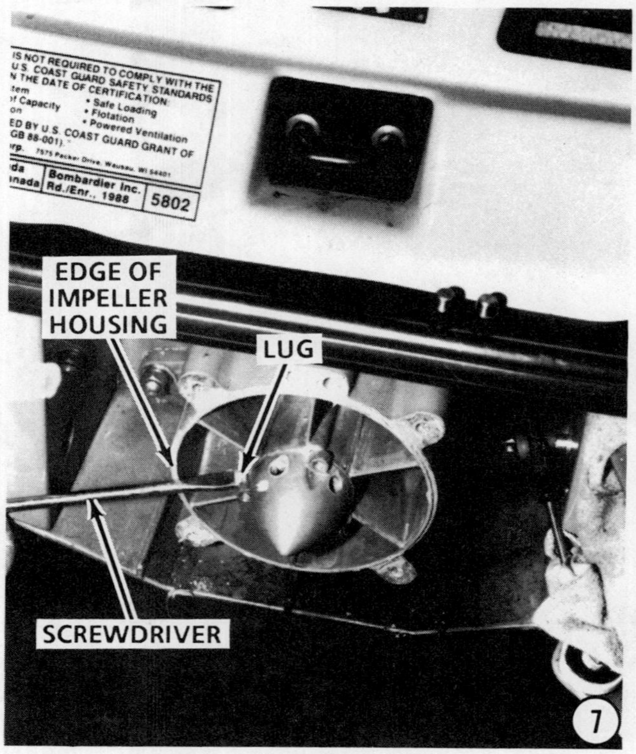

7

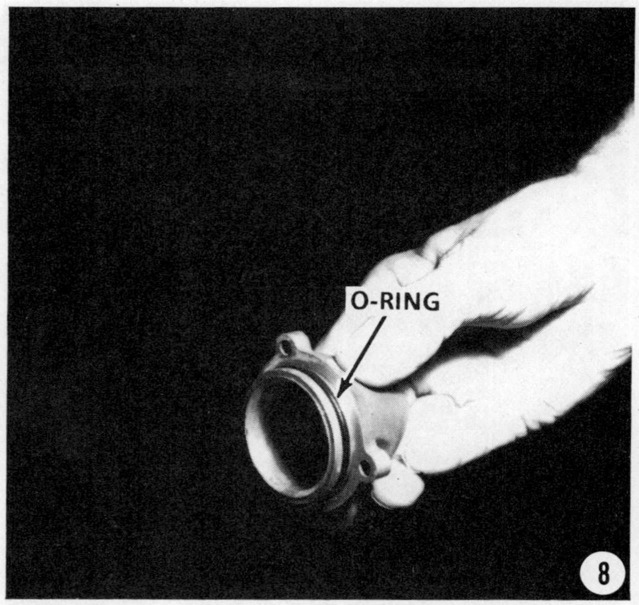

8

7- With the screwdriver behind the end cap lug, use the edge of the impeller housing as a pivot point and pry the end cap free of impeller housing.

8- Remove the end cap, and then remove and discard the O-ring.

9- Loosen the two clamps on the boot around the driveshaft.

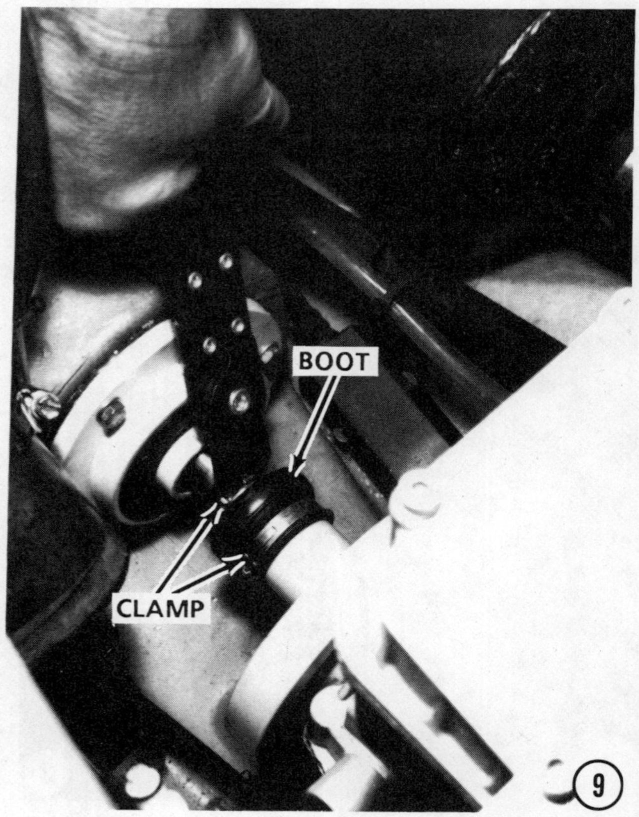

9

10- Remove the four nuts and washers securing the impeller housing to the mounting studs on the craft hull.

Jet Pump Puller Tool Fabrication

The impeller housing can be pulled from the hull with the use of two home-made tools. Each is fabricated from a metal bar strap approximately 16" long by one inch wide. For adequate strength, the bar should be 1/2" thick -- (400mm x25mmx 12mm). Drill two 3/8" (10mm) holes through each bar in the position shown in the accompanying illustration. Wrap tape around the ends of the bars to prevent scratching the hull surface.

11- Use the bolts removed in the previous step to secure the two fabricated bars across the impeller housing against the hull. Tighten the bolts alternately and evenly to separate the impeller housing from the hull.

12- Withdraw the impeller housing from the hull. The driveshaft may or may not come away with the housing.

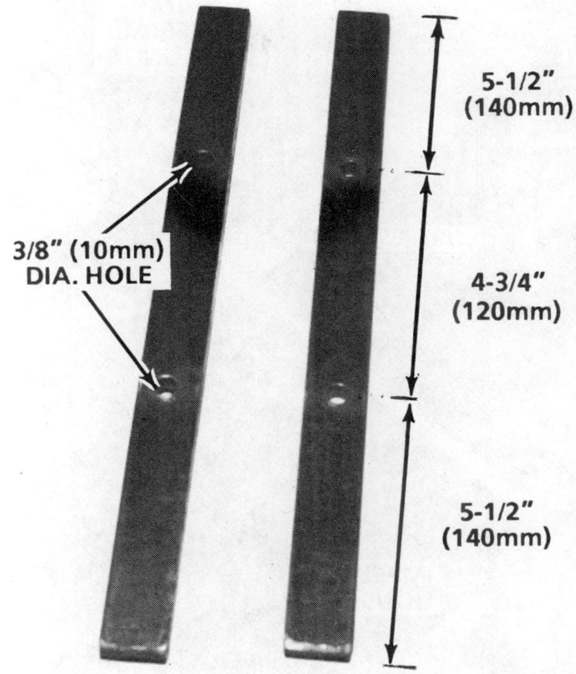

If special manufacturer's tools are not available, a set of homemade tools may be easily fabricated using the dimensions given. The holes drilled are used to pull the jet pump from the hull, as explained in the text.

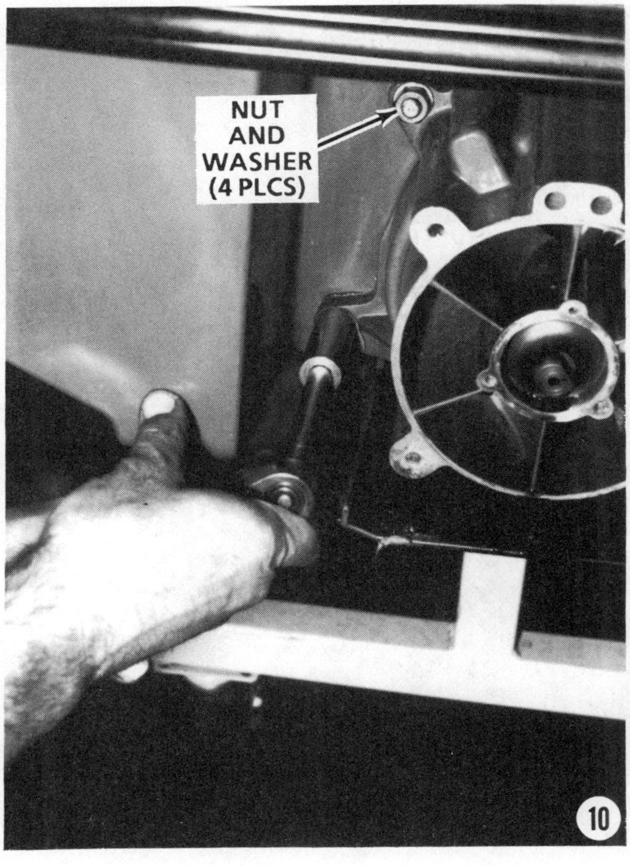

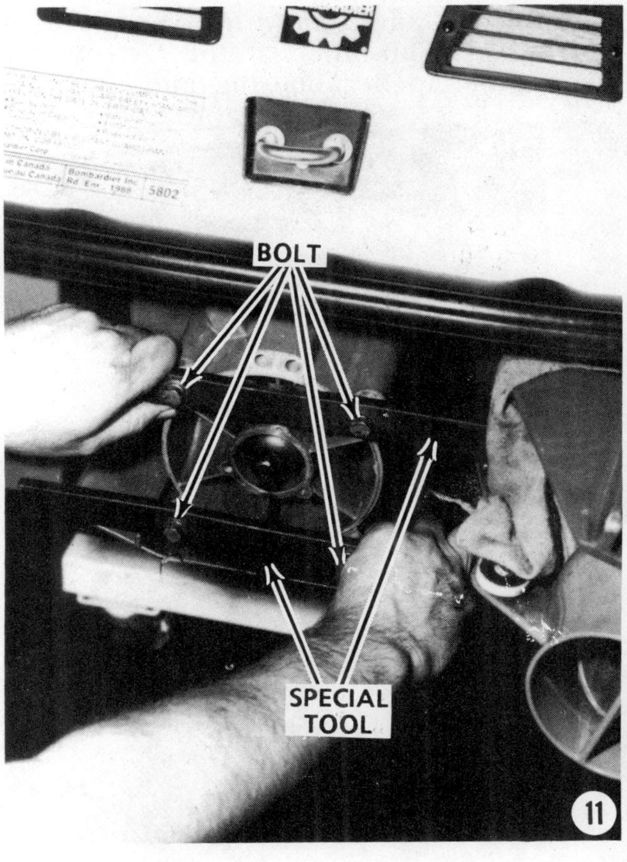

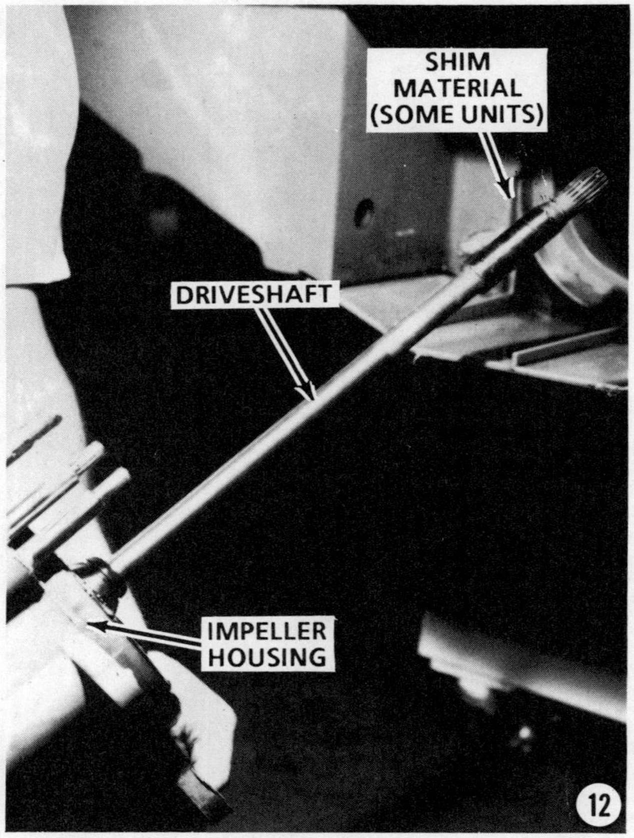

SPECIAL SHIM MATERIAL WORDS

Some jet pump may have a piece of shim material installed at the factory, between the impeller housing and the hull.

Look for and **SAVE** any shim material found at this location.

If the driveshaft came out with the impeller housing, skip the following step and proceed directly to Step 14. If the driveshaft remained in the hull, perform the next step.

13- Pull the driveshaft from the seal carrier.

SPECIAL DRIVESHAFT WORDS

Do not attempt to pull the driveshaft free of the impeller housing without first prying the boot free of the housing. Such action will surely damage the boot. If the boot is stuck, apply liquid soap at the sealing surfaces around the driveshaft and impeller housing. Allow the liquid soap to "seep" in for a few minutes, and then try to pull the driveshaft free again.

14- Attempt to pull the driveshaft free of the impeller housing. If the attempt fails, hold the impeller housing by the driveshaft just a few inches above the workbench. Strike the impeller housing lugs alternately and evenly with a soft head mallet to free the shaft.

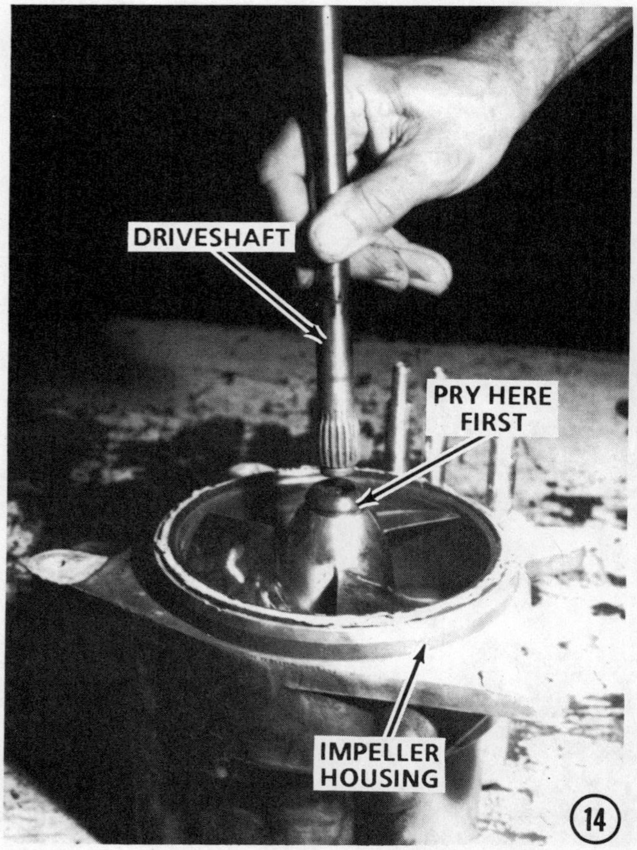

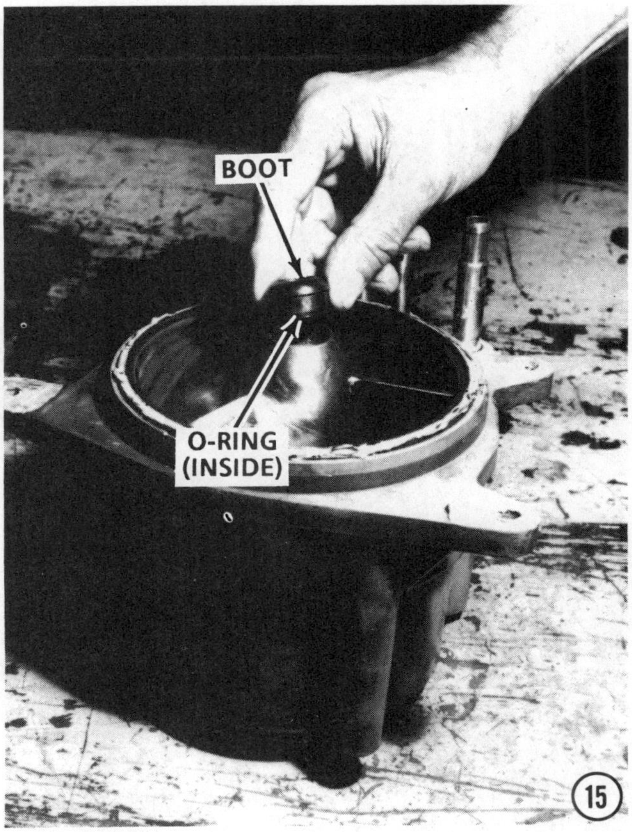

15- Remove the boot from the impeller housing. Remove and discard the **O**-ring inside the boot.

16- Remove the threaded dampener from the forward end of the driveshaft.

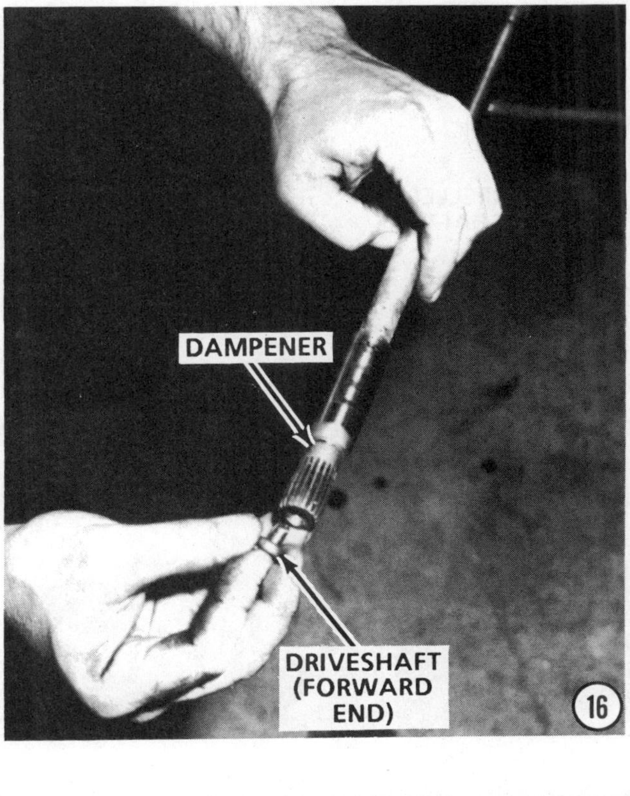

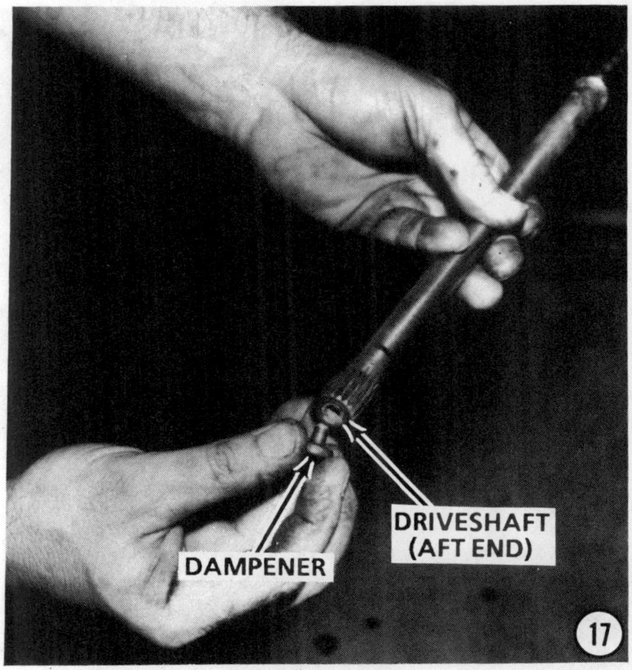

17- Remove another threaded dampener from the aft end of the driveshaft.

Impeller Removal

18- Clamp the impeller shaft in a vise equipped with soft jaws, with the housing upward over the vise and the vise jaws making contact with the two flats on the end of the impeller shaft.

Obtain Impeller Remover P/N 295-000-001. Insert the tool into the impeller splines.

When the impeller is secured in a vise, to disassemble the impeller housing, soft jaws, even a couple pieces of wood will do, ***MUST*** *be used to clamp on the shaft flats.* ***NEVER*** *attempt to clamp the shaft without a cushion between the shaft and the vise.*

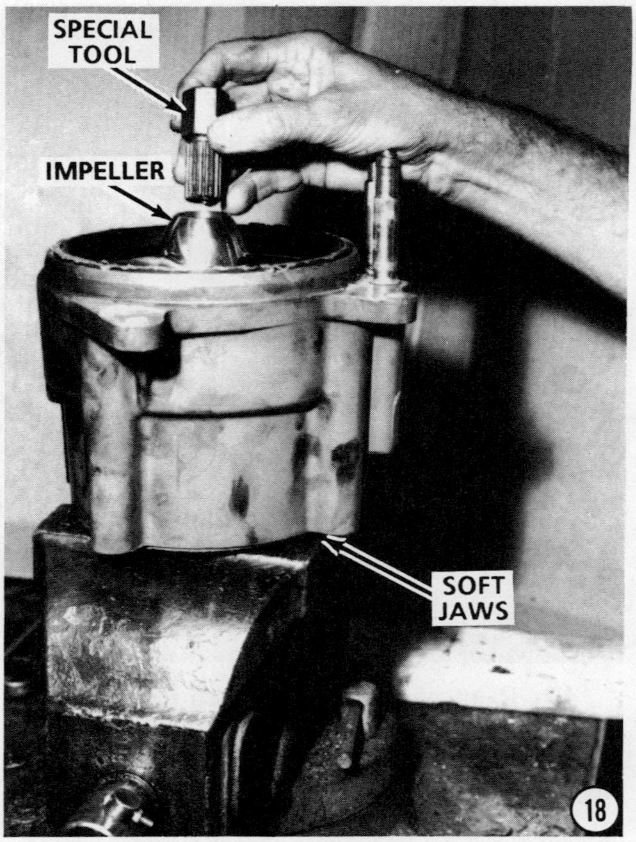

19- Place a suitable socket and breaker bar over the special tool and "break" the impeller free of the shaft.

20- Continue to unthread the impeller from the shaft. Release the impeller shaft from the vise.

21- Remove the impeller shaft and thrust bearing from the housing.

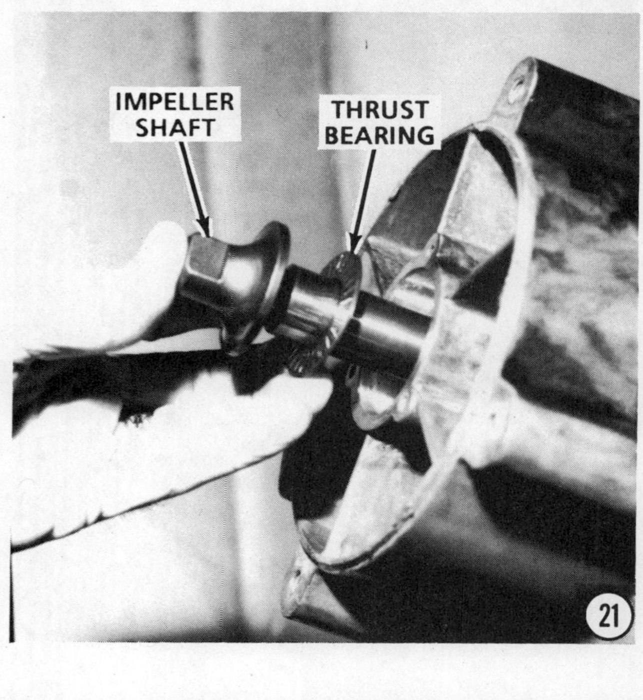

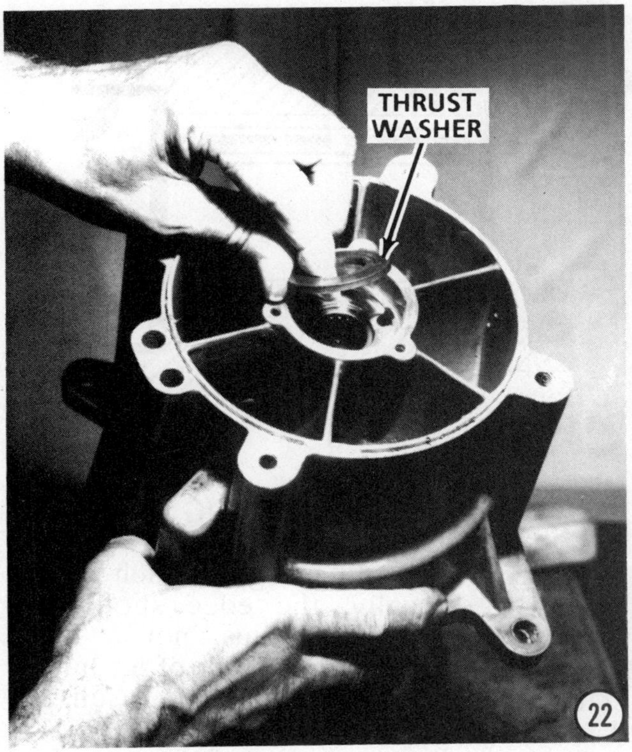

22- Lift out the thrust washer from the forward end of the impeller housing.

23- Pry out the oil seal from the forward end of the impeller housing. If the needle bearing behind the seal is no longer fit for service, obtain a slide hammer with a puller jaw attachment. Hook the end of the jaws behind the caged needle bearing and work the bearing free of the housing.

Wear Ring Removal

24- Secure the impeller housing in a vise equipped with soft jaws. Observe the slot in the housing directly across from the cooling water fitting. This slot is machined into the housing to avoid damage to the housing when the wear ring is cut through. Cut the wear ring in line with the slot provided using a small jig saw or low clearance hack saw. After the ring is cut, insert a screwdriver blade under a cut end. Work the ring up and out of the housing.

25- If the needle bearing inside the impeller housing is no longer fit for service, use the same slide hammer and puller jaws and pull the bearing free of the housing.

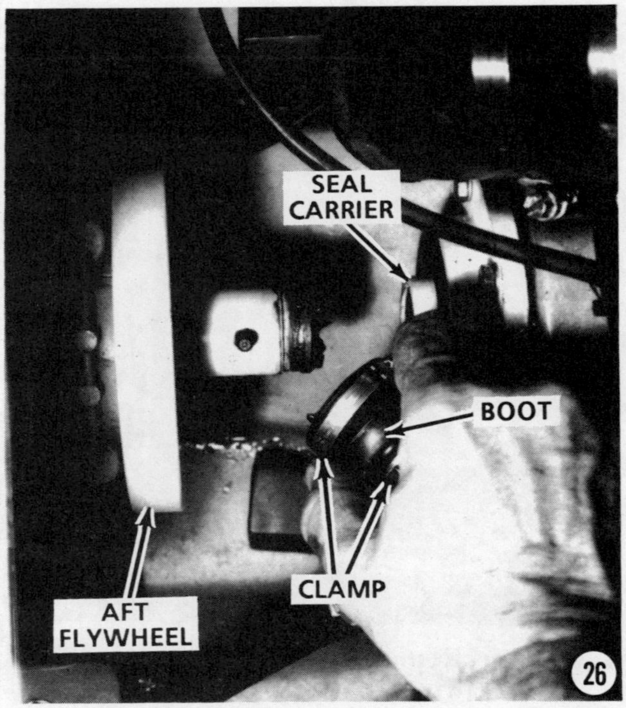

Seal Carrier Service

26- Loosen the two hose clamps on the boot between the seal carrier and the aft flywheel. Remove the boot between the seal carrier and the aft flywheel.

27- Remove the six bolts and washers securing the seal carrier to the hull, and then remove the carrier.

28- Pry both oil seals out of the carrier. Use a suitable driver and drive the caged needle bearing free of the carrier.

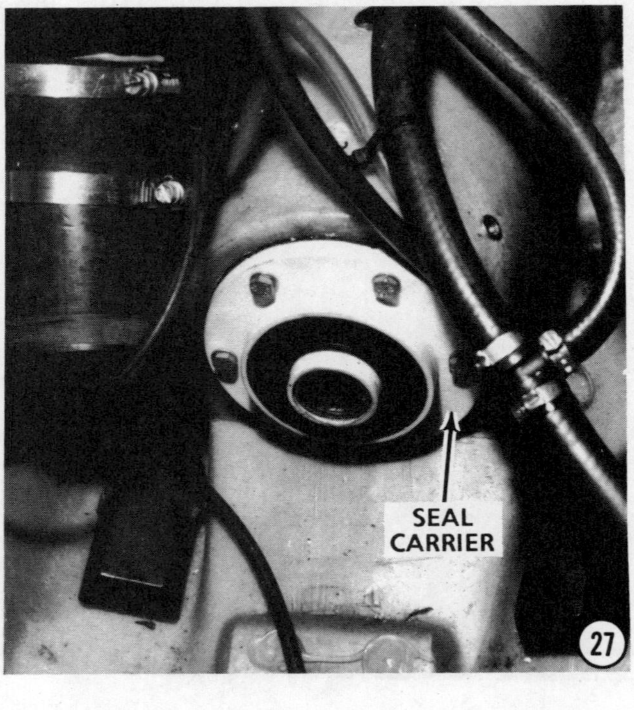

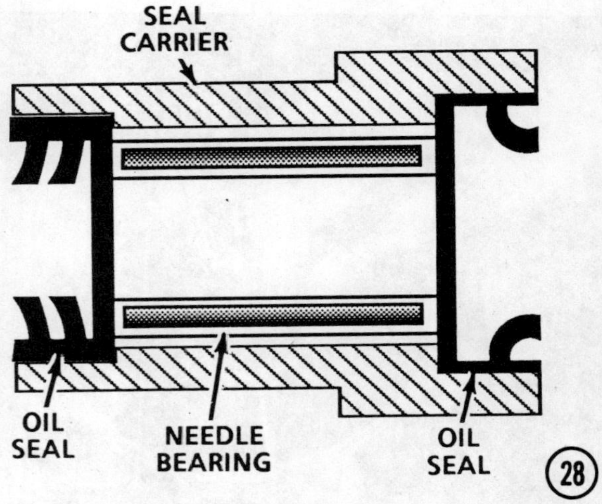

CLEANING AND INSPECTING

Inspect the leading edge of each impeller blade. Ideally, the edge should be rounded to a 1/32" radius, not a knife edge. The cross-section shape of an impeller blade is similar to the shape of an airplane wing.

If the craft is consistently operated in shallow water with a sandy or "pebbly" bottom, the leading edge of the impeller blades will become rounded or will become chipped with pieces missing.

Very little material may be removed from the impeller blades during reconditioning work. Therefore, if the edge is rounded off or corroded, a file can be carefully used to "dress" the edge.

If an impeller blade is chipped more than 1/4", the impeller should be replaced.

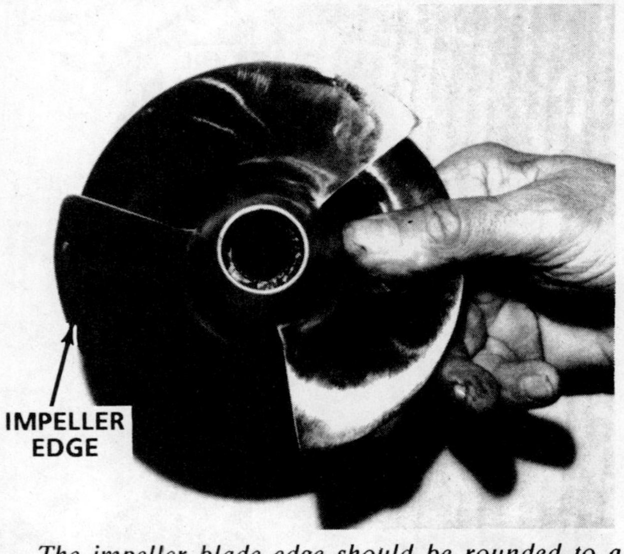

The impeller blade edge should be rounded to a 1/32" radius. The text explains how this is to be done.

Any chip less than 1/4" can be evenly filed away along the length of the blade. Whatever work is done to one blade, must be done to all blades in order to maintain impeller balance. If the impeller becomes excessively out of balance, a definite vibration condition will be quite noticeable, as the craft moves through the water. Such vibration will also cause premature bearing and shaft wear.

Measure the outside diameter of the impeller. If the blades of the impeller are worn, reducing the impeller diameter close to or beyond the service limit, 5.446" (138.33mm), the impeller must be replaced. If the wear ring was not removed, inspect the interior surface. If the surface of the wear ring has deep scratches, the ring should be replaced.

Driveshaft Runout

Set the long drive shaft on two "V" blocks -- one at each end. Position a dial indicator gauge over the center portion of the shaft, and measure the runout. The maximum allowable runout is 0.020" (0.5mm). Excessive driveshaft runout causes damage to driveshaft splines, impeller, flywheel, and the seal carrier.

Check to be sure the rock grate fins are not bent or missing. Such a condition might allow a large rock to pass through and cause serious damage to the impeller.

Inspect the vanes in the impeller housing. File the blades to a sharp edge on the forward end. Be sure to keep all the edges the same distance from the front of the housing.

Use a scraper to remove all traces of silicone sealer from both surfaces of the

In order to remove the wear ring, the ring must be cut through using a small jig saw or a low clearance hacksaw. Notice the deep scratches in the ring surface -- requiring the ring to be replaced.

The vanes in this impeller housing were "chewed away" by the impeller, when the impeller worked loose on the shaft.

impeller housing, the hull/ride shoe, and the nozzle. Final traces can be removed with a steel brush mounted in an electric drill. A clean air tight seal is important when the parts are assembled, otherwise air could be drawn in and cause cavitation -- damage to the impeller and loss of power.

Check the water passages in the impeller housing -- three openings between two of the vanes and two on an outside lug. Check the two oil passages in the aft impeller housing hub. Blow low compressed air through each passage to verify the passage is not restricted or blocked.

Inspect the splines on both ends of the driveshaft, the impeller, and the engine flywheel. If the original impeller shaft is to be used again, clean the threads with Loctite Safety Solvent 755.

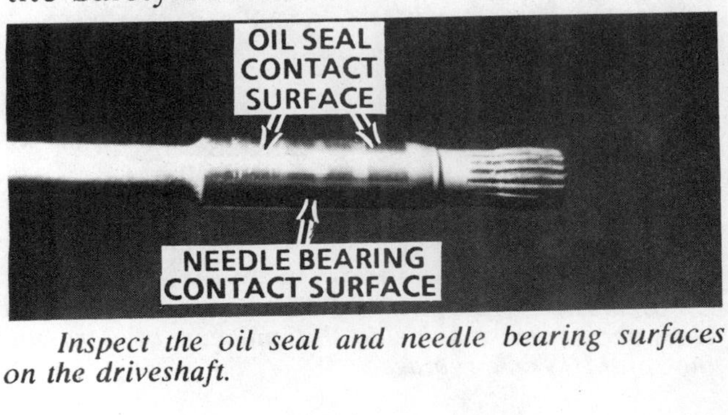

Inspect the oil seal and needle bearing surfaces on the driveshaft.

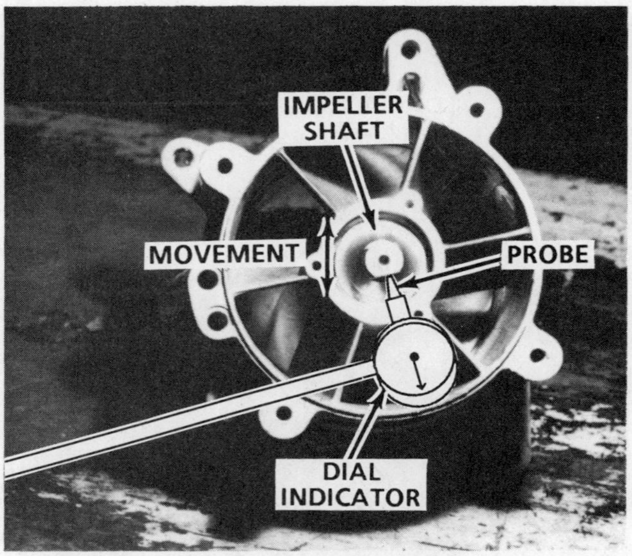

Setup to measure the radial "play" of the impeller shaft. The setup to measure the end "play" is similar, except the dial indicator probe rests against the end of the shaft.

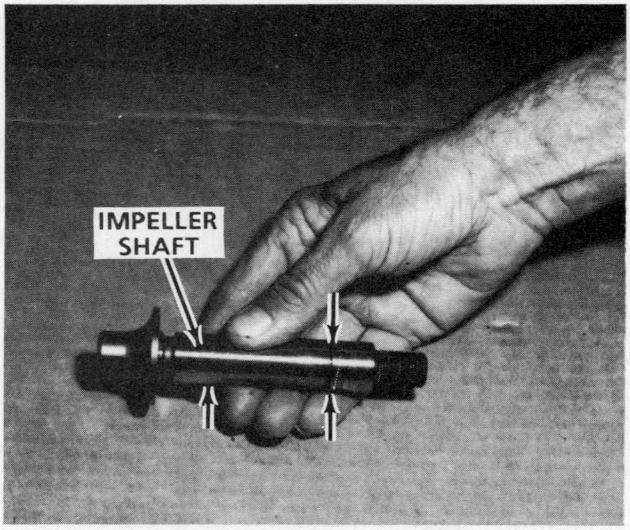

Measure the diameter of the bearing contact surface, as shown.

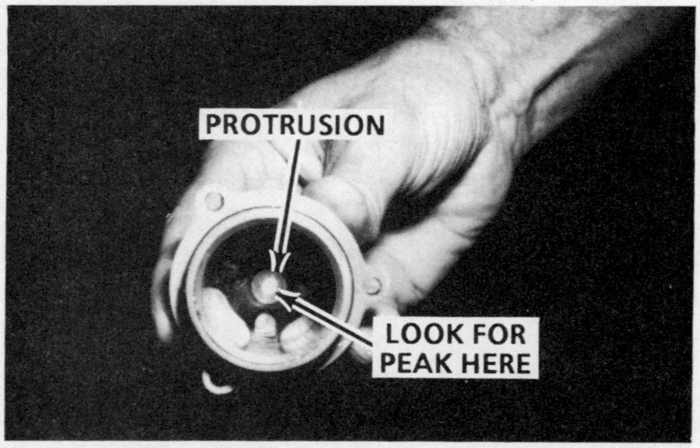

Excessive end "play" of the impeller shaft indicates the protrusion inside the end cap is worn and the cap MUST be replaced.

Impeller Shaft
Radial "Play"

A minimum amount of radial "play" in the impeller shaft is necessary to ensure long life for the jet pump. To check for radial "play", insert the impeller shaft into the impeller housing, without the impeller installed, but with both needle bearings in place.

Clamp the impeller housing lug in a vise equipped with soft jaws to prevent housing movement. Make a dial indicator gauge setup with the gauge probe making contact with one of the flats on the end of the impeller shaft. Now, move the shaft up and down and make a note of the maximum deflection indicated by the indicator gauge.

If the deflection is more than 0.002" (0.05mm), the needle bearings may be worn or the impeller shaft may be worn. Use a micrometer and measure the diameter of the impeller shaft at the needle bearing surface. If the diameter is less than 0.8753" (22.23mm), the shaft is worn and **MUST** be replaced. A diameter greater than the figure specified, means the needle bearing is worn and **MUST** be replaced.

GOOD SHOP WORDS

Good shop practice dictates the shaft, the two needle bearings, the thrust washer, and the thrust bearing, be replaced, if excessive "play" is detected.

Impeller Shaft
End "Play"

To check the end "play" of the impeller shaft, use the same setup as for checking the radial "play", except this time, the impeller is installed on the shaft and the end cap is installed on the housing.

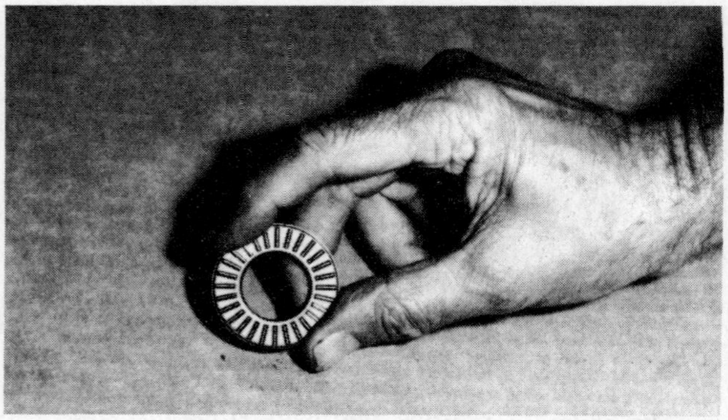

Inspect the needle bearings in the thrust bearing. If the bearings show signs of wear, the bearing and thrust washer should be replaced.

Position the probe of the dial indicator on the end of the impeller shaft -- on the same axis as the shaft. Move the shaft in and out of the housing. Make a note of the maximum deflection registered by the dial indicator. If the deflection is greater than 0.21" (0.54mm), the protrusion inside the end cap is worn and the cap **MUST** be replaced.

JET PUMP ASSEMBLING

The following procedures cover assembly of all parts to the jet pump. The procedures outline detailed instructions to assemble virtually all parts of the jet pump. As the work progresses, if a particular item was not removed or disassembled -- simply skip the step involved for the part and proceed directly to the work required.

Seal Carrier Assembling

1- Make a suitable setup in an arbor press and install a new needle bearing into the seal carrier. Use a suitable driver which presses on the needle bearing cage -- not the needle bearing. The bearing is correctly installed when each side is flush with shoulders inside the carrier.

Pack both seal lips with Lithium lubricant. Apply a coating of Loctite 515 around the circumference of each seal. Install the single lip seal on the engine side of the carrier with the seal lip facing **OUTWARD**. Install the double lip seal on the jet pump side of the carrier with the lip of the seal facing **OUTWARD.** The seals are correctly installed when the seal surface is flush with the housing.

2- Install the seal carrier into the hull. Apply Loctite 242 onto the threads of the securing bolts. Install the bolts and washers. Tighten the bolts to a torque value of 18 ft lb (25Nm).

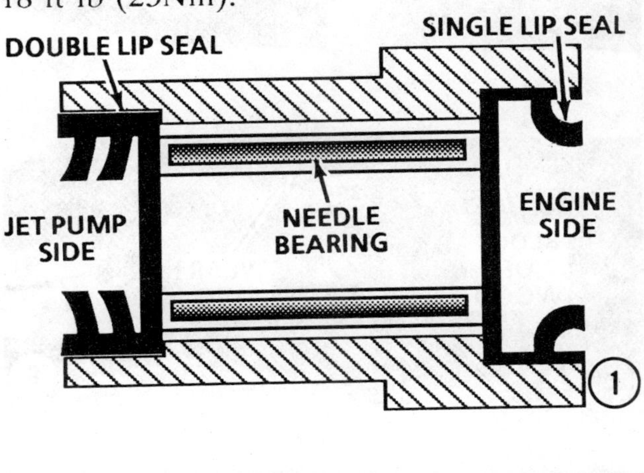

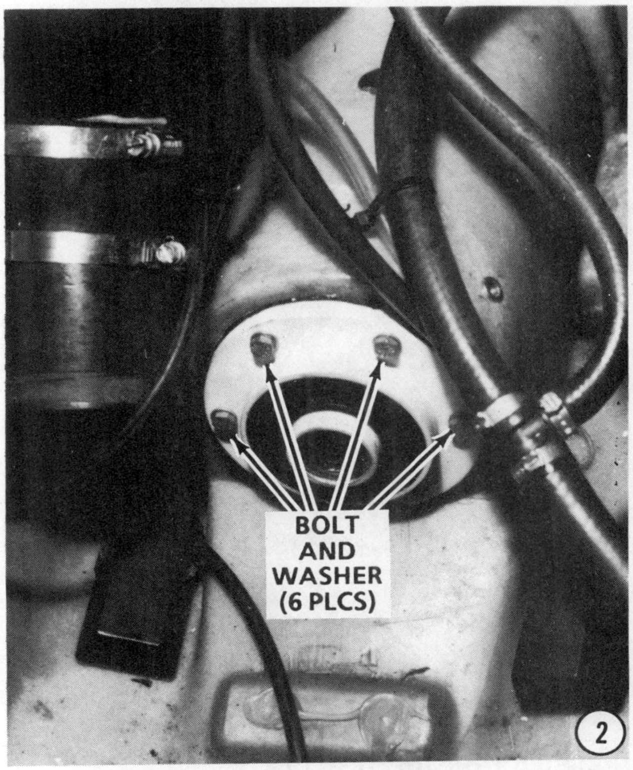

3- Slide a clamp onto each end of the boot, and then install the boot between the aft flywheel and the seal carrier. Tighten the clamps just **SNUG**. These clamps must not be permanently tightened until after the driveshaft has been installed, or the engine will be misaligned.

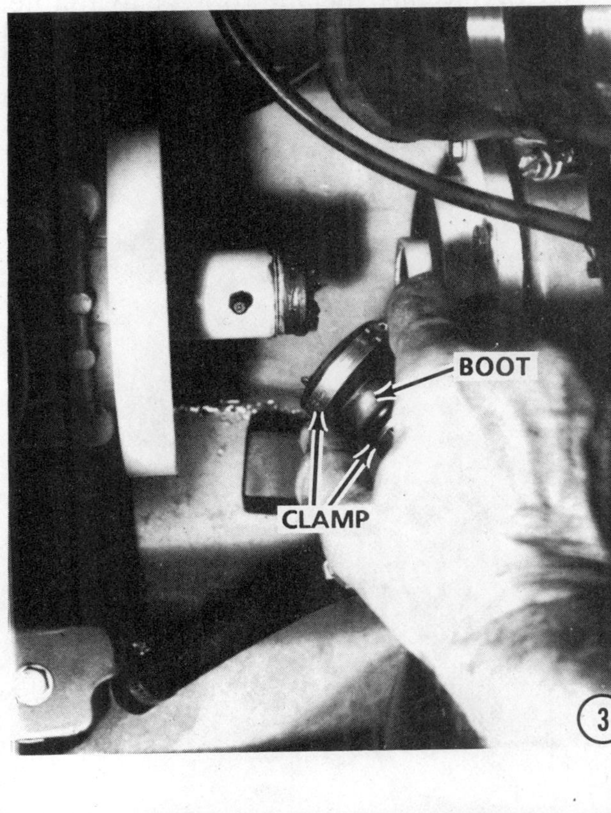

Impeller Housing Bearing and Seal Installation

4- Obtain Bearing Installer P/N 295-000-014. This tool will properly position the bearing within the impeller housing hub. If this tool is not available, a second but more difficult method of installation is presented at the end of this step.

Assuming the special tool is available: Position a new bearing, with the stamped surface facing **OUTWARD,** into the forward housing hub. Insert the tool into the bearing, using a rotating motion. Press the bearing in, using an arbor press, until a shoulder on the tool makes contact with the housing hub. The manufacturer strongly recommends the use of an arbor press -- and further states the bearing should **NEVER** be driven into place with a hammer.

If the special tool is not available: Obtain a suitable driver which contacts the bearing cage only -- not the needle bearing. Using an arbor press, work the bearing into place, with the stamped surface facing **OUTWARD**, until a clearance of 0.060-0.100" (1.5-2.5mm), exists between the stamped face of the bearing and the oil seal seat.

Pack the seal lip with Lithium lubricant. Coat the outside of the seal with Loctite 515. Install the seal using the special tool or suitable driver, with the seal lip facing **OUTWARD**, until the seal seats inside the housing.

5- Position a new needle bearing, with the stamped surface **OUTWARD**, into the aft housing hub. Insert the same tool used in the previous step, into the bearing, using a rotating motion. Using an arbor press, move the bearing, until the proper shoulder on the tool makes contact with housing hub surface.

The manufacturer strongly recommends the use of an arbor press -- and further states the bearing should **NEVER** be driven into place with a hammer.

If the special tool is not available, obtain a suitable driver which contacts only

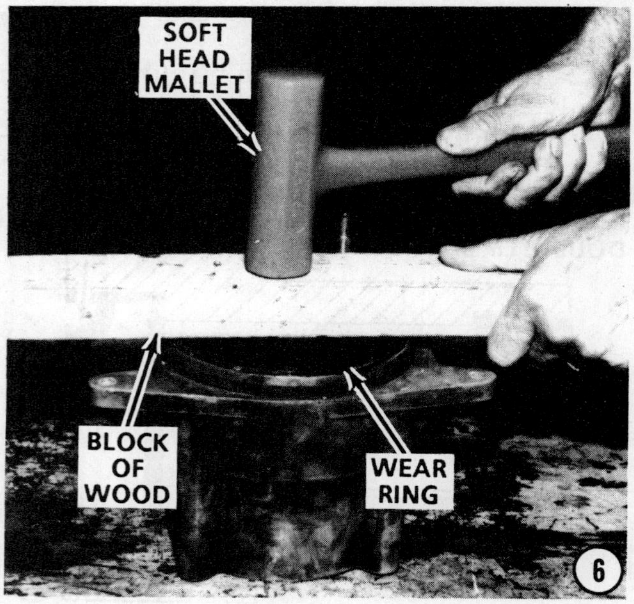

the bearing cage -- **NOT** the needle bearing. Make a suitable setup in the arbor press and press the bearing into the housing hub, with the stamped surface facing **OUTWARD**. The bearing is fully seated when the clearance between the stamped bearing surface and the thrust washer seat is 0.080-0.120" (2-3mm).

The manufacturer strongly recommends the use of an arbor press -- and further states the bearing should **NEVER** be driven into place with a hammer.

Wear Ring Installation

6- Cover the exterior surface of the wear ring with a coating of liquid soap. If the replacement ring has a lip on one face, position this lip to be on the outside of the housing.

If the ring is to be pressed in using an arbor press, obtain a steel plate approximately 7" x 7" x 1/4" thick (180mm x 180mm x 6mm). Fit the ring squarely in place over the housing. Place the plate over the ring and press the wear ring into the housing until it seats in the bottom of the housing.

If an arbor press is not available, use a 2" x 4" x 12" piece of wood. Fit the ring squarely into the housing, and then place the piece of wood over the ring. Now, work the ring into the housing by striking the wood with a hefty hammer, rotating the piece of wood 90o, striking again. Continue rotating the wood and striking with the hammer, until the wear ring seats in the housing.

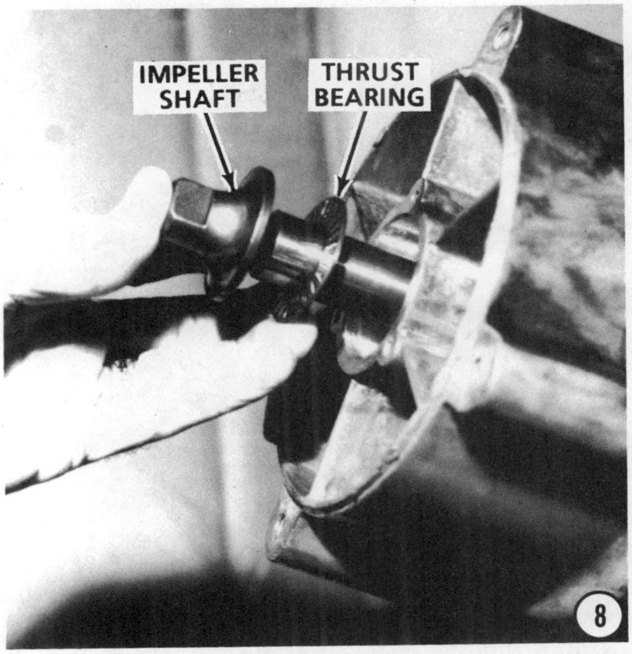

Impeller Shaft Installation

7- Coat both surfaces of the thrust washer with two-stroke engine oil, and then insert the washer into the aft impeller housing hub.

8- Coat both surfaces of the thrust bearing with two-stroke engine oil. Slide the bearing over the impeller shaft, and then insert the shaft and bearing into the aft housing hub. **TAKE CARE** not to damage the oil seal in the housing hub as the shaft

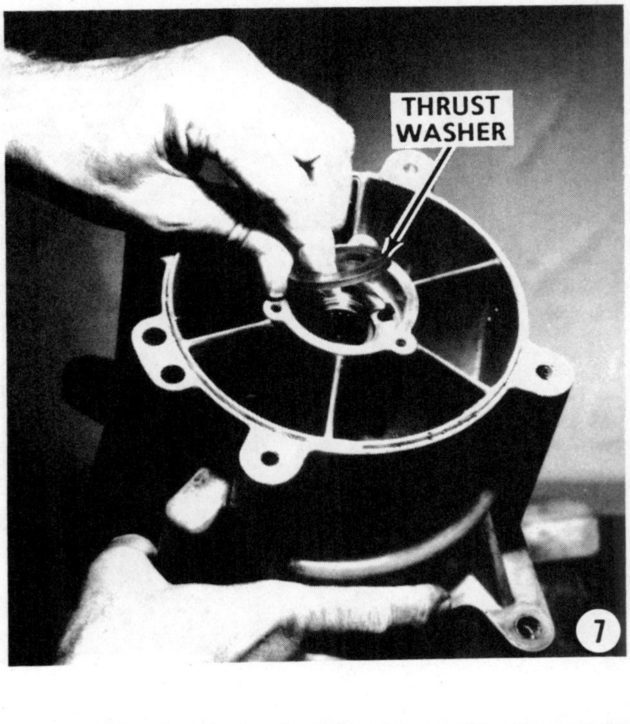

moves through the seal and into place in the impeller housing hub.

Impeller Installation

9- Clamp the end of the impeller shaft with the housing installed on the shaft, in a vise equipped with soft jaws. Check to be sure the impeller shaft splines are clean. Apply Loctite Primer N to the shaft threads. Allow the primer to dry for about twenty minutes. Apply Loctite 271 Red over the primer. The primer reduces the curing time of the Loctite 271 to six hours -- the time required for the 271 to reach maximum effectiveness. If the Loctite Primer N is not used, the 271 has a curing time of about 24 hours.

Make a final check to ensure the thrust washer and bearing are correctly seated in the hub, before the impeller is installed. Thread the impeller onto the impeller shaft.

The vanes in this impeller housing were damaged when the impeller worked loose on the shaft and moved forward.

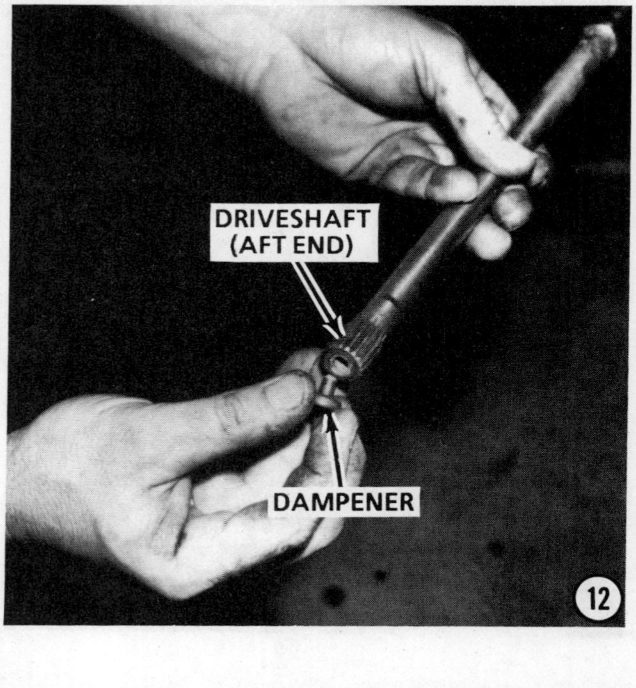

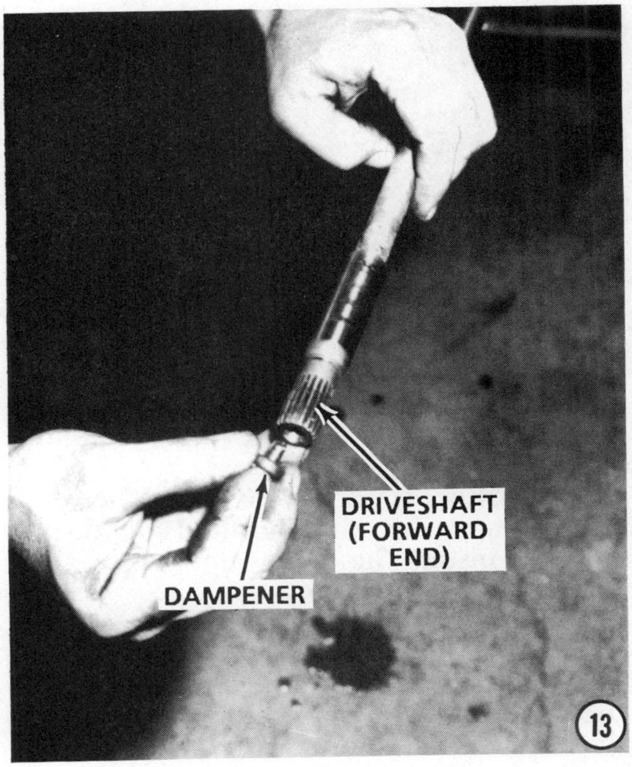

10- Slide Impeller Remover, P/N 295-000-001, into the impeller, with the splines of the tool and the splines of the impeller indexed.

11- Use the correct size socket over the tool, along with a torque wrench, and tighten the impeller to a torque value of 52 ft lb (70Nm).

Driveshaft Assembling

12- Install and tighten the threaded dampener into the aft end of the driveshaft.

13- Install and tighten another dampener into the forward end of the driveshaft.

14- Lay down a bead of silicone sealer around the perimeter of the impeller housing, looping around the three water passages and all the bolt holes.

15- Dip the driveshaft boot in a container of liquid soap, and then work it into the impeller.

16- Slide a new O-ring onto the splines at the aft end of the driveshaft. Apply a

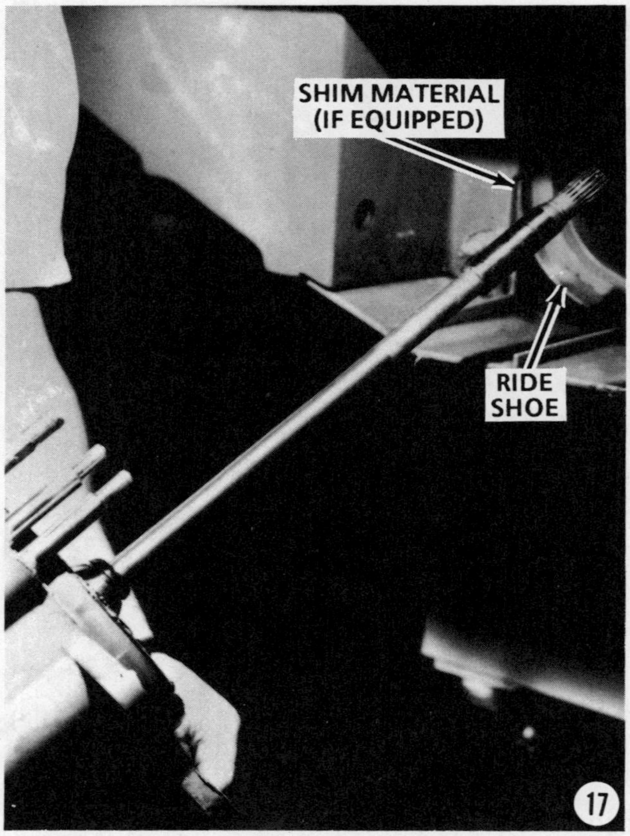

17

coating of synthetic lubricant onto the O-ring and splines. Work the splines through the boot and into the impeller, with the splines of the shaft indexed with the splines in the impeller. To prevent folding the lip of the seal backward, while the driveshaft is being installed, move the shaft back-and-

18

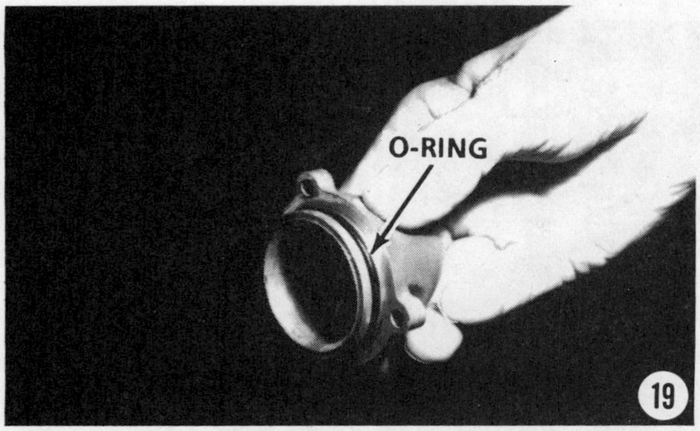

19

forth as the splines pass through the seal. Push the O-ring down the driveshaft and work it into the boot.

17- If shim material was found between the hull and the impeller housing during removal, install the same amount of shim material against the hull surface. Some models do not have shim material at this location. Insert the driveshaft through the ride shoe, seal carrier, and into the aft flywheel. Rotate the driveshaft until the splines of the shaft index with the splines in the flywheel. Install the impeller housing against the hull/shim material.

18- Apply a coating of Loctite 242 to the hull studs. Install the washers, lockwashers and nuts. Tighten the nuts alternately and evenly to a torque value of 26 ft lb (35Nm).

20

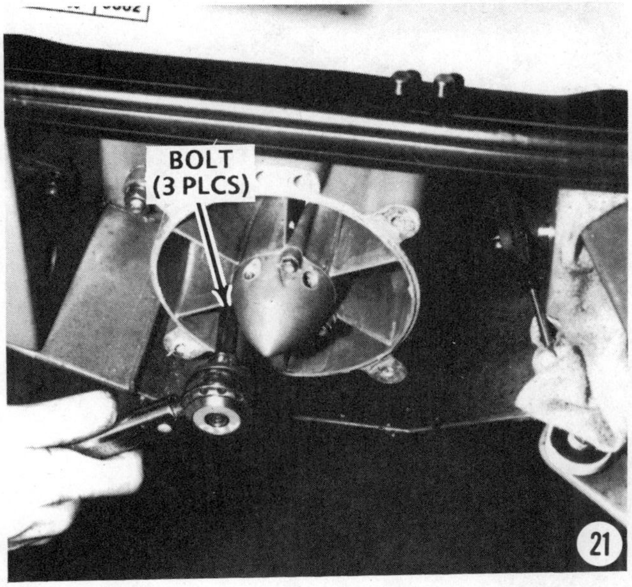

19- Coat a new end cap **O**-ring with Loctite PST 567, and then install the ring into the cap groove.

20- Install the end cap over the impeller housing, with the three bolt holes aligned and the two plug holes facing **UPWARD**, as shown.

21- Coat the threads of the three attaching bolts with Loctite 242. Install the bolts and tighten them alternately and evenly to a torque value of 62 in lb (7Nm).

22- Remove both plugs from the end cap reservoir. The starboard hole is the "fill hole" and the port hole is the "full hole". Fill the reservoir through the "fill hole" with Sea-Doo synthetic oil -- approximately 2.2 ounces, (65ml), or until the oil begins to exit through the "full hole".

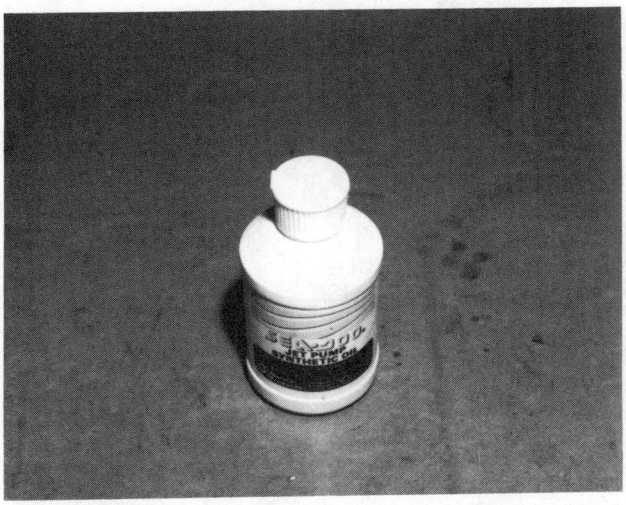

*Bombardier Sea-Doo Jet Pump Synthetic Oil is the **ONLY** lubricant recommended by the manufacturer for use in the jet pump.*

LUBRICANT CAUTION

The manufacturer strongly states: Sea-Doo synthetic oil must **NEVER** be mixed with any other type of oil, especially mineral based oil.

Apply Loctite 242 to the plug threads. Install the plugs and tighten them securely.

23- Release the bungee cord holding the nozzle and nozzle housing to one side.

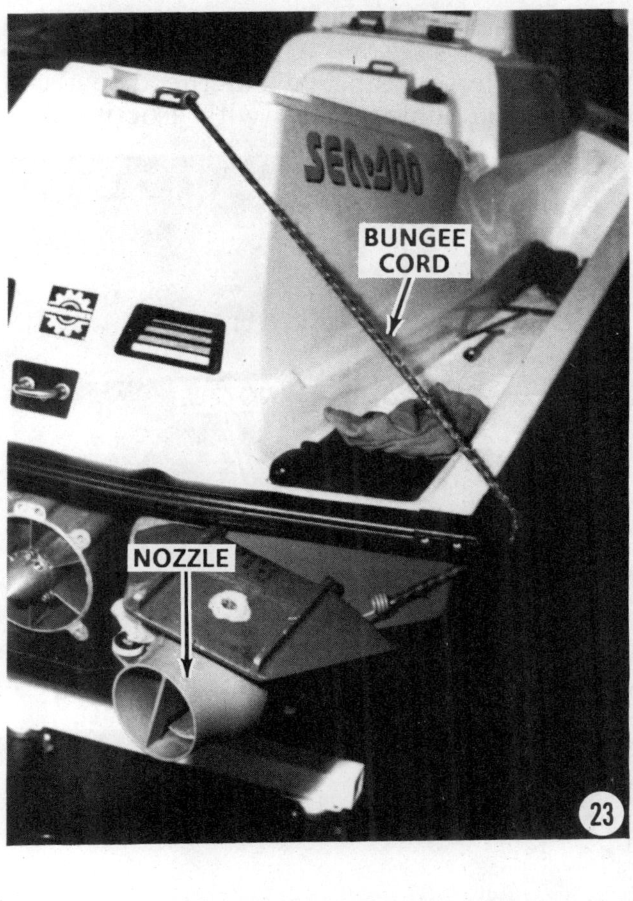

24- Install two new O-rings in the water passages in the top of the impeller housing.

25- Install the nozzle, with the steering nozzle and steering cable attached, onto the impeller housing. Coat the threads of the four attaching bolts with Loctite 242.

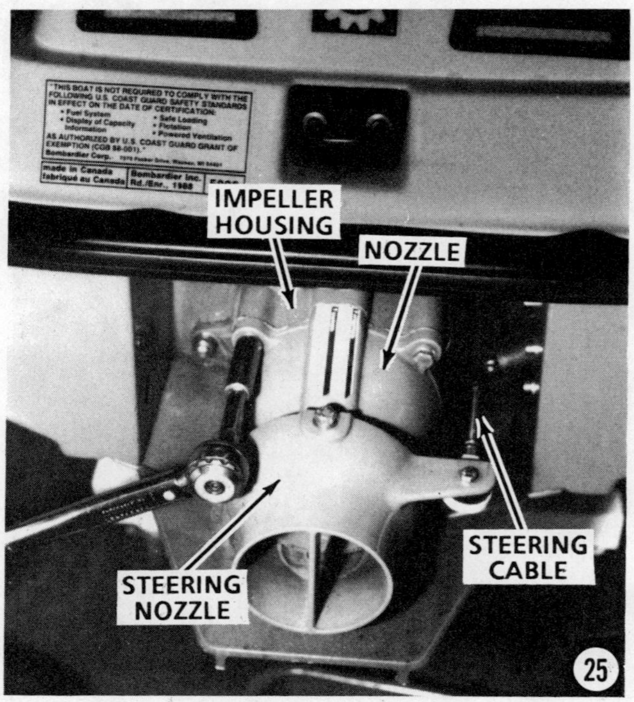

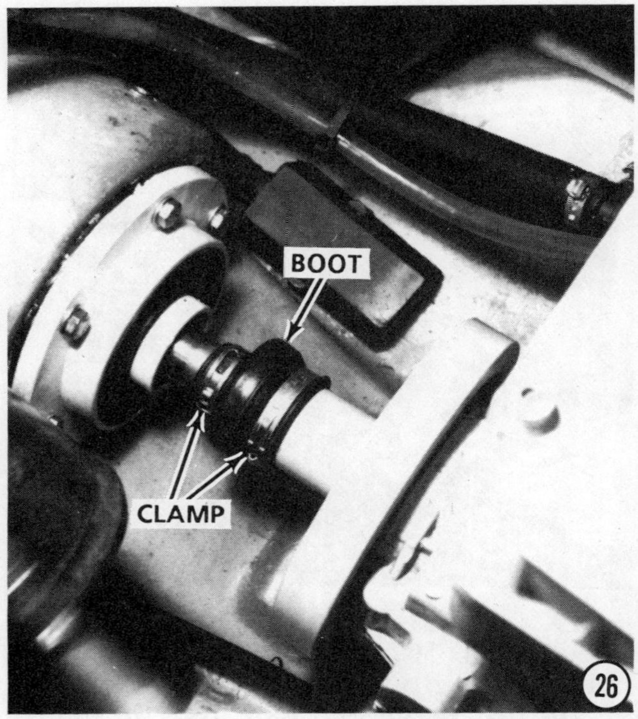

Install and tighten the bolts alternately and evenly to a torque value of 18 ft lb (25Nm). If the steering cable needs adjustment, see Chapter 11 -- **CONTROL ADJUSTMENTS**.

26- Tighten the two clamps on the boot between the aft flywheel and the seal carrier.

27- Slide clamps over the three hoses and then work the hoses onto the jet pump fittings. Tighten the clamps securely.

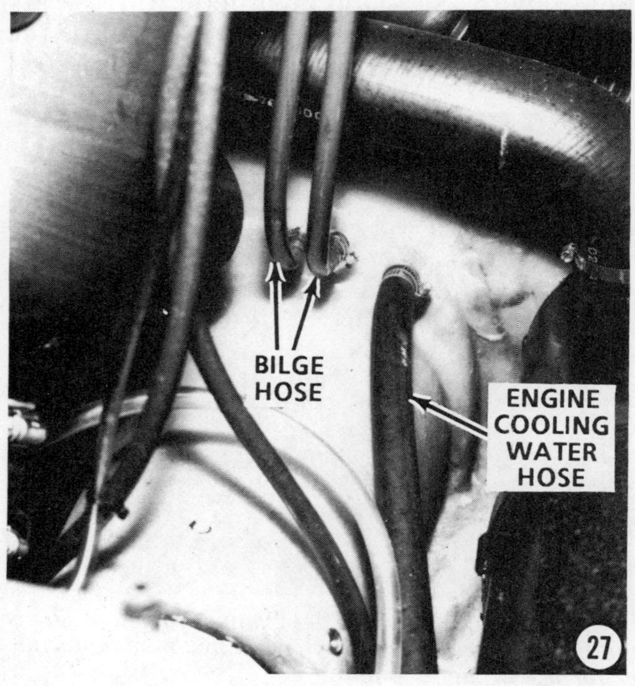

11
CONTROL ADJUSTMENTS

11-1 INTRODUCTION

Control adjustments on Bombardier Sea-Doo personal watercraft are limited to changing the length of the steering cable and the reverse shift cable. Normally, adjustments are not necessary for extended periods of time, unless the craft has experienced an accident or the control cable has been disconnected to service other components.

Detailed procedures to adjust the steering cable and the reverse shift cable are presented in this short chapter. Adjustment of the throttle cable and the choke cable is thoroughly covered -- in detail -- within the pages of Chapter 6 -- **FUEL.**

The "business" end of a Bombardier Sea-Doo personal watercraft, with the steering cable hookup to the steering nozzle clearly visible.

11-2 STEERING CABLE ADJUSTMENT

The steering nozzle axis must be positioned exactly parallel to the axis of the craft to prevent the craft from veering to port or to starboard when the handlebar is positioned for "dead ahead".

Adjustment

1- Measure the distance from the bottom of the handlebar grip to the foot rest, as shown. Position the handlebar with equal vertical distances on both sides -- the "dead ahead" position. Perform steps 2, 3, and 4 **WITHOUT** disturbing the handlebar position.

2- Place a straight edge -- a piece of wood will do nicely -- against the nozzle -- parallel to the transom. Hold the piece of wood in place, as shown, and measure the

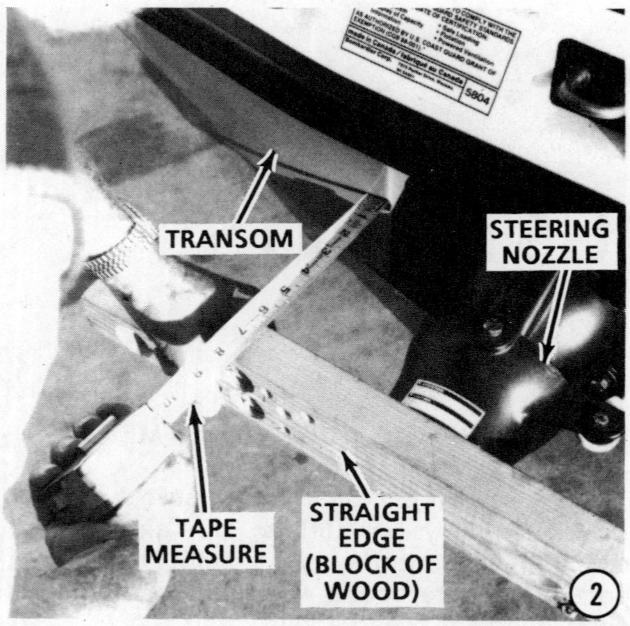

distance, on one side, from the straight edge piece of wood to the transom. Record or remember the dimension.

3- Keep the straight edge in place against the nozzle and measure the distance between the straight edge and the transom on the other side of the nozzle.

4- If the distances measured in Steps 2 and 3 are not identical, adjust the two nuts on the cable at the steering bracket to bring the straight edge an equal distance from the transom on both sides of the nozzle.

5- Turn the handlebar and swing the steering nozzle hard over to port. If the clearance between the outlet nozzle and the steering nozzle is less than 0.005" (0.13mm), adjust the nuts on the steering cable bracket until this minimum clearance is obtained. **DO NOT** exceed this clearance on one side, because the added clearance on one side of the nozzle will take away clearance on the other side. Repeat this procedure with the handlebars turned in the opposite direction -- hard over to starboard -- swinging the nozzle hard over to starboard.

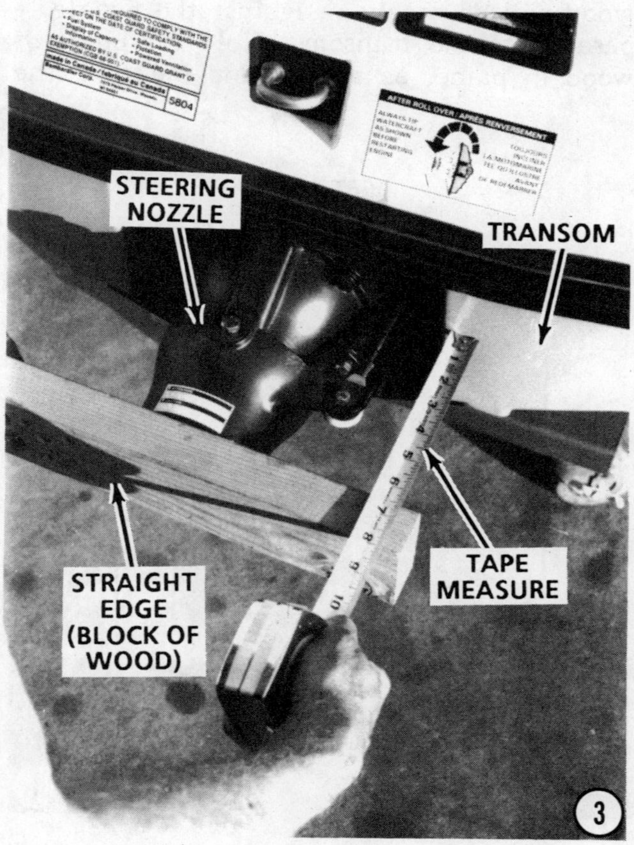

SPECIAL WORDS

A final check **MUST** be made with the craft in a body of water or in a test tank to **ENSURE** proper cable adjustment. Start the engine and allow it to warm to operating temperature.

CAUTION

Water must circulate through the jet pump — to and from the engine, anytime the engine is operating. Circulating water will

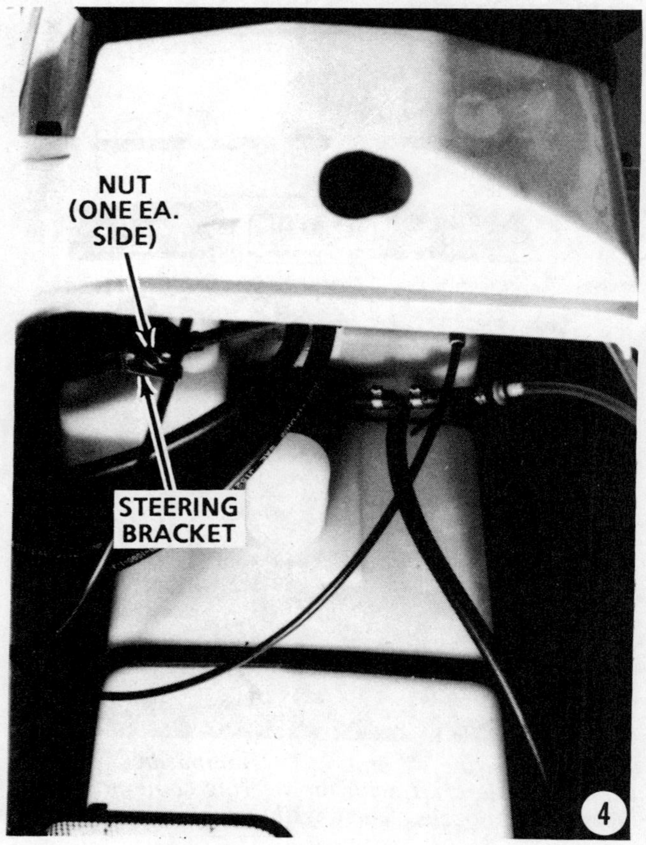

prevent overheating — which could cause damage to moving engine parts and possible engine seizure.

With the craft in the water; with the engine operating; and holding the handlebar in the "dead ahead" position; advance the throttle and observe if the bow tends to veer to port or to starboard. If the bow appears to remain relatively on a true course, the adjustment is satisfactory -- adjustment task completed.

Plastic Handlebar Cover

The plastic handlebar cover fits very snugly over the two foam inserts. If the old cover has become tattered and torn, cut it clear with a pair of scissors. To ease installation of a new cover, first soak it in hot water for, say 1/4 hour or until it seems to be a bit pliable. The cover could also be heated with a hair dryer. If a dryer is used, keep the dryer moving slowly over the entire surface of the cover. Once the cover is ready, work quickly before it has a chance to cool. Close the zipper.

Handlebar Grips

A grip which has worked loose on the handlebar is a nuisance and presents a definite danger to the operator. The old grip, if still in good condition may be reinstalled using the proper adhesion techniques, or a new set may be installed. Do not plan on using the craft following installation of new grips for at least **TWO** full days. This 48-hour period is absolutely necessary for the sealant to cure for maximum effectiveness.

Remove the old grip -- cut it away -- if necessary. Clean all traces of old sealant from the knurled section of the handlebar end. If the old grip is to be used, clean all traces of old sealant from inside the bore of the grip.

Apply Sikaflex primer 260 P/N 293-530-013 to the knurled portion of the handlebar and onto the inside bore of the grip. Allow the primer to cure for **AT LEAST** one full hour. Next, apply Sikaflex sealant 221 over the primer on the bar and the grip. Slide the grip over the end of the handlebar as far as possible. Now, wait at least **TWO** full days before putting the watercraft back into action on the water.

11-3 REVERSE CAPABILITY

The Bombardier Sea-Doo, Model GT, is equipped with reverse capability. As mentioned in Chapter 1 -- **DESCRIPTION AND OPERATION** — "reverse" is accomplished by a gate installed in such a manner to swing down over the steering nozzle. The gate is

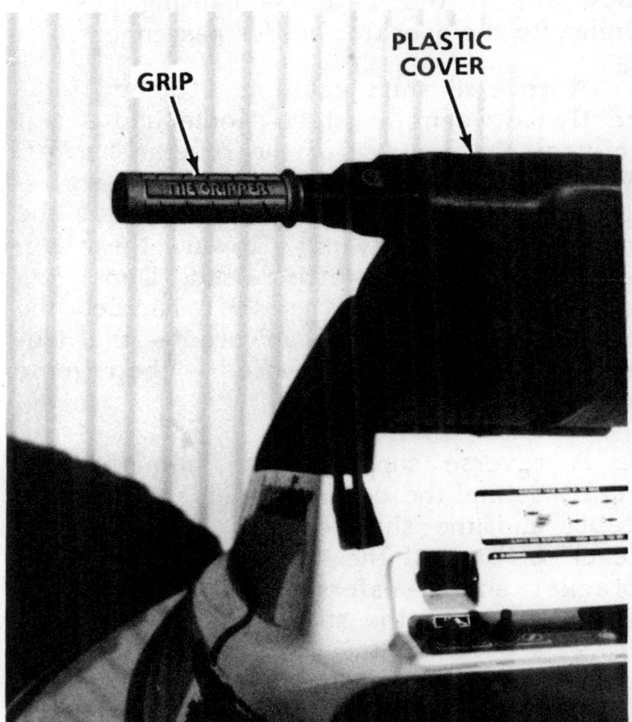

The plastic handlebar cover takes a "beating" from the sun, water, and operator abuse. Specific, detailed, and critical instructions to replace the cover are given in the text.

controlled with a cable. When the gate is in the fully raised position, allowing all water from the nozzle to be discharged sternward -- the craft moves forward.

When the gate is swung down just part way, some of the discharged water moves sternward and the other portion of water is deflected downward and forward under the craft. Positioning the gate in such a manner places the jet pump in **NEUTRAL**, because the water moving aft equals the water moving forward and the craft is static ("dead in the water").

When the gate is moved to the full down position, virtually all discharged water from the nozzle is deflected down and forward -- the craft moves sternward.

The position of the reverse bucket determines the direction of thrust from water expelled from the nozzle. If the bucket is not adjusted properly, forward motion of the craft will be seriously restricted because some of the expelled water will be forced back under the craft --as for going astern.

More importantly, improper adjustment of the reverse bucket can cause the bucket to be unintentionally lowered -- throwing the craft into reverse -- suddenly raising the bow out of the water -- causing possible injury to the operator and/or passengers.

A reverse shift cable is attached indirectly between the shift handle and a ball joint on the portside of the reverse bucket. When the shift handle is pulled back -- toward the operator -- the aft end of the cable is pulled forward -- toward the craft, and the reverse bucket is raised. Depending on the position of the shift handle, the bucket will be partially lowered -- the neutral position, or fully lowered -- the reverse position.

A reverse support plate, mounted directly behind the control panel, supports the cable and the shift handle. A triangular lever pivots at the base of the support bracket and transfers the motion of the shift handle to the shift cable. A locking lever, secured to the shift handle shaft, indents with two rectangular notches cut in the plate to hold the forward and reverse positions of the shift handle. Reverse shift cable adjustment is made at the point where the cable is supported -- at the rear of the plate.

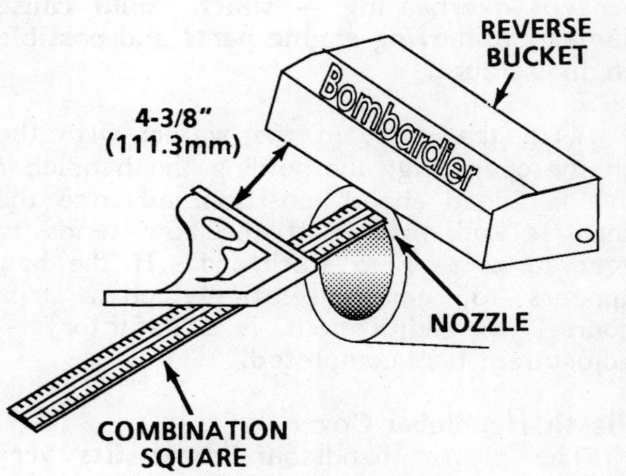

*Simple line drawing to indicate placement of the square against the end face of the nozzle when making the **NEUTRAL** position adjustment, as shown in the drawing below.*

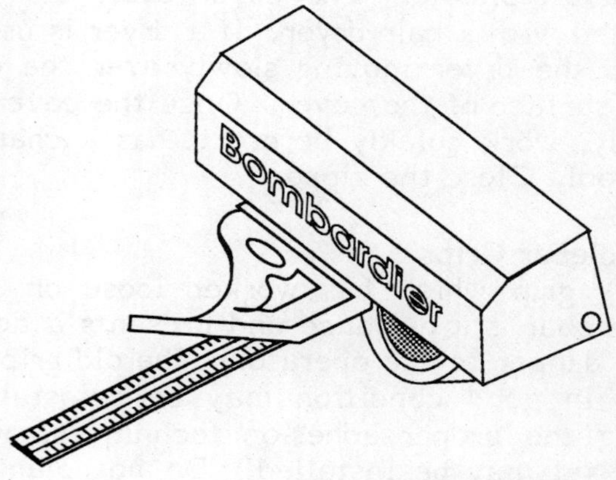

*When making the reverse bucket adjustment for the **NEUTRAL** position, a combination square is used to set the outer lower face edge of the bucket the required distance from the nozzle. The return springs must be unhooked for this adjustment.*

Two return springs, one on each side of the reverse bucket, provide tension on the bucket to keep it in the forward -- up -- position.

REVERSE CABLE ADJUSTMENT

Obtain a combination square (set square), and set it to 4-3/8" (111.3mm). Unhook both return springs from the reverse bucket. Set the shift handle in the **NEUTRAL** position. Place the handlebar in the "dead ahead" position, as described in the steering cable adjustment -- the back face of the nozzle parallel with the transom. Hold the square up to the top edge of the nozzle, under the reverse bucket, as shown. The lower edge of the bucket must

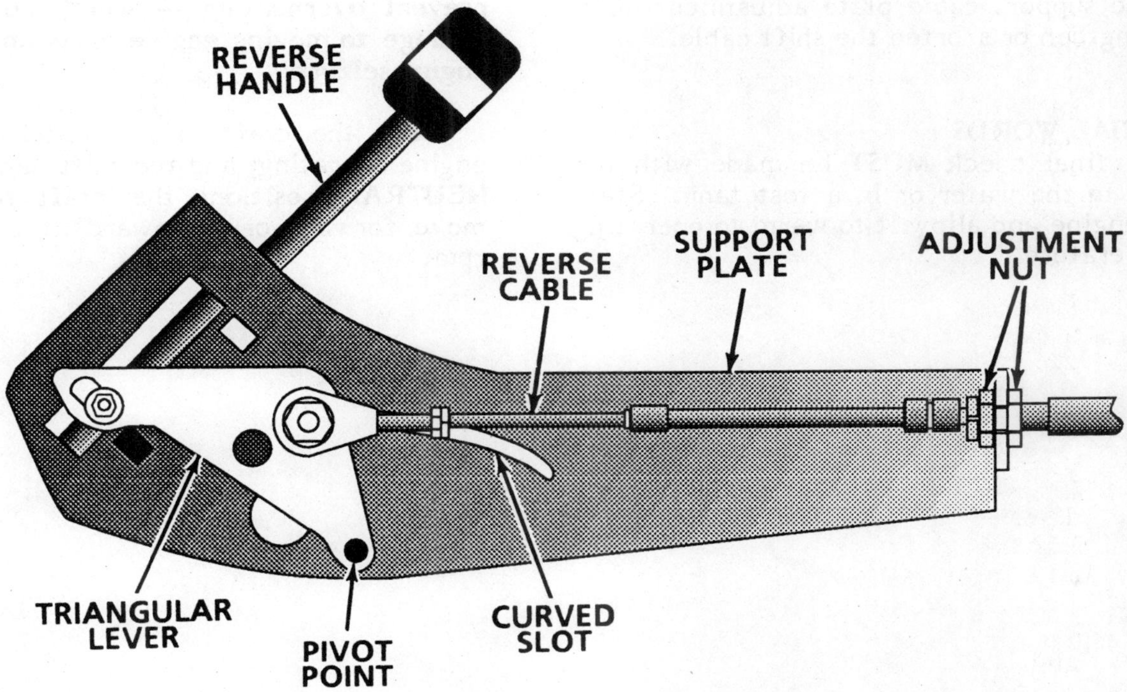

*Line drawing to illustrate the portside of the reverse bracket, with the handle in the **NEUTRAL** position.*

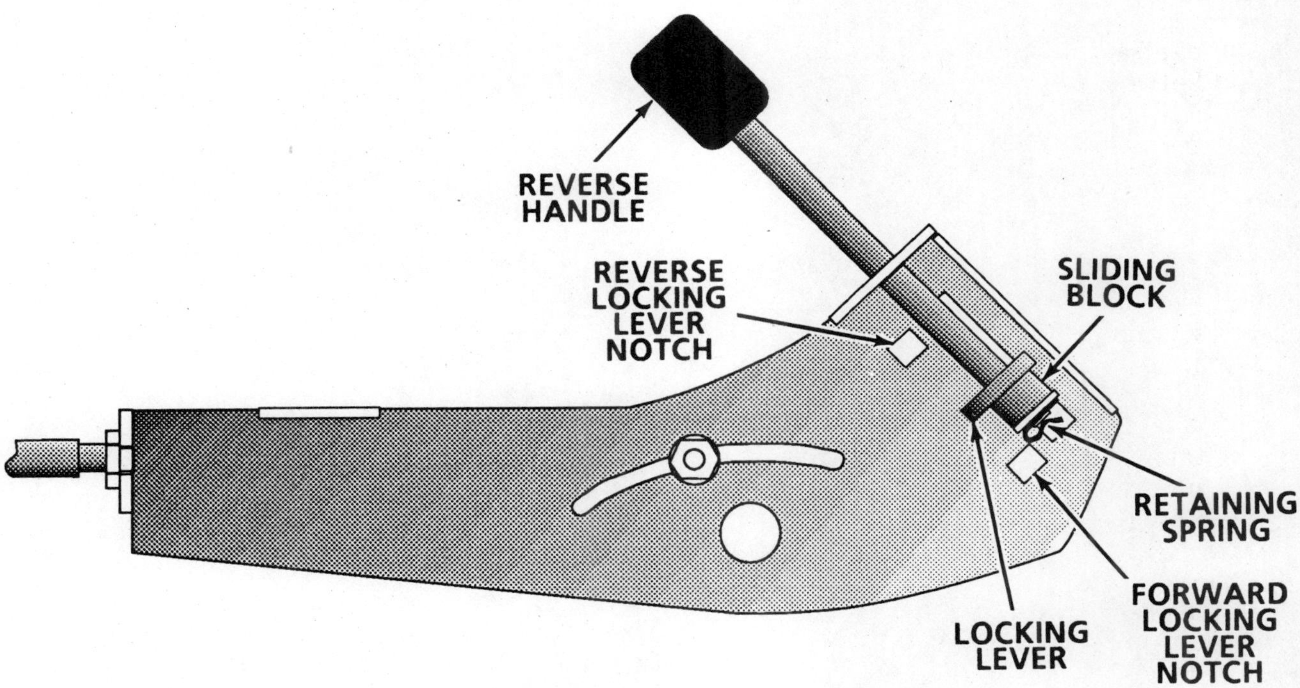

*Line drawing to illustrate the starboard side of the reverse bracket, with the handle in the **NEUTRAL** position.*

be between 4-3/8" to 4-7/8" (111.3-111.7mm) from the upper edge of the nozzle. If the bucket is positioned incorrectly, make the necessary adjustment at the support cable plate adjustment nuts, to lengthen or shorten the shift cable.

SPECIAL WORDS

A final check **MUST** be made with the craft in the water or in a test tank. Start the engine and allow it to warm to operating temperature.

CAUTION

Water must circulate through the jet pump -- to and from the engine, anytime the engine is operating. Circulating water will prevent overheating — which could cause damage to moving engine parts and possible engine seizure.

With the craft in the water; with the engine operating; and the shift handle in the **NEUTRAL** position; the craft **MUST** not move forward or sternward at any engine rpm.

APPENDIX

METRIC CONVERSION CHART

LINEAR

inches	X 25.4	= millimetres (mm)
feet	X 0.3048	= metres (m)
yards	X 0.9144	= metres (m)
miles	X 1.6093	= kilometres (km)
inches	X 2.54	= centimetres (cm)

AREA

$inches^2$	X 645.16	= $millimetres^2$ (mm^2)
$inches^2$	X 6.452	= $centimetres^2$ (cm^2)
$feet^2$	X 0.0929	= $metres^2$ (m^2)
$yards^2$	X 0.8361	= $metres^2$ (m^2)
acres	X 0.4047	= hectares (10^4 m^2) (ha)
$miles^2$	X 2.590	= $kilometres^2$ (km^2)

VOLUME

$inches^3$	X 16387	= $millimetres^3$ (mm^3)
$inches^3$	X 16.387	= $centimetres^3$ (cm^3)
$inches^3$	X 0.01639	= litres (l)
quarts	X 0.94635	= litres (l)
gallons	X 3.7854	= litres (l)
$feet^3$	X 28.317	= litres (l)
$feet^3$	X 0.02832	= $metres^3$ (m^3)
fluid oz	X 29.60	= millilitres (ml)
$yards^3$	X 0.7646	= $metres^3$ (m^3)

MASS

ounces (av)	X 28.35	= grams (g)
pounds (av)	X 0.4536	= kilograms (kg)
tons (2000 lb)	X 907.18	= kilograms (kg)
tons (2000 lb)	X 0.90718	= metric tons (t)

FORCE

ounces - f (av)	X 0.278	= newtons (N)
pounds - f (av)	X 4.448	= newtons (N)
kilograms - f	X 9.807	= newtons (N)

ACCELERATION

$feet/sec^2$	X 0.3048	= $metres/sec^2$ (m/S^2)
$inches/sec^2$	X 0.0254	= $metres/sec^2$ (m/s^2)

ENERGY OR WORK (watt-second - joule - newton-metre)

foot-pounds	X 1.3558	= joules (j)
calories	X 4.187	= joules (j)
Btu	X 1055	= joules (j)
watt-hours	X 3500	= joules (j)
kilowatt - hrs	X 3.600	= megajoules (MJ)

FUEL ECONOMY AND FUEL CONSUMPTION

miles/gal	X 0.42514	= kilometres/litre (km/l)

Note:
235.2/(mi/gal) = litres/100km
235.2/(litres/100 km) = mi/gal

LIGHT

footcandles	X 10.76	= $lumens/metre^2$ (lm/m^2)

PRESSURE OR STRESS (newton/sq metre - pascal)

inches HG (60 F)	X 3.377	= kilopascals (kPa)
pounds/sq in	X 6.895	= kilopascals (kPa)
inches H_2O (60° F)	X 0.2488	= kilopascals (kPa)
bars	X 100	= kilopascals (kPa)
pounds/sq ft	X 47.88	= pascals (Pa)

POWER

horsepower	X 0.746	= kilowatts (kW)
ft-lbf/min	X 0.0226	= watts (W)

TORQUE

pound-inches	X 0.11299	= newton-metres (N·m)
pound-feet	X 1.3558	= newton-metres (N·m)

VELOCITY

miles/hour	X 1.6093	= kilometres/hour (km/h)
feet/sec	X 0.3048	= metres/sec (m/s)
kilometres/hr	X 0.27778	= metres/sec (m/s)
miles/hour	X 0.4470	= metres/sec (m/s)

TEMPERATURE

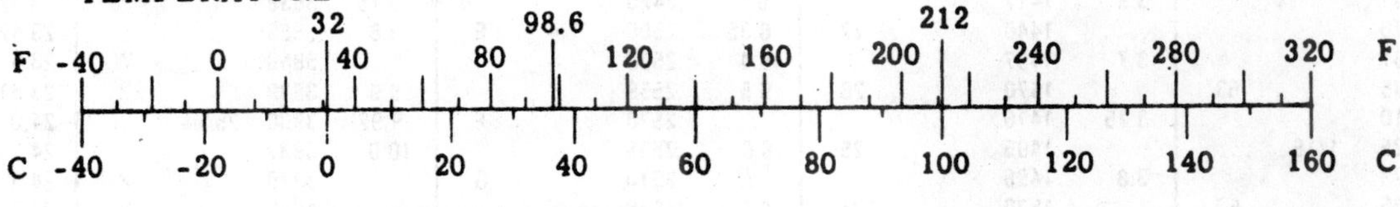

Celsius = 0.556 (°F -32) F = (1.8°C) +32

DRILL SIZE CONVERSION CHART

SHOWING MILLIMETER SIZES, FRACTIONAL AND DECIMAL INCH SIZES AND NUMBER DRILL SIZES

Milli-Meter	Dec. Equiv.	Frac-tional	Num-ber	Milli-Meter	Dec. Equiv.	Frac-tional	Num-ber	Milli-Meter	Dec. Equiv.	Frac-tional	Num-ber	Milli-Meter	Dec. Equiv.	Frac-tional	Num-ber	Milli-Meter	Dec. Equiv.	Frac-tional
.1	.0039			1.75	.0689			....	.1570		22	6.8	.2677			10.72	.4219	27/64
.15	.0059			....	.0700		50	4.0	.1575			6.9	.2716			11.0	.4330	
.2	.0079			1.8	.0709			....	.1590		21	....	.2720		I	11.11	.4375	7/16
.25	.0098			1.85	.0728			....	.1610		20	7.0	.2756			11.5	.4528	
.3	.0118			....	.0730		49	4.1	.1614			....	.2770		J	11.51	.4531	29/64
....	.0135		80	1.9	.0748			4.2	.1654			7.1	.2795			11.91	.4687	15/32
.35	.0138			....	.0760		48	....	.1660		19	....	.2811		K	12.0	.4724	
....	.0415		79	1.95	.0767			4.25	.1673			7.14	.2812	9/32	....	12.30	.4843	31/64
.39	.0156	1/64	....	1.98	.0781	5/64	....	4.3	.1693			7.2	.2835			12.5	.4921	
.4	.0157			....	.0785		47	....	.1695		18	7.25	.2854			12.7	.5000	1/2
....	.0160		78	2.0	.0787			4.37	.1719	11/64	....	7.3	.2874			13.0	.5118	
.45	.0177			2.05	.0807			....	.1730		17	....	.2900		L	13.10	.5156	33/64
....	.0180		77	....	.0810		46	4.4	.1732			7.4	.2913			13.49	.5312	17/32
.5	.0197			....	.0820		45	....	.1770		16	....	.2950		M	13.5	.5315	
....	.0200		76	2.1	.0827			4.5	.1771			7.5	.2953			13.89	.5469	35/64
....	.0210		75	2.15	.0846			....	.1800		15	7.54	.2968	19/64	....	14.0	.5512	
.55	.0217			....	.0860		44	4.6	.1811			7.6	.2992			14.29	.5624	9/16
....	.0225		74	2.2	.0866			....	.1820		14	....	.3020		N	14.5	.5709	
.6	.0236			2.25	.0855			4.7	.1850		13	7.7	.3031			14.68	.5781	37/64
....	.0240		73	....	.0890		43	4.75	.1870			7.75	.3051			15.0	.5906	
....	.0250		72	2.3	.0905			4.76	.1875	3/16	....	7.8	.3071			15.08	.5937	19/32
.65	.0256			2.35	.0925			4.8	.1890		12	7.9	.3110			15.48	.6094	39/64
....	.0260		71	....	.0935		42	....	.1910		11	7.94	.3125	5/16	....	15.5	.6102	
....	.0280		70	2.38	.0937	3/32	....	4.9	.1929			8.0	.3150			15.88	.6250	5/8
.7	.0276			2.4	.0945			....	.1935		10	....	.3160		O	16.0	.6299	
....	.0292		69	....	.0960		41	....	.1960		9	8.1	.3189			16.27	.6406	41/64
.75	.0295			2.45	.0964			5.0	.1968			8.2	.3228			16.5	.6496	
....	.0310		68	....	.0980		40	....	.1990		8	....	.3230		P	16.67	.6562	21/32
.79	.0312	1/32	....	2.5	.0984			5.1	.2008			8.25	.3248			17.0	.6693	
.8	.0315			....	.0995		39	....	.2010		7	8.3	.3268			17.06	.6719	43/64
....	.0320		67	....	.1015		38	5.16	.2031	13/64	....	8.33	.3281	21/64	....	17.46	.6875	11/16
....	.0330		66	2.6	.1024			....	.2040		6	8.4	.3307			17.5	.6890	
.85	.0335			....	.1040		37	5.2	.2047			....	.3320		Q	17.86	.7031	45/64
....	.0350		65	2.7	.1063			....	.2055		5	8.5	.3346			18.0	.7087	
.9	.0354			....	.1065		36	5.25	.2067			8.6	.3386			18.26	.7187	23/32
....	.0360		64	2.75	.1082			5.3	.2086			....	.3390		R	18.5	.7283	
....	.0370		63	2.78	.1094	7/64	....	....	.2090		4	8.7	.3425			18.65	.7344	47/64
.95	.0374			....	.1100		35	5.4	.2126			8.73	.3437	11/32	....	19.0	.7480	
....	.0380		62	2.8	.1102			...	.2130		3	8.75	.3445			19.05	.7500	3/4
....	.0390		61	....	.1110		34	5.5	.2165			8.8	.3465			19.45	.7656	49/64
1.0	.0394			....	.1130		33	5.56	.2187	7/32	....	....	.3480		S	19.5	.7677	
....	.0400		60	2.9	.1141			5.6	.2205			8.9	.3504			19.84	.7812	25/32
....	.0410		59	....	.1160		32	....	.2210		2	9.0	.3543			20.0	.7874	
1.05	.0413			3.0	.1181			5.7	.2244			....	.3580		T	20.24	.7969	51/64
....	.0420		58	....	.1200		31	5.75	.2263			9.1	.3583			20.5	.8071	
....	.0430		57	3.1	.1220			....	.2280		1	9.13	.3594	23/64	....	20.64	.8125	13/16
1.1	.0433			3.18	.1250	1/8	....	5.8	.2283			9.2	.3622			21.0	.8268	
1.15	.0452			3.2	.1260			5.9	.2323			9.25	.3641			21.04	.8218	53/64
....	.0465		56	3.25	.1279			....	.2340		A	9.3	.3661			21.43	.8437	27/32
1.19	.0469	3/64	...	....	.1285		30	5.95	.2344	15/64	....	....	.3680		U	21.5	.8465	
1.2	.0472			3.3	.1299			6.0	.2362			9.4	.3701			21.83	.8594	55/64
1.25	.0492			3.4	.1338			....	.2380		B	9.5	.3740			22.0	.8661	
1.3	.0512			....	.1360		29	6.1	.2401			9.53	.3750	3/8	....	22.23	.8750	7/8
....	.0520		55	3.5	.1378			....	.2420		C	....	.3770		V	22.5	.8858	
1.35	.0513			....	.1405		28	6.2	.2441			9.6	.3780			22.62	.8906	57/64
....	.0550		54	3.57	.1406	9/64	....	6.25	.2460		D	9.7	.3819			23.0	.9055	
1.4	.0551			3.6	.1417			6.3	.2480			9.75	.3838			23.02	.9062	29/32
1.45	.0570			....	.1440		27	6.35	.2500	1/4	E	9.8	.3858			23.42	.9219	59/64
1.5	.0591			3.7	.1457			6.4	.2520			....	.3860		W	23.5	.9252	
....	.0595		53	....	.1470		26	6.5	.2559			9.9	.3839			23.81	.9375	15/16
1.55	.0610			3.75	.1476			....	.2570		F	9.92	.3906	25/64	....	24.0	.9449	
1.59	.0625	1/16	....	....	.1495		25	6.6	.2598			10.0	.3937			24.21	.9531	61/64
1.6	.0629			3.8	.1496			....	.2610		G	....	.3970		X	24.5	.9646	
....	.0635		52	....	.1520		24	6.7	.2638			....	.4040		Y	24.61	.9687	31/32
1.65	.0649			8.9	.1535			6.75	.2657	16/64	....	10.32	.4062	13/32	...	25.0	.9843	
1.7	.0669			....	.1540		23	6.75	.2657			....	.4130		Z	25.03	.9844	63/64
....	.0670		51	3.97	.1562	5/32	....	....	.2660		H	10.5	.4134			25.4	1.0000	1

ENGINE SPECIFICATIONS
ALL MODELS

Engine type	Rotax 587, water cooled, two-stroke
Induction type	Rotary valve
Exhaust system	Water cooled, water injected
Starting system	Electric
Lubrication	
Fuel/oil mixture	Variable ratio oil injection
Oil injection pump	Gear driven Mikuni
Oil type	Sea-Doo injection oil
No. of cylinders	Two
Standard bore	2.992" (76.00mm)
1st. oversize bore	3.002" (76.25mm)
2nd. oversize bore	3.012" (76.50mm)
Stroke	2.519" (64.00mm)
Displacement	35.43 cu. in. (580.7cc)
Spark plug make	NGK
Spark plug type	BR7ES
Spark plug gap	0.024" (0.60mm)
Ignition timing	0.086" (2.18mm) {17-21 degrees}
RPM limiter range	6300-6700 rpm
Low speed mixture screw adj.	1-1/8 turn out
High speed mixture screw adj.	1/4 turn out (sealed)
Idle speed	1300-1500 rpm

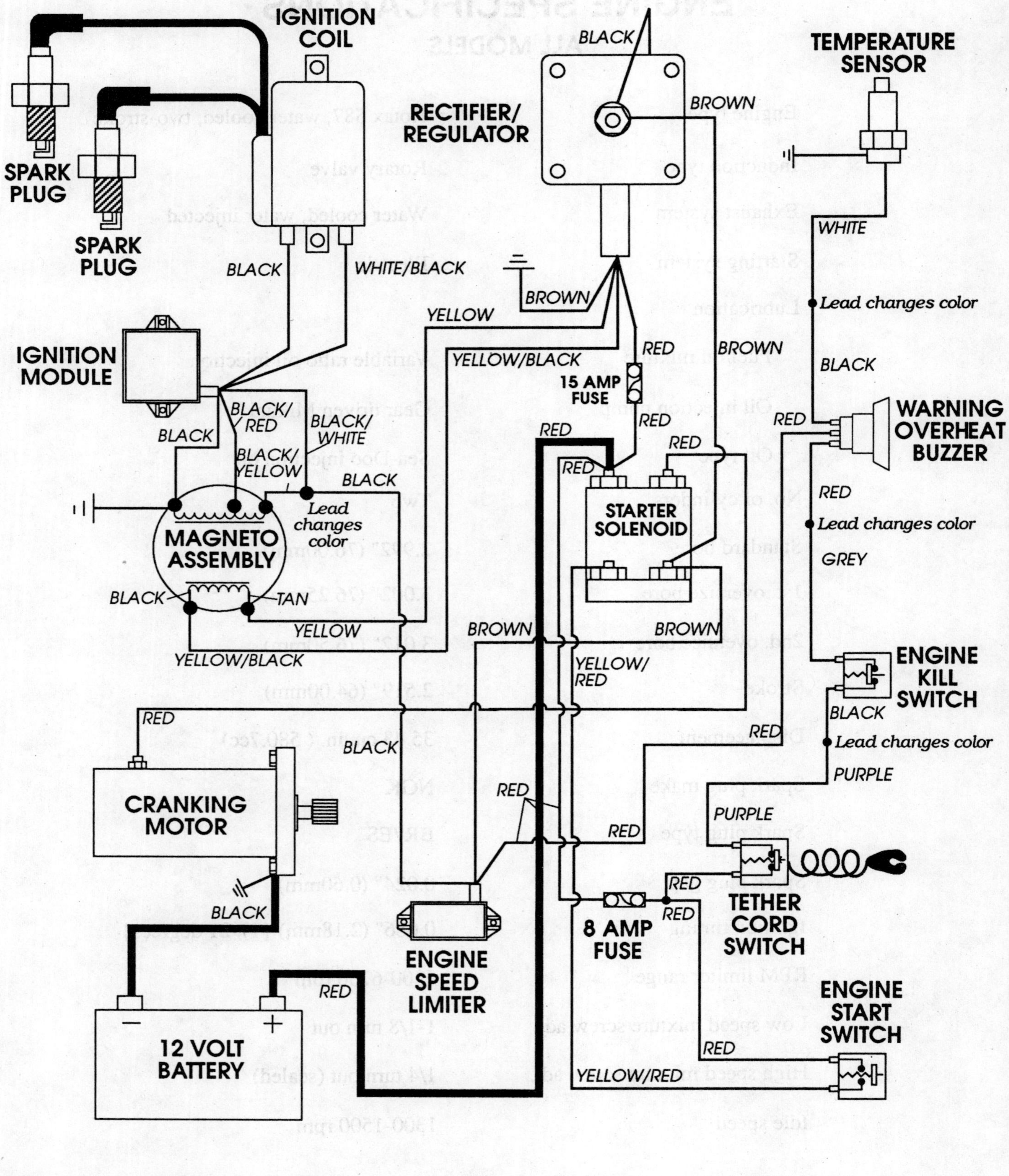

Wire Identification -- All Bombardier Sea-Doo models ***WITHOUT*** *tachometer and oil/fuel gauge.*

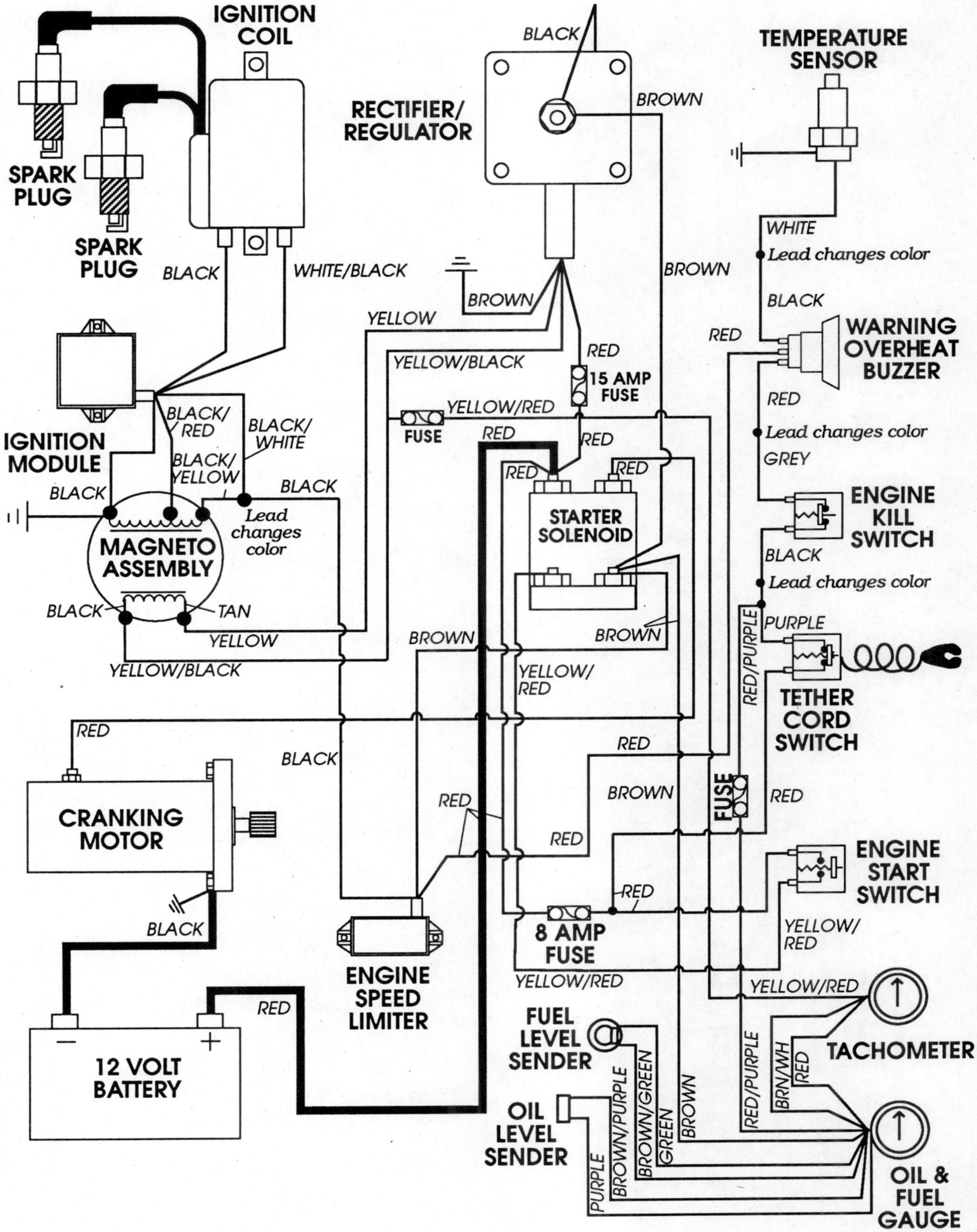

Wire Identification -- All Bombardier Sea-Doo models ***WITH*** *tachometer and oil/fuel gauge.*

DEGREE WHEEL FOR ROTARY VALVE TIMING

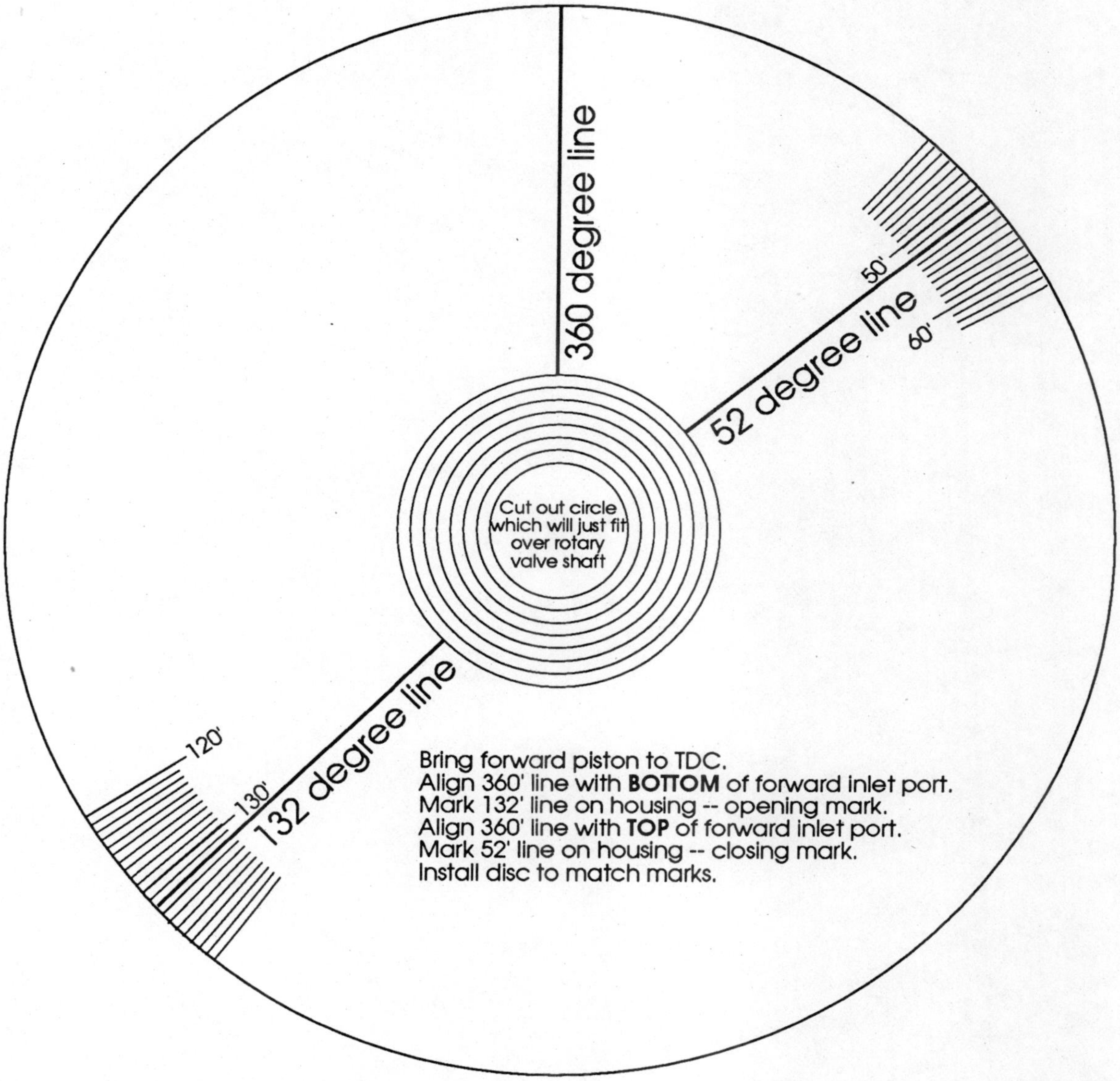

Bring forward piston to TDC.
Align 360' line with **BOTTOM** of forward inlet port.
Mark 132' line on housing -- opening mark.
Align 360' line with **TOP** of forward inlet port.
Mark 52' line on housing -- closing mark.
Install disc to match marks.

DEGREE WHEEL FOR ROTARY VALVE TIMING

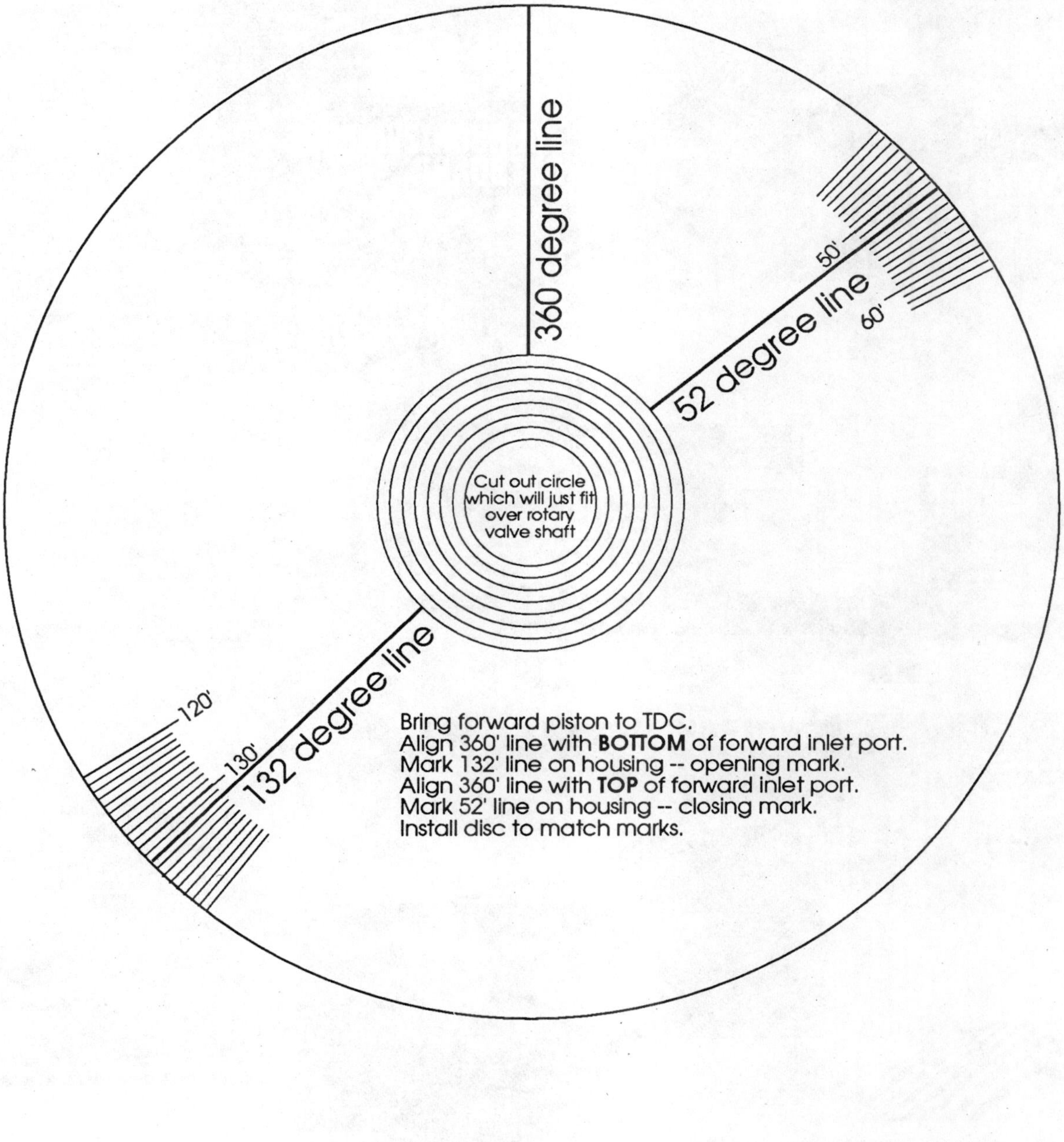

Other Seloc Marine Manuals

New titles are constantly being produced and the updating work on existing manuals never ceases.

All manuals contain complete detailed instructions, specifications, and wiring diagrams.